Soft Computing for Biological Systems

Hemant J. Purohit • Vipin Chandra Kalia
Ravi Prabhakar More
Editors

Soft Computing for Biological Systems

Springer

Editors
Hemant J. Purohit
Environmental Biotechnology
and Genomics Division
CSIR-National Environmental
Engineering Research Institute
(NEERI)
Nagpur, Maharashtra, India

Vipin Chandra Kalia
Microbial Biotechnology and Genomics
CSIR-Institute of Genomics and Integrative
Biology (IGIB)
Delhi University Campus
Delhi, India

Ravi Prabhakar More
ADBS, Lab 18, Neural Stem Cell Program
TIFR-National Centre for Biological Sciences
Bangalore, Karnataka, India

ISBN 978-981-10-7454-7 ISBN 978-981-10-7455-4 (eBook)
https://doi.org/10.1007/978-981-10-7455-4

Library of Congress Control Number: 2018931480

Printed on acid-free paper

This Springer imprint is published by the registered company Springer Nature Singapore Pte Ltd. part of Springer Nature.
The registered company address is: 152 Beach Road, #21-01/04 Gateway East, Singapore 189721, Singapore

Dedicated to our mentors
and
inspiration – the respected Mr. Dashrath
Manjhi:
The Mountain Man

Preface

The biological systems and their functions are driven by information stored in the genetic material, the DNA, and their expression is driven by different factors. The active units of these DNA sequences are genes, which also interact with each other to define a condition-specific expression. The soft computing approaches recognize the different patterns in DNA sequence and assign them biological relevance with available information. At times these patterns not only help in the classification of but also predict functionally active domains. These approaches are equally helpful in predicting protein-protein interaction. To understand any stressed scenario, there is need to predict gene networks by applying tools which can suggest differential gene expressions. The issue extends these tools in a wide range of models from bacteria to human cancers. We wish to present the status of diverse possibilities and our views and opinions to finally provide mankind with novel, innovative, and long-lasting strategies, in the book entitled *Soft Computing for Biological Systems*. The book provides insights into bioinformatics tools for neural networks, metagenomics data analysis, genetic barcoding, machine learning, and diagnostic predictions. The well-illustrated articles written by the experts in the area provide information on thrust scientific R&D areas and their future perspectives for the prospective researchers and graduate students – future of the scientific society. This book has reached its completion primarily due to the sincere efforts of the dedicated academic experts – to share their vision and wisdom. This collection of chapters has been presented in a manner which can benefit the curious minds of the society. We are indebted to all the people, whose invaluable contributions brought this book to fruition.

Delhi, India Vipin Chandra Kalia

Contents

About the Editors

Dr. Vipin Chandra Kalia is presently working as emeritus scientist. He has been the chief scientist and the deputy director at Microbial Biotechnology and Genomics, CSIR-Institute of Genomics and Integrative Biology, Delhi. He is a professor in AcSIR who obtained his M.Sc. and Ph.D. in genetics from Indian Agricultural Research Institute, New Delhi. He has been elected as (1) fellow of the National Academy of Sciences (FNASc), (2) fellow of the National Academy of Agricultural Sciences (FNAAS), and (3) fellow of the Association of Microbiologists of India (FAMSc). His main areas of research are microbial biodiversity, bioenergy, biopolymers, genomics, microbial evolution, quorum sensing, quorum quenching, drug discovery, and antimicrobials. He has published more than 100 papers in scientific journals such as (1) *Nature Biotechnology*, (2) *Biotechnology Advances*, (3) *Trends in Biotechnology*, (4) *Annual Review of Microbiology*, (5) *Critical Reviews in Microbiology*, (6) *Bioresource Technology*, (7) *PLOS ONE*, (8) *BMC Genomics*, (9) *International Journal of Hydrogen Energy*, and (10) *Gene*. He has authored 14 book chapters. His works have been cited 4100 times with an h-index of 35 and an i10-index of 76 (http://scholar.google.co.in/citations?hl=en&user=XaUw-VIAAAAJ). He has edited seven books: (i) *Quorum Sensing vs Quorum Quenching: A Battle with No End in Sight* (2015), (http://link.springer.com/book/10.1007/978-81-322-1982-8), (ii) *Microbial Factories: Biofuels, Waste Treatment – Vol. 1* (2015) (http://link.springer.com/book/10.1007%2F978-81-322-2598-0), (iii) *Microbial Factories: Biodiversity, Biopolymers, Bioactive Molecules – Vol. 2* (2015) (http://link.springer.com/book/10.1007%2F978-81-322-2595-9), (iv) *Waste Biomass Management: A Holistic Approach* (2017) (http://www.springer.com/in/book/9783319495941), (v) *Drug Resistance in Bacteria, Fungi, Malaria, and Cancer* – Editors: Arora, Gunjan, Sajid, Andaleeb, Kalia, Vipin Chandra (Eds.) (http://www.springer.com/in/book/9783319486826), (vi) *Microbial Applications* – Vol. 1 Kalia, V. (Ed), Kumar, P. (Ed) (2017) (http://www.springer.com/in/book/9783319526652), and (vii) *Microbial Applications – Vol. 2* Kalia, V. (Ed) (2017) (http://www.springer.com/in/book/9783319526683). He is presently the editor-in-chief of the *Indian Journal of Microbiology* (2013–2021) and editor of (1) *Journal of Microbiology and Biotechnology* (Korea), (2) *International Scholarly Res. Network Renewable Energy*, (3) *Dataset Papers in Microbiology*, and (4) *PLOS ONE*. He is a life member of the following scientific societies: (1) the Society of

Biological Chemists of India; (2) Society for Plant Biochemistry and Biotechnology, India; (3) Association of Microbiologists of India; (4) Indian Science Congress Association; (5) BioEnergy Council of India; and (6) Biotech Research Society of India (BRSI). He can be contacted at vckalia@igib.res.in; vc_kalia@yahoo.co.in

Dr. Hemant J. Purohit is head of Environmental Biotechnology and Genomics Division, National Environmental Engineering Research Institute (CSIR), Nagpur. He is also a professor in AcSIR (Academy of Scientific and Innovative Research), New Delhi. He completed his PhD from Nagpur University in 1986. He has been involved in designing a strategy for capturing microbial diversity by interfacing culturable and DNA fingerprinting tools; developing genomics-based monitoring tools for EIA and bio-remediation process; studying stress-dependent microbial response using dynamic gene expression and its application in bioprocess optimization; developing better insights into microbial capacities for utilization of organics through genome sequence analysis, etc. He has been project coordinator for a number of high-value projects. He has 225 publications to his credit. His Google scholar citations is 4711 (as of June 5, 2017), and he has an h-index of 38 and i10-index of 111. He has supervised 25 PhD students and more than 100 MSC student dissertations. He is a highly distinguished scientist. He is a recipient of a number of prestigious awards, including Fogarty International Exchange Program Fellowship; Commonwealth Fellowship, Department of Biochemistry, University of Hull, UK; CSIR Research Fellowships (JRF and SRF); etc.

Dr. Ravi Prabhakar More is a bioinformatician at the National Centre for Biological Sciences in Bangalore, Karnataka. He completed his Ph.D. from Swami Ramanand Teerth Marathwada University (SRTMU) and CSIR-NEERI, India, in 2015. He has developed signature (regular expression)-based DNA BarID and matK-QR classifier software for the identification of bacteria and plant species. He has been involved in next-generation sequencing (NGS), whole-genome sequencing (WGS), and exome sequencing (WES) data analysis for the identification of genes responsible for human brain disorders and in developing automated bioinformatics pipeline by using Perl and shell scripts on high-performance computing (HPC) for NGS data mining; he has worked on insect transcriptomics and phylogenetics, bacterial genomics, metagenomics, and DNA barcoding. He has published 11 scientific papers in reputed journals. He has worked at the National Institute for Basic Biology, Okazaki, Japan, and developed combined supervised approach (naïve Bayesian and homology) for detecting horizontal gene transfer in microbes. He is a recipient of awards of "Outstanding Project Personnel" for contribution in R&D activity, CSIR-NEERI, Nagpur, India, in the year 2014.

1 Current Scenario on Application of Computational Tools in Biological Systems

Hemant J. Purohit, Hitesh Tikariha, and Vipin Chandra Kalia

Abstract

The uncertainties and complexities of biological system challenge analytical approach and process of understanding. The wet lab experiments supported by soft algorithms find a way to resolve these scenarios. In the last decade, the biological analytical approach has found tremendous shift in data generation and analysis capacities. From sequencing of DNA and RNA to prediction of 3D structure and function of protein, there are a wide array of soft tools to make the job of exploring a system lot easier. This development eases our understanding of gene networks, plasticity and pattern of gene expression at gene to epigenomic level. In this book, we attempted to document selected areas of biological system and their advances, which will be frontier areas.

Keywords

Databases · Epigenome · Gene networks · Omic tools · Plasticity · Signatures

1.1 Introduction

The biological research has seen rapid progress through the use of computational tools for understanding physiological events. However, with the advent of next generation sequencing, there has been an explosive generation of data at different

H. J. Purohit (✉) · H. Tikariha
Environmental Biotechnology and Genomics Division, CSIR-National Environmental Engineering Research Institute (NEERI), Nagpur, Maharashtra, India
e-mail: hj_purohit@neeri.res.in; hemantdrd@hotmail.com

V. C. Kalia
Microbial Biotechnology and Genomics, CSIR-Institute of Genomics and Integrative Biology (IGIB), Delhi University Campus, Delhi, India
e-mail: vckalia@igib.in; vc_kalia@yahoo.co.in

H. J. Purohit et al. (eds.), *Soft Computing for Biological Systems*,
https://doi.org/10.1007/978-981-10-7455-4_1

levels of cellular organisation. A deeper understanding of protein expression profiles further supported this phenomenon. This has brought the data generated by biological systems into the domain of the big data analysis. The soft computing and artificial intelligence have become a prerequisite for the field of biological research to unfold system phenomenon. Bioinformatics tools have now become an essential hand for every section of biological data not only for handling and processing but also for validating the wet laboratory experiments. The omics era actually has now started emerging out of its lag phase. The progress of every laboratory is based on how intelligently they are harnessing the analytical tools for shaping the log phase trend of their physiological understanding. Keeping all these ingredients in the mind, this book opens up the current recipes of biological data de-codification. It is an attempt to focus on a few key areas and define their present status. The different areas challenge the readers to exploit a diversity of tools for applications in biological systems.

1.2 Protein Structure Prediction and Interaction

From the protocol of protein assay to chromatography and finally to NMR, now time has brought the reliability and rapidity in understanding the same information by in silico protein structure and function prediction (Tikariha et al. 2016). Even for an unknown protein, the implementation starts with the determination of its primary sequence. There has been an intense shift in the simple prediction of the secondary and tertiary structure of a protein from geometrical-based programming to new machine learning algorithms. Spencer et al. (2015) have given a vivid detail on ab initio protein secondary and tertiary prediction with the help of deep learning network. The concept eliminates the need for large protein structure database with known predicted proteins. This concept along with the incorporation of dihedral angle, torsion angle, solvent accessible surface area, positions and interactions of hydrogen bonds data can make the structure prediction a piece of cake (Heffernan et al. 2015). Even protein sequence and PDB database are on the rise, which will add to our knowledge on protein folding. This database can help in training the programs, which can help them in predicting the folding pattern and hence proposing the structure of an unknown protein. Thus, to get a vivid insight, one of the chapters gives an idea about the application of machine learning advancement in protein structure prediction.

The prediction of protein 3D structure is followed by the challenge of unearthing its interaction with other molecules, such as DNA and mRNA, or even with another protein. This part of the study holds immense potential for application in cellular pharmacology and drug discovery. Majorly there are three methods to study protein-protein interactions (PPIs) such as (1) residue coupling, (2) prediction of binding surface patch and (3) assembly prediction (Keskin et al. 2016). Based on this information, dozens of tools to analyse interfacial changes and calculate residues physicochemical changes have been developed. The database is also being constructed where one can look for curated PPIs such as CORUM, HIPPIE, IntAct, SPIKE, etc. Exploration of an interaction of peptide chain is also on the rise, and there is a huge market build-up on using peptide as a therapeutic agent (Nevola and Giralt 2015). Research avenues are also being built for modulation of PPI using small inhibitory molecules (Arkin et al. 2014). Computational tools can modulate the dynamics of protein structure and its behaviour in a particular solvent system. Integration of this data with the thermodynamics of molecule interactions and characteristics of amino acid residues involved in the interactions can simulate the interactive behaviour of two protein molecules. Protein interactions with smaller molecules can also assist in deciphering the signalling cascade; and so by understanding this, one can precisely regulate/modulate the machinery inside a cell. One can also detect crucial amino acid residue involved in PPI. An evolutionary biologist can also seek for changes in protein-protein interactions, which can be responsible for metabolic and phenotypic changes (Bartlett et al. 2016). Seeking this past trend and huge market potential, this chapter aims to provide a deep insight on PPIs.

Drug designing strategies are driven by the protein interactions with various other molecules, in which the target locations are very specific and are with minimum free energy levels. Data mining and drug discovery have been rising in the field of pharmacology in recent years (Lavecchia 2015). Machine learning systems mimic from nature's own cellular system, where a molecule can play multiple roles, and now this drives drug designing forward to poly-pharmacology (Lavecchia and Cerchia 2016). In silico analyses are carried out in designing multi-target drug relying on a huge database of ligands and protein 3D structure, docking dynamics and pattern-based designing of the molecule. This not only ensures the drug design but also its delivery to the site of action that is also a major concern for its effectiveness. The whole effectiveness of drug relies on the intracellular transporter system (Nigam 2015). Here algorithms on basis of nature of molecule, transporter protein and interaction between them can predict how well the drug can find its way into the cell and carry out their action. We have discussed the potential of drug transporters system in one of the chapters.

1.3 Emerging Areas in Tool Development

With the advent of sequencing technologies, there has been a progressive rise in computational tools (Kalia 2015; Koul et al. 2015; Yu et al. 2015; Ambardar et al. 2016; Koul and Kalia 2016; Kalia et al. 2017; Kumar et al. 2015, 2017; Meza-Lucas et al. 2016). From pairing and assembling, the sequence reads to their annotation as genes with different algorithms are becoming faster and accurate. In this, the foremost approach is to design a robust multiple sequence alignment (MSA) program. MSA is a key step for functional annotation, phylogenetic studies and a necessity for comparative genomics and metagenomics (Pooja et al. 2015). Most of the MSA tools such as CLUSTAL, MUSCLE, K-align and a lot more are based on de novo assembly and pairwise alignment by tree construction. These programs are good at handling a small set of sequences, but they become redundant while handling thousands of dataset. For overcoming this deficiency, new tools have been designed such as HAlign, a fast multiple similar DNA/RNA sequence alignment (Zou et al. 2015), and PASTA, ultralarge MSA (Mirarab et al. 2015) which can resolve this issue. Even tools like GUIDANCE2 are introduced to detect unreliable alignment regions in MSA (Sela et al. 2015). The hardware driver limitations are also being resolved, which can be seen in the development of GPU named CUDA ClustalW v1.0, and these will accelerate the computation of large datasets (Hung et al. 2015). We have dealt in detail the development in the domain of MSA and its application in the sequence alignment.

Next-generation sequencing has brought the computational biology to a new level. The databases for DNA, mRNA and proteins are growing geometrically. The repositories such as NCBI, EMBL, IMG, MG-RAST, SILVA and RCSB PDB are among the most exhaustively used databases. The Web has a large number of repositories and analysis pipeline for each separate domain such as CRCDA for cancer, Cas-Analyzer, Omics Pipe, etc. (Fisch et al. 2015; Thangam and Gopal 2015; Park et al. 2017). With NGS, the cost is going down, and there has been a tremendous amount of metagenome data generation. It is thus demanding new tools for accurate and reliable processing of this huge datasets. The attention has now been laid on the interpretation of this data rather than functional and taxonomical categorisation. Machine learning techniques, deep neural network generation and highly sophisticated statistical analysis are being used to understand this data. Integrated approach has been wired to connect all the analysis pipeline. A programming language such as Pearl, Python, R and Ruby are extensively used for investigation of NGS data. Nowadays, the Python and R have become two hands for interpretation of complex biological system and aid in connecting new links between the large and different datasets. The demands put regular pressure on program developers for updating the algorithms; recently QIIME pipeline was updated with the incorporation of PhyloToAST, which boosts its species-level classification and gives more elaborative visual evaluation (Dabdoub et al. 2016).

Transcriptome analysis has even brought the sequencing and analysis of miRNA, piRNA and IncRNAs possible, which is a big deal for disease diagnostic especially in the case of cancer. Extraction of secondary data from sequenced and annotated primary data is now becoming a remarkable strategy. Genome construction from metagenome is a new technique developed recently employing the process of binning, coverage, reassembly and curation (Sangwan et al. 2016). Tools like CheckM are devised to check the quality of reconstructed genomes (Parks et al. 2015). Apart from the reconstruction of the genome, the scheme is being designed to understand community-level talks, gene transfer and resistance development (More et al. 2014; Kapley et al. 2015). Thus to make the reader aware of this vast area, a chapter has been dedicated to bioinformatics tools for NGS data analysis.

Genomic tools are not limited to sequence identification or characterisation but can be implemented as pattern search algorithms to generate signatures, which can be utilised as biomarkers for diagnostic purposes (Porwal et al. 2009; Bhushan et al. 2015). A genomic biomarker can be used as both prognostic and predictive biomarker. Due to its high sensitivity and high specificity, the medical industry is looking for the discovery of such biomarker for every type of diseases (Kalia and

Kumar 2015; Kalia et al. 2015, 2016; Kekre et al. 2015; Kumar et al. 2016; Lee et al. 2016; Puri et al. 2016). Cancer is one of the deadly diseases and hard to diagnose at initial stages and opens a wide door for exploration of the genomic marker. A whole bunch of biomarkers discovered till date for head and neck cancer have been presented in a recent review (Kang et al. 2015). The promising nature of biomarker application has provoked us to include a chapter on the use of genomic biomarker in the case of oral cancer.

1.4 Gene Networks and Plasticity

Cells represent a collection of very well-coordinated and synchronised interactions and movement of every molecule residing in it. This is due to inherited intelligence cell carries for regulating expression of genes for every desirable event. Understanding this network of genes and how they regulate various machineries of cells by modulating itself is a challengeable task. Exploration of gene network involves the study of their expression pattern. The biological phenomenon evolved over a period of time, with one gene, one expression and identified physiology to a now collection of genes but even with the most sophisticated tools not completely understood till date. Gene Expression Atlas, Gene Expression Commons, CODEX and many more single gene expression databases are being created of which BloodSpot is the recent one which provides the tree-based relationship between different gene expression profiles present in the database (Bagger et al. 2016). As genes are differentially expressed in diverse conditions, it provides the plasticity to the gene networking; the wide range of data need to be generated to predict even an interaction of a single gene that behaves as a node in a network. A database on gene plasticity named ImmuSort is already being released, which provides an electric sorting system for immune cells (Wang et al. 2015). Thus from different expression profiles of a single gene to linking its connection with the expression profile of another gene requires a network-based analysis and a mammoth database.

Artificial neural networks are a set of models designed to classify and predict the outcome from a provided data; hence they are widely used algorithms in gene network prediction. Feedforward neural network, radial basis function network, modular neural network and physical neural network are the general types of neural network that are routinely applied in such analyses. Implying the data within a given set of conditions, the network is designed to calculate the expression behaviour for a set of genes. We have discussed the beneficial role of above study in diagnostic prediction by the aid of gene expression profile and artificial neural network.

The array of the genetic circuit in the cells can be grouped into various categories and modules specialised to carry out a specific task. Carving out this module of gene network could render the task easier for decoding the process associated with it. We are mostly interested in a specific set of the gene network, which we can modulate in a way that achieves a specific task such as understanding the response of a signal cascade when osmotic stress is faced by the cell. Exploration of each module can give an idea of complete genetic web collectively working in the cell. Realising the core importance of this idea, we have added a chapter on soft computing approaches to extract biologically significant gene network modules that presents how through computational convergence one can study such network module and function carried out by each module separately.

Not only understanding of gene network is essential but we should also know how we can create a network. Mapping of a network relies on data used for its creation which in our case would be gene expression profiles. This requires a series of expression data of every single gene than stacking them upon one another in time series or imposed variant conditions and in the last layering and connecting the links between each gene involved in the network. Either the supervised or unsupervised model can be used for creating a network. In a recent paper, authors describe the use of both the approaches for the creation of gene regulatory network (Huynh-Thu and Sanguinetti 2015). Single cell network synthesis toolkit has been used to identify an interconnected network of 20 transcription factors in human blood cell (Moignard et al. 2015). So to get acquainted about such emerging topic, we have incorporated a chapter which deals with the construction of gene network.

1.5 Epigenome: Emerging Area

All the current techniques target the pattern of the four nucleotides, thereby predicting its functions, but in the case of eukaryotes, this scenario changes. The methylation pattern under the epigenetic tag governs which gene will get expressed and which does not. Epigenomic research targets such molecule which can alter the expression pattern of the genes in a chromosome. The epigenetic study has two core areas – DNA methylation and histone modification. Methylation of DNA is usually on CpG islands and follows a particular pattern used to deduce the expression profile of gene under study. Techniques like methylation array detect DNA methylation, whereas ChIP sequencing determines a modification in histone. Both the tools have helped in generating an epigenetic map of the human chromosome. The epigenomic study is particularly interesting as it delivers the regulatory expression channel of gene thereby influencing the phenotypic expression. The cross-links through which epigenetic action is controlled by environmental factors are also a great issue of interest. Lots of epigenome-wide-associated studies have linked diets, smoking, stress, etc. to changes in genotypic and phenotypic variation in human (Lee et al. 2015; Provençal and Binder 2015). Realising such rising trend in the area of epigenomic, we have included a chapter on Module-Based Knowledge Discovery for Multiple-Cytosine-Variant Methylation Profile.

1.6 Expanding the Domain of Computational Statistical Analysis

With the expansion of biological data, lots of statistical tools have been developed to sort, group, analyse and predict the outcome from the data. Statistic combined with appropriate programming language results in more analytical approach and visually enhanced result. Along with the application of computational tools, various modelling techniques are also being integrated to understand the pest population dynamics (Whish et al. 2015; Gilioli et al. 2016). With so much focus on the application of computational statistic in the field of biotechnology, we introduce our reader to the domain of agriculture with such analysis. This chapter describes the role of various computing tools and techniques based on background statistical analysis for studying pest population dynamics.

1.7 Pattern Recognition/Barcoding/Diagnostics

Identification of species and determining its role in an environment are crucial step in ecological discernment. DNA barcoding is one of the emerging genomic tools to tackle this problem. Based on consensus pattern of a sequence, it aids in the identification of species (Kalia and Kumar 2015; Kalia et al. 2015, 2016; Kekre et al. 2015; Kumar et al. 2016; Lee and Rho 2016). DNA barcoding applies to all domain of life for their classification. A great deal of DNA barcode application till date has been broadly reviewed recently (Kress et al. 2015). Barcoding also allows revealing the diversity pattern of flora and fauna thereby producing a species map for niche/habitat (More and Purohit 2016; More et al. 2016).

The great nature of DNA barcoding is that it is applicable to every organism with minor modification. The initial step in barcoding is deducing the signature sequence in the species. After the identification, it can be used to tag every other species which have an exact signature (Porwal et al. 2009; Kalia et al. 2011; Bhushan et al. 2013). Thus the nature and location of code vary from species to species and organism to organism. The DNA barcode has a great role in conservative biology as it can help in tracking the species of interest. To open up the reader more about the application of DNA barcoding, we have discussed thoroughly the fish DNA barcoding as a model. This chapter also covers up the various bioinformatics tools and techniques deployed in generating a DNA barcode for a given species.

Earlier when DNA barcode was introduced, it was limited to eukaryotic organisms due to high mutation rate in prokaryotes and absence of mitochondrial or plastid DNA, which have rich consensus region. But now this scenario is changing, and bacterial DNA barcode is being introduced in recent years along with the introduction of meta-barcoding. Recent publications on marine benthic meta-barcoding have already laid down this trend (Leray and Knowlton 2015). In upcoming years we can expect the rise of meta-barcoding along with the metagenomics. For providing a complete package of tools and software used in bacteria DNA barcoding and analysis, reader can refer to a later chapter.

Pattern- and network-based computational analyses are not only limited to the microorganism or medical biology, but it has an expanded horizon in plant biology too. Earlier it was concentrated to the regime of plant classical genetics and breeding but gradually arose with plant genomics. The surge in plant genomics can be seen with the recent introduction of PLAZA 3.0 which is a server assisting in comparative plant genomics (Proost et al. 2015). Genomic analysis has already been extended to the study of metabolite-based quantitative trait loci. Identification of metabolites is one of the highlighted areas in plant metabolomics. Luo in 2015 discussed the genome-wide association studies based on metabolite. Not only genetic trait but analysis of phenotypic trait in plant biology is a keen area. The various repositories have been created to store phenotypic data for a selected plant species, e.g. MaizeGDB, Ephesis databases, etc. This has laid down the incorporation of microarray, metabolomics, sequencing and proteomics data in a single platform for understanding the link between phenotypic expression, genetic makeup and environmental factors. This has arisen the need for handling ample amount of data synchronising it with metadata (Krajewski et al. 2015). Modelling framework is also being applied in plant biology for better resolution of its cellular event (Boudon et al. 2015). Observing a high trend in the application of computational tools in the subject of plant biology, a vivid description of the integration of computational approach in plant biology and also its field application has been discussed in this book.

With metadata, biological systems are challenging the scientific community with its complexity. Covering different emerging disciplines in biology where computational approach is essential or playing an essential role has been discussed in this book, which will surely give the reader a new paradigm in their analytical processes.

References

Ambardar S, Gupta R, Trakroo D, Lal R, Vakhlu J (2016) High throughput sequencing: an overview of sequencing chemistry. Indian J Microbiol 56:394–404. https://doi.org/10.1007/s12088-016-0606-4

Arkin MR, Tang Y, Wells JA (2014) Small-molecule inhibitors of protein-protein interactions: progressing toward the reality. Chem Biol 21:1102–1114. https://doi.org/10.1016/j.chembiol.2014.09.001

Bagger FO, Sasivarevic D, Sohi SH, Laursen LG, Pundhir S, Sønderby CK, Winther O, Rapin N, Porse BT (2016) BloodSpot: a database of gene expression profiles and transcriptional programs for healthy and malignant haematopoiesis. Nucleic Acids Res 44(D1):D917–D924. https://doi.org/10.1093/nar/gkv1101

Bartlett M, Thompson B, Brabazon H, Del Gizzi R, Zhang T, Whipple C (2016) Evolutionary dynamics of floral homeotic transcription factor protein–protein interactions. Mol Biol Evol 33:1486–1501. https://doi.org/10.1093/molbev/msw031

Bhushan A, Joshi J, Shankar P, Kushwah J, Raju SC, Purohit HJ, Kalia VC (2013) Development of genomic tools for the identification of certain *Pseudomonas* up to species level. Indian J Microbiol 53:253–263. https://doi.org/10.1007/s12088-013-0412-1

Bhushan A, Mukherjee T, Joshi J, Shankar P, Kalia VC (2015) Insights into the origin of *Clostridium botulinum* strains: evolution of distinct restriction endonuclease sites in *rrs* (16S rRNA gene). Indian J Microbiol 55:140–150. https://doi.org/10.1007/s12088-015-0514-z

Boudon F, Chopard J, Ali O, Gilles B, Hamant O, Boudaoud A, Traas J, Godin C (2015) A computational framework for 3D mechanical modeling of plant morphogenesis with cellular resolution. PLoS Comput Biol 11:e1003950. https://doi.org/10.1371/journal.pcbi.1003950

Dabdoub SM, Fellows ML, Paropkari AD, Mason MR, Huja SS, Tsigarida AA, Kumar PS (2016) PhyloToAST: bioinformatics tools for species-level analysis and visualization of complex microbial datasets. Sci Rep 6. https://doi.org/10.1038/srep29123

Fisch KM, Meißner T, Gioia L, Ducom JC, Carland TM, Loguercio S, Su AI (2015) Omics pipe: a community-based framework for reproducible multi-omics data analysis. Bioinformatics 31:1724–1728. https://doi.org/10.1093/bioinformatics/btv061

Gilioli G, Pasquali S, Marchesini E (2016) A modelling framework for pest population dynamics and management: an application to the grape berry moth. Ecol Model 320:348–357

Heffernan R, Paliwal K, Lyons J, Dehzangi A, Sharma A, Wang J, Sattar A, Yang Y, Zhou Y (2015) Improving prediction of secondary structure, local backbone angles, and solvent accessible surface area of proteins by iterative deep learning. Sci Rep 5:11476. https://doi.org/10.1038/srep11476

Hung CL, Lin YS, Lin CY, Chung YC, Chung YF (2015) CUDA ClustalW: an efficient parallel algorithm for progressive multiple sequence alignment on multi-GPUs. Comput Biol Chem 58:62–68. https://doi.org/10.1016/j.compbiolchem.2015.05.004

Huynh-Thu VA, Sanguinetti G (2015) Combining tree-based and dynamical systems for the inference of gene regulatory networks. Bioinformatics 31:1614–1622. https://doi.org/10.1093/bioinformatics/btu863

Kalia VC (2015) Let's explore the latent features of genes to identify bacteria. J Mol Genet Med 9: e105. https://doi.org/10.4172/1747-0862.1000E105

Kalia VC, Kumar P (2015) Genome wide search for biomarkers to diagnose *Yersinia* infections. Indian J Microbiol 55:366–374. https://doi.org/10.1007/s12088-015-0552-6

Kalia VC, Mukherjee T, Bhushan A, Joshi J, Shankar P, Huma N (2011) Analysis of the unexplored features of *rrs* (16S rDNA) of the genus *Clostridium*. BMC Genomics 12:18. https://doi.org/10.1186/1471-2164-12-18

Kalia VC, Kumar P, Kumar R, Mishra A, Koul S (2015) Genome wide analysis for rapid identification of *Vibrio* species. Indian J Microbiol 55:375–383. https://doi.org/10.1007/s12088-015-0553-5

Kalia VC, Kumar R, Kumar P, Koul S (2016) A genome-wide profiling strategy as an aid for searching unique identification biomarkers for *Streptococcus*. Indian J Microbiol 56:46–58. https://doi.org/10.1007/s12088-015-0561-5

Kalia VC, Kumar R, Koul S (2017) In silico analytical tools for phylogenetic and functional bacterial genomics. In: Arora G, Sajid A, Kalia VC (eds) Drug resistance in bacteria, fungi, malaria and cancer. Springer, Cham, pp 339–355. https://doi.org/10.1007/978-3-319-48683-3_15. ISBN: 978-3-319-48682-6

Kang H, Kiess A, Chung CH (2015) Emerging biomarkers in head and neck cancer in the era of genomics. Nat Rev Clin Oncol 12:11–26. https://doi.org/10.1038/nrclinonc.2014.192

Kapley A, Liu R, Jadeja NB, Zhang Y, Yang M, Purohit HJ (2015) Shifts in microbial community and its correlation with degradative efficiency in a wastewater treatment plant. Appl Biochem Biotechnol 176:2131–2143. https://doi.org/10.1007/s12010-015-1703-2

Kekre A, Bhushan A, Kumar P, Kalia VC (2015) Genome wide analysis for searching novel markers to rapidly identify *Clostridium* strains. Indian J Microbiol 55:250–257. https://doi.org/10.1007/s12088-015-0535-7

Keskin O, Tuncbag N, Gursoy A (2016) Predicting protein–protein interactions from the molecular to the proteome level. Chem Rev 116:4884–4909. https://doi.org/10.1021/acs.chemrev.5b00683

Koul S, Kalia VC (2016) Comparative genomics reveals biomarkers to identify *Lactobacillus* species. Indian J Microbiol 56:253–263. https://doi.org/10.1007/s12088-016-0605-5

Koul S, Kumar P, Kalia VC (2015) A unique genome wide approach to search novel markers for rapid identification of bacterial pathogens. J Mol Genet Med 9:194. https://doi.org/10.4172/1747-0862.1000194

Krajewski P, Chen D, Ćwiek H, van Dijk AD, Fiorani F, Kersey P, Klukas C, Lange M, Markiewicz A, Nap JP, van Oeveren J, Pommier C, Scholz U, van Schriek M, Usadel B, Weise S (2015) Towards recommendations for metadata and data handling in plant phenotyping. J Exp Bot 66:5417–5427. https://doi.org/10.1093/jxb/erv271

Kress WJ, García-Robledo C, Uriarte M, Erickson DL (2015) DNA barcodes for ecology, evolution, and conservation. Trends Ecol Evol 30:25–35

Kumar A, Mohanty NN, Chacko N, Yogisharadhya R, Shivachandra SB (2015) Structural features of a highly conserved Omp16 protein of *Pasteurella multocida* strains and comparison with related peptidoglycan-associated lipoproteins (PAL). Indian J Microbiol 55:50–56. https://doi.org/10.1007/s12088-014-04896-1

Kumar R, Koul S, Kumar P, Kalia VC (2016) Searching biomarkers in the sequenced genomes of *Staphylococcus* for their rapid identification. Indian J Microbiol 56:64–71. https://doi.org/10.1007/s12088-016-0565-9

Kumar R, Koul S, Kalia VC (2017) Exploiting bacterial genomes to develop biomarkers for identification. In: Arora G, Sajid A, Kalia VC (eds) Drug resistance in bacteria, fungi, malaria and cancer. Springer, Cham, pp 357–370. https://doi.org/10.1007/978-3-319-48683-3_16. ISBN: 978–3–319-48682-6

Lavecchia A (2015) Machine-learning approaches in drug discovery: methods and applications. Drug Discov Today 20:318–331. https://doi.org/10.1016/j.drudis.2014.10.012

Lavecchia A, Cerchia C (2016) *In silico* methods to address polypharmacology: current status, applications and future perspectives. Drug Discov Today 21:288–298. https://doi.org/10.1016/j.drudis.2015.12.007

Lee S, Rho JY (2016) Development of a specific diagnostic system for detecting *Turnip Yellow Mosaic Virus* from Chinese cabbage in Korea. Indian J Microbiol 56:103–107. https://doi.org/10.1007/s12088-015-0557-1

Lee KW, Richmond R, Hu P, French L, Shin J, Bourdon C, Reischl E, Waldenberger M, Zeilinger S, Gaunt T, McArdle W, Ring S, Woodward G, Bouchard L, Gaudet D, Smith GD, Relton C, Paus T, Pausova Z (2015) Prenatal exposure to maternal cigarette smoking and DNA methylation: epigenome-wide association in a discovery sample of adolescents and replication in an independent cohort at birth through 17 years of age. Environ Health Perspect 123:193. https://doi.org/10.1289/ehp.1408614.

Lee S, Kim CS, Shin YG, Kim JH, Kim YS, Jheong WH (2016) Development of nested PCR-based specific markers for detection of peach rosette mosaic virus in plant quarantine. Indian J Microbiol 56:108–111. https://doi.org/10.1007/s12088-015-0548-2

Leray M, Knowlton N (2015) DNA barcoding and metabarcoding of standardized samples reveal patterns of marine benthic diversity. Proc Natl Acad Sci 112:2076–2081. https://doi.org/10.1073/pnas.1424997112

Luo J (2015) Metabolite-based genome-wide association studies in plants. Curr Opin Plant Biol 24:31–38. https://doi.org/10.1016/j.pbi.2015.01.006

Meza-Lucas A, Pérez-Villagómez M, Martínez-López JP, García-Rodea R, Martínez-Castelán MG, Escobar-Gutiérrez A, de la Rosa-Arana JL, Villanueva-Zamudio A (2016) Comparison of DOT-ELISA and Standard-ELISA for detection of the *Vibrio cholerae* toxin in culture supernatants of bacteria isolated from human and environmental samples. Indian J Microbiol 56:379–382. https://doi.org/10.1007/s12088-016-0596-2

Mirarab S, Nguyen N, Guo S, Wang LS, Kim J, Warnow T (2015) PASTA: ultra-large multiple sequence alignment for nucleotide and amino-acid sequences. J Comp Biol 22:377–386. https://doi.org/10.1089/cmb.2014.0156

Moignard V, Woodhouse S, Haghverdi L, Lilly AJ, Tanaka Y, Wilkinson AC, Buettner F, Macaulay IC, Jawaid W, Diamanti E, Nishikawa S, Piterman N, Kouskoff V, Theis FJ, Fisher J, Göttgens B (2015) Decoding the regulatory network of early blood development from single-cell gene expression measurements. Nat Biotechnol 33:269–276. https://doi.org/10.1038/nbt.3154

More RP, Purohit HJ (2016) The identification of discriminating patterns from 16S rRNA gene to generate signature for *Bacillus* genus. J Comp Biol 23:651–661. https://doi.org/10.1089/cmb.2016.0002

More RP, Mitra S, Raju SC, Kapley A, Purohit HJ (2014) Mining and assessment of catabolic pathways in the metagenome of a common effluent treatment plant to induce the degradative capacity of biomass. Bioresour Technol 153:137–146. https://doi.org/10.1016/j.biortech.2013.11.065

More RP, Mane RP, Purohit HJ (2016) matK-QR classifier: a patterns based approach for plant species identification. BioData Min 9:39. https://doi.org/10.1186/s13040-016-0120-6

Nevola L, Giralt E (2015) Modulating protein–protein interactions: the potential of peptides. Chem Commun 51:3302–3315. https://doi.org/10.1039/c4cc08565e

Nigam SK (2015) What do drug transporters really do? Nat Rev Drug Discov 14:29–44. https://doi.org/10.1038/nrd4461

Park J, Lim K, Kim JS, Bae S (2017) Cas-analyzer: an online tool for assessing genome editing results using NGS data. Bioinformatics 33:286–288. https://doi.org/10.1093/bioinformatics/btw561

Parks DH, Imelfort M, Skennerton CT, Hugenholtz P, Tyson GW (2015) CheckM: assessing the quality of microbial genomes recovered from isolates, single cells, and metagenomes. Genome Res 25:1043–1055. https://doi.org/10.1101/gr.186072.114

Pooja S, Pushpanathan M, Jayashree S, Gunasekaran P, Rajendhran J (2015) Identification of periplasmic a-amlyase from cow dung metagenome by product induced gene expression profiling (Pigex). Indian J Microbiol 55:57–65. https://doi.org/10.1007/s12088-014-0487-3

Porwal S, Lal S, Cheema S, Kalia VC (2009) Phylogeny in aid of the present and novel microbial lineages: diversity in *Bacillus*. PLoS One 4:e4438. https://doi.org/10.1371/journal.pone.0004438

Proost S, Van Bel M, Vaneechoutte D, Van de Peer Y, Inzé D, Mueller-Roeber B, Vandepoele K (2015) PLAZA 3.0: an access point for plant comparative genomics. Nucleic Acids Res 43 (D1):D974–D981. https://doi.org/10.1093/nar/gku986

Provençal N, Binder EB (2015) The effects of early life stress on the epigenome: from the womb to adulthood and even before. Exp Neurol 268:10–20. https://doi.org/10.1016/j.expneurol.2014.09.001

Puri A, Rai A, Dhanaraj PS, Lal R, Patel DD, Kaicker A, Verma M (2016) An *in silico* approach for identification of the pathogenic species, *Helicobacter pylori* and its relatives. Indian J Microbiol 56:277–286. https://doi.org/10.1007/s12088-016-0575-7

Sangwan N, Xia F, Gilbert JA (2016) Recovering complete and draft population genomes from metagenome datasets. Microbiome 4:8. https://doi.org/10.1186/s40168-016-0154-5

Sela I, Ashkenazy H, Katoh K, Pupko T (2015) GUIDANCE2: accurate detection of unreliable alignment regions accounting for the uncertainty of multiple parameters. Nucleic Acids Res 43 (W1):W7–W14. https://doi.org/10.1093/nar/gkv318

Spencer M, Eickholt J, Cheng J (2015) A deep learning network approach to ab initio protein secondary structure prediction. IEEE/ACM Trans Comp Biol Bioinform 12:103–112. https://doi.org/10.1109/TCBB.2014.2343960

Thangam M, Gopal RK (2015) CRCDA – Comprehensive resources for cancer NGS data analysis. Database 2015:bav092. https://doi.org/10.1093/database/bav092

Tikariha H, Pal RR, Qureshi A, Kapley A, Purohit HJ (2016) *In silico* analysis for prediction of degradative capacity of *Pseudomonas putida* SF1. Gene 591:382–392. https://doi.org/10.1016/j.gene.2016.06.028

Wang P, Yang Y, Han W, Ma D (2015) ImmuSort, a database on gene plasticity and electronic sorting for immune cells. Sci Rep 5:10370. https://doi.org/10.1038/srep10370

Whish JP, Herrmann NI, White NA, Moore AD, Kriticos DJ (2015) Integrating pest population models with biophysical crop models to better represent the farming system. Environ Model Softw 72:418–425

Yu S, Peng Y, Zheng Y, Chen W (2015) Comparative genome analysis of *Lactobacillus casei*: insights into genomic diversification for niche expansion. Indian J Microbiol 55:102–107. https://doi.org/10.1007/s12088-014-0496-2

Zou Q, Hu Q, Guo M, Wang G (2015) HAlign: fast multiple similar DNA/RNA sequence alignment based on the centre star strategy. Bioinformatics 31(15):2475–2481. https://doi.org/10.1093/bioinformatics/btv177

Diagnostic Prediction Based on Gene Expression Profiles and Artificial Neural Networks

2

Eugene Lin and Shih-Jen Tsai

Abstract

Recent advances in scientific research point out that diagnostic prediction represents a novel paradigm because of the decreased expense and the expanded productivity of multi-omics technologies such as gene expression profiling. In order to evaluate a mammoth amount of biomarkers produced by high-throughput technologies, machine learning and predictive approaches such as artificial neural network (ANN) algorithms have widely been utilized to assess disease mechanisms and intervention outcomes. In this chapter, we first illustrated ANN algorithms for establishing biomarkers in diagnostic prediction studies. We then surveyed a variety of diagnostic prediction applications for numerous diseases and treatments with consideration of ANN algorithms and gene expression profiling. Finally, we outlined their limitations and future directions. Future work in diagnostic prediction studies promises to lead to innovative ideas related to disease prevention and drug responsiveness in light of multi-omics technologies as well as machine learning and predictive algorithms.

Keywords

Gene expression · Artificial neural networks · Machine learning · Chemotherapy · Schizophrenia

E. Lin
Institute of Biomedical Sciences, China Medical University, Taichung, Taiwan

Department of Electrical Engineering, University of Washington, Seattle, WA, USA

TickleFish Systems Corporation, Seattle, WA, USA

S.-J. Tsai (✉)
Department of Psychiatry, Taipei Veterans General Hospital, Taipei, Taiwan

Division of Psychiatry, National Yang-Ming University, Taipei, Taiwan
e-mail: tsai610913@gmail.com

H. J. Purohit et al. (eds.), *Soft Computing for Biological Systems*,
https://doi.org/10.1007/978-981-10-7455-4_2

2.1 Introduction

In this chapter, we briefly describe some key emerging diagnostic prediction studies for various diseases and treatments of significance for public health with consideration of gene expression profiles and machine learning algorithms such as artificial neural network (ANN) models (Lin and Tsai 2011). This review is not intended as a comprehensive survey of all possible diagnostics applications studied in the literature.

First, we described machine learning and predictive algorithms such as ANN models that have been widely used in the research community for pinpointing biomarkers as well as for associating with diseases and drug responses in the diagnostic prediction studies. Furthermore, we surveyed some potential biomarkers that were investigated in the diagnostic prediction studies using gene expression profiles and ANN algorithms and were reported to be linked with disease status or drug efficacy. Moreover, we highlighted the limitations and future outlook regarding the diagnostic prediction studies in terms of gene expression profiles as well as machine learning and predictive algorithms. In future work, replication studies with extensive and independent cohorts will be indispensable in order to establish the characteristics of the potential biomarkers identified in the diagnostic prediction studies in disease diagnosis as well as treatment response (Lin 2012; Lin and Tsai 2012).

2.2 Machine Learning and Artificial Neural Networks

Machine learning and predictive methods contain computer algorithms which are able to naturally perceive complicated patterns based on empirical data (Kononenko 2001; Lane et al. 2012; Lin and Tsai 2016c). The objective of machine learning and predictive algorithms is to facilitate computer algorithms to gain from data of the past or present and then make decisions or predictions for unrecognized forthcoming circumstances by utilizing that knowledge (Landset et al. 2015; Lin and Tsai 2016c). In the general terms, the workflow (as shown in Fig. 2.1) for a machine learning and predictive algorithm incorporates three phases including construct the model from pattern inputs, appraise and refine the model, and then establish the model into construction in prediction-making (Landset et al. 2015). In other words, machine learning and predictive algorithms for classification

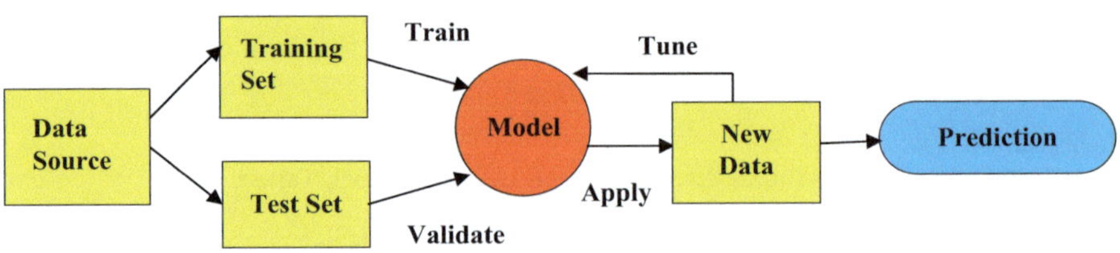

Fig. 2.1 Machine learning workflow

applications such as medical diagnosis or diagnostic prediction are procedures for adopting the best assumption from a set of alternatives that are qualified for a set of observations (Witten and Frank 2005). The strengths of machine learning and predictive algorithms for classification, including nonlinearity, fault tolerance, and real-time operation, make them suitable for complicated applications (Lane et al. 2012).

ANN models, such as multilayer feedforward neural networks, can be frequently utilized to solve complicated applications in classification and predictive modeling due to the fact that ANN algorithms possess the benefits of fault tolerance, nonlinearity, integrality, and real-time operations (Lin et al. 2006; Kung and Hwang 1998). A multilayer feedforward neural network is one category of ANN algorithms where networks between entities construct no directed cycles (Bishop 1995). In other words, a loop or cycle does not exist in the network because the data only relays in an onward order from the input entities, by means of the hidden entities (if any), and then to the output entities.

Moreover, from an algorithmic point of view, the primary operation of this ANN is separated into the learning and retrieving stages (Kung and Hwang 1998). In the learning stage of this ANN, the back-propagation algorithm (Rumelhart et al. 1996) is adopted for the learning scheme. Furthermore, in the retrieving stage, this ANN repeats through all the panels to achieve the retrieval response at the output panel in keeping with the inputs of test patterns. On the other hand, from a structural point of view, this ANN is an iterative and spatial neural network that possesses numerous panels of hidden neuron groups among the input and output neuron panels (Kung and Hwang 1998).

The ANN models can be executed using favored machine learning tools such as R (the R Project for Statistical Computing; http://www.r-project.org/) or the Waikato Environment for Knowledge Analysis (WEKA) software (Witten and Frank 2005). However, popular open-source machine learning tools including R and WEKA are not originally constructed and implemented for large-scale data (Landset et al. 2015). To effortlessly design and adopt for big data, there are assorted machine learning tools, such as Mahout (http://mahout.apache.org/), MLlib (https://spark.apache.org/mllib/), H2O (http://h2o.ai/), and SAMOA (https://github.com/samoa-moa/samoa-moa), available to run in a distributed environment (Lin and Tsai 2016c).

2.3 Gene Expression Profile

Noncoding RNAs, such as long noncoding RNAs and small noncoding RNAs, are distinct from their complement messenger RNAs (mRNAs) because the sequence of nucleotides in noncoding RNAs encodes no proteins (Nagano and Fraser 2011; Lin and Tsai 2016a). While long noncoding RNAs represent transcripts with more than 200 nucleotides in length, small noncoding RNAs, such as the microRNAs, are smaller than 200 nucleotides in length.

The microRNAs govern gene expression by regulating mRNA translation, stability, and degradation (Dwivedi 2014; Lin and Tsai 2016a). The characteristics of mRNAs, microRNAs, and long noncoding RNAs in examining disease pathogenesis and in keeping track of response to treatment for human disease are developing rapidly. Future work will be conducted to assess whether gene expression profiling including mRNAs, microRNAs, and long noncoding RNAs may be established as potential biomarkers with respect to human disease and therapeutic responses (Lin and Tsai 2016a).

2.4 Gene Expression Profile Studies with ANN

Table 2.1 summarizes the relevant diagnostic prediction studies by using gene expression profile and ANN models. This is by no means a comprehensive survey of all probable diagnostic prediction studies discovered so far. Nonetheless, a growing body of studies has been investigated when scientists remain to pay much attention to diagnostic prediction research.

2.4.1 Cancer

There were a variety of diagnostic prediction studies for cancer research using ANN models and gene expression profiling. First, Pass et al. (2004) trained a three-layer ANN model based on the expression value of differentially regulated genes and derived a set of 27 genes that distinguishes good-risk and poor-risk surgically

Table 2.1 Diagnostic prediction studies of gene expression profiling for various diseases and treatments of significance using artificial neural networks

Disease/treatment	Results	References
Malignant pleural mesothelioma	Achieved 76% accuracy	Pass et al. (2004)
Neuroblastoma	Achieved 88% accuracy	Wei et al. (2004)
Astrocytic brain tumors	Identified an optimum set of 37 genes	Petalidis et al. (2008)
Breast cancer	Reduced a 70-gene signature to nine genes	Lancashire et al. (2010)
Schizophrenia	Achieved 87.9% accuracy	Takahashi et al. (2010)
Diffuse large B-cell lymphoma	Achieved 93% accuracy	Mehridehnavi and Ziaei (2013)
Luminal A-like breast cancer	Revealed ten microRNAs for further analysis	McDermott et al. (2014)
Childhood sarcomas	Showed strong connection links on certain genes	Tong et al. (2014)
Chemotherapy in non-small cell lung cancer	Achieved 65.71% accuracy	Chen et al. (2015)

treated patients with malignant pleural mesothelioma. A rare and aggressive cancer called malignant pleural mesothelioma usually evolves in the thin row of tissue neighboring the lungs known as the pleura. Of the 27 genes revealed to be significant, 18 have been intensely investigated in the literature, and few have been linked with malignant pleural mesothelioma (Pass et al. 2004).

Secondly, Wei et al. (2004) utilized gene expression profiles from cDNA microarrays to forecast the outcome and extract a minimal gene set in patients with neuroblastoma by using ANN models. Neuroblastoma is the most common cancer in childhood and in infancy. They suggested that the top 24 ANN-ranked clones, which represented 19 unique genes as a minimal gene set, resulted in the minimal classification error. Wei et al. (2004) also indicated that ANN models can predict additional patients according to their survival status based on either all genes or in particular the 19 genes.

Thirdly, Petalidis et al. (2008) assessed whether molecular signatures can define survival prognostic subclasses of astrocytic tumors by using gene expression data from 65 highly annotated tumors and a simple ANN model in the form of a single-layer perceptron. Astrocytic tumors are the most common type of cancer in the brain. They analyzed the ANN model to optimize leave-one-out cross-validation runs, which resulted in an optimum set of 37 genes. Petalidis et al. (2008) selected two genes of special interest, *PEA15* and *ADM*, for further analysis in their study.

In addition, Lancashire et al. (2010) leveraged a previously published dataset of breast cancer and applied an ANN approach to identify an optimal gene expression signature for predicting the outcome of patients with breast cancer. Lancashire et al. (2010) found that only nine genes were needed to forecast metastatic spread with sensitivity of 98% by utilizing an ANN algorithm implemented especially for the optimal biomarker subgroups in gene expression data.

Moreover, Mehridehnavi and Ziaei (2013) utilized ANN models to find the most significant genes and classify patients with diffuse large B-cell lymphoma, which is a cancer of B cells, on the basis of their gene expression profiles. Diffuse large B-cell lymphoma is a form of white blood cell responsible for generating antibodies. Mehridehnavi and Ziaei (2013) used the signal-to-noise ratio as a major approach to reduce the number of genes from 4026 to 2 most significant genes. By using two most significant genes to train the ANN model, their results showed that the training and testing errors were 0% and 7%, respectively (Mehridehnavi and Ziaei 2013).

Furthermore, based on a cDNA microarray dataset, Tong et al. (2014) utilized ANN models to find the potential gene-gene interactions among previously determined biomarkers in children sarcomas, which are a rare kind of cancer arising from transformed cells of mesenchymal origin. Their analysis revealed that seven key genes including *FCGRT*, *FNDC5*, *GATA2*, *HLA-DPB1*, *MT1L*, *OLFM1*, and *TNNT1* had significant associations (Tong et al. 2014).

Finally, McDermott et al. (2014) employed ANN models and microarray profiling to pinpoint circulating microRNAs that were expressed in a differential manner among individuals with luminal A-like breast cancer in comparison to those without luminal A-like breast cancer. They found 76 microRNAs with differential expression in subjects with luminal A-like breast cancer and also identified 10 microRNAs for further analysis using ANN models (McDermott et al. 2014).

2.4.2 Chemotherapy

The use of genetic information and other biomarkers has played a major role in better predicting patients' responses to targeted therapy. For example, adjuvant chemotherapy for non-small cell lung cancer can be used after surgery to put an end to recurrence or metastases. Unfortunately, not every patient is suitable for treatment. Chen et al. (2015) aimed to construct prediction models to recognize who was suitable for adjuvant chemotherapy in subjects with non-small cell lung cancer. Their analysis showed that the best ANN model achieved 65.71% accuracy with two genes such as *DUSP6* and *LCK*.

2.4.3 Schizophrenia

Schizophrenia is a chronic and severe mental disorder that affects social behavior, beliefs, and thinking for a person (Liou et al. 2012; Lin and Tsai 2016b). Takahashi et al. (2010) used an ANN algorithm to assess whether the gene expression signature in whole blood consists of sufficient information to segregate patients with schizophrenia. They singled out 14 probes as predictors for differential diagnosis of schizophrenia with the quality filtering and stepwise forward selection methods. The ANN model was then constructed with the selected probes, and it carried out 91.2% accuracy in the training data and 87.9% accuracy in the testing data (Takahashi et al. 2010).

2.5 Perspectives

Several limitations exist with respect to the aforementioned diagnostic prediction studies. Firstly, studies with limited sample size did not warrant well-defined results (Lin and Lane 2015). Secondly, researchers often investigate all of the available algorithms because the only sure way to find the very best algorithm is to try all of them (Lin and Tsai 2016c; Lin and Lane 2017).

Besides ANN models, there are a variety of machine learning tools we can use to analyze gene expression profiling data in diagnostic prediction studies. Some of the best-known machine learning and predictive algorithms encompass naive Bayes (Domingos and Pazzani 1997), C4.5 decision tree (Quinlan 1993), ANNs (Lin et al. 2006; Kung and Hwang 1998; Bishop 1995; Rumelhart et al. 1996), support vector machine (SVM) (Vapnik 1995), k-means (Lloyd 1982), k-nearest neighbors (kNN) (Altman 1992), and regression (Friedman et al. 2010; Zou and Hastie 2005). These classifiers are usually adopted for comparison owing to the fact that these methods possess a diversity of capacities with distinctively representational models, such as probabilistic models for naive Bayes, decision tree models for the C4.5 algorithm, and regression models for SVM (Hewett and Kijsanayothin 2008).

For instance, Table 2.2 summarizes the relevant diagnostic prediction studies by using gene expression profile and a variety of machine learning models. In order to

Table 2.2 Diagnostic prediction studies of gene expression profiling for various diseases and treatments of significance using a variety of machine learning algorithms

Disease/treatment	Results	References
Breast cancer	Identified 21 most-associated genes	Chou et al. (2013)
Colorectal tumors	Achieved 99% accuracy	Chu et al. (2014)
Colon cancer	Achieved 91% accuracy	Hu et al. (2015)

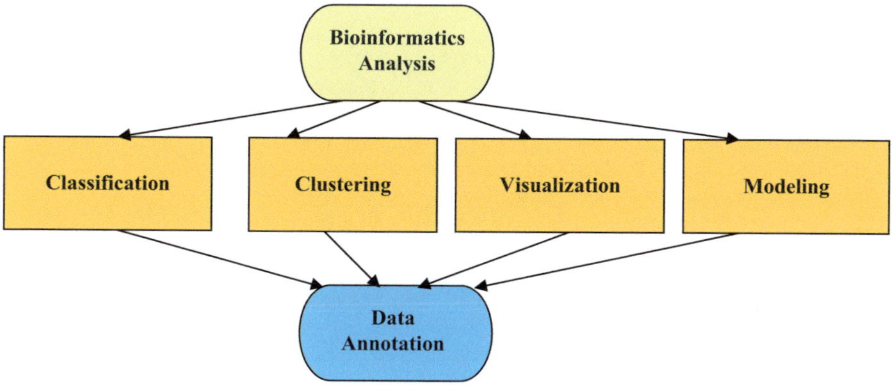

Fig. 2.2 Bioinformatics tools for analyzing and visualizing the relationship between gene expression data and human diseases

predict breast cancer recurrence, Chou et al. (2013) employed gene expression profiling of breast cancer survivability and three methods including logistic regression, decision tree, and ANN models. Their analysis indicated 21 genes closely relevant to breast cancer recurrence (Chou et al. 2013). In addition, in order to screen for the variations in gene expression between colorectal tumors and normal mucosa tissues, Chu et al. (2014) employed four methods, including ANN, prediction analysis of microarray, classification and regression trees (CART), and C5.0 algorithms. Colorectal cancer is a cancer that starts in the colon or rectum. Chu et al. (2014) adopted a two-tier genetic screen to reduce the number of candidate significant genes, and the ANN model achieved the best classification performance, with an average 99% test accuracy. Moreover, based on gene expression data, Hu et al. (2015) classified colon cancer subjects treated with elective standard oncological resection into two groups such as relapse and no relapse by using ANN, Kohonen neural network, and SVM models. The Kohonen neural network model achieved the best classification performance, with an average 91% test accuracy (Hu et al. 2015).

In future work, a bioinformatics pipeline can be used to provide a thorough evaluation and validate whether the findings are replicated in diagnostic prediction studies. Figure 2.2 shows a bioinformatics pipeline for analyzing and visualizing gene expression profiling data in diagnostic prediction studies. Additionally, we could investigate potential biomarkers by using a custom data mining pipeline so that genetic networks would be illustrated at the genome level.

Acknowledgments The authors extend their sincere thanks to Vita Genomics, Inc. and SBIR grants (S099000280249-154) from the Department of Economic Affairs in Taiwan for funding this research.

References

Altman NS (1992) An introduction to kernel and nearest-neighbor nonparametric regression. Am Stat 46:175–185

Bishop CM (1995) Neural networks for pattern recognition. Clarendon Press, Oxford

Chen YC, Chang YC, Ke WC, Chiu HW (2015) Cancer adjuvant chemotherapy strategic classification by artificial neural network with gene expression data: an example for non-small cell lung cancer. J Biomed Inform 56:1–7. https://doi.org/10.1016/j.jbi.2015.05.006

Chou HL, Yao CT, Su SL, Lee CY, Hu KY, Terng HJ, Shih YW, Chang YT, Lu YF, Chang CW, Wahlqvist ML, Wetter T, Chu CM (2013) Gene expression profiling of breast cancer survivability by pooled cDNA microarray analysis using logistic regression, artificial neural networks and decision trees. BMC Bioinform 14:100. https://doi.org/10.1186/1471-2105-14-100

Chu CM, Yao CT, Chang YT, Chou HL, Chou YC, Chen KH, Terng HJ, Huang CS, Lee CC, Su SL, Liu YC, Lin FG, Wetter T, Chang CW (2014) Gene expression profiling of colorectal tumors and normal mucosa by microarrays meta-analysis using prediction analysis of microarray, artificial neural network, classification, and regression trees. Dis Markers 2014:634123. https://doi.org/10.1155/2014/634123

Domingos P, Pazzani M (1997) On the optimality of the simple Bayesian classifier under zero-one loss. Mach Learn 29:103-137

Dwivedi Y (2014) Emerging role of microRNAs in major depressive disorder: diagnosis and therapeutic implications. Dialogues Clin Neurosci 16:43–61

Friedman J, Hastie T, Tibshirani R (2010) Regularization paths for generalized linear models via coordinate descent. J Stat Softw 33:1–22

Hewett R, Kijsanayothin P (2008) Tumor classification ranking from microarray data. BMC Genomics 9:S21. https://doi.org/10.1186/1471-2164-9-S2-S21

Hu HP, Niu ZJ, Bai YP, Tan XH (2015) Cancer classification based on gene expression using neural networks. Genet Mol Res 14:17605–17611. https://doi.org/10.4238/2015.December.21.33

Kononenko I (2001) Machine learning for medical diagnosis: history, state of the art and perspective. Artif Intell Med 23:89–109

Kung SY, Hwang JN (1998) Neural networks for intelligent multimedia processing. Proc IEEE 86:1244–1272

Lancashire LJ, Powe DG, Reis-Filho JS, Rakha E, Lemetre C, Weigelt B, Abdel-Fatah TM, Green AR, Mukta R, Blamey R, Paish EC, Rees RC, Ellis IO, Ball GR (2010) A validated gene expression profile for detecting clinical outcome in breast cancer using artificial neural networks. Breast Cancer Res Treat 120:83–93. https://doi.org/10.1007/s10549-009-0378-1

Landset S, Khoshgoftaar TM, Richter AN, Hasanin T (2015) A survey of open source tools for machine learning with big data in the Hadoop ecosystem. J Big Data 2:24

Lane HY, Tsai GE, Lin E (2012) Assessing gene-gene interactions in pharmacogenomics. Mol Diagn Ther 16:15–27. https://doi.org/10.2165/11597270-000000000-00000

Lin E (2012) Novel drug therapies and diagnostics for personalized medicine and nanomedicine in genome science, nanoscience, and molecular engineering. Pharm Regul Aff Open Access 1: e116

Lin E, Lane HY (2015) Genome-wide association studies in pharmacogenomics of antidepressants. Pharmacogenomics 16:555–566. https://doi.org/10.2217/pgs.15.5

Lin E, Lane HY (2017) Machine learning and systems genomics approaches for multi-omics data. Biomarker Res 5:2. https://doi.org/10.1186/s40364-017-0082-y

Lin E, Tsai SJ (2011) Gene-gene interactions in a context of individual variability in antipsychotic drug pharmacogenomics. Curr Pharmacogenomics Pers Med 9:323–331

Lin E, Tsai SJ (2012) Novel diagnostics R&D for public health and personalized medicine in Taiwan: current state, challenges and opportunities. Curr Pharmacogenomics Pers Med 10:239–246

Lin E, Tsai SJ (2016a) Genome-wide microarray analysis of gene expression profiling in major depression and antidepressant therapy. Prog Neuro-Psychopharmacol Biol Psychiatry 64:334–340. https://doi.org/10.1016/j.pnpbp.2015.02.008

Lin E, Tsai SJ (2016b) Genetics and suicide. In: Courtet P (ed) Understanding suicide – risk assessment, prevention, and treatment. Springer, Cham

Lin E, Tsai SJ (2016c) Machine learning and predictive algorithms for personalized medicine: from physiology to treatment. In: Turnbull A (ed) Personalized medicine. Nova Science Publishers, New York

Lin E, Hwang Y, Wang SC, Gu ZJ, Chen EY (2006) An artificial neural network approach to the drug efficacy of interferon treatments. Pharmacogenomics 7:1017–1024. https://doi.org/10.2217/14622416.7.7.1017

Liou YJ, Bai YM, Lin E, Chen JY, Chen TT, Hong CJ, Tsai SJ (2012) Gene-gene interactions of the INSIG1 and INSIG2 in metabolic syndrome in schizophrenic patients treated with atypical antipsychotics. Pharmacogenomics J 12:54–61. https://doi.org/10.1038/tpj.2010.74

Lloyd SP (1982) Least squares quantization in PCM. IEEE Trans Inf Theory (Special Issue on Quantization) IT-28:129–137

McDermott AM, Miller N, Wall D, Martyn LM, Ball G, Sweeney KJ, Kerin MJ (2014) Identification and validation of oncologic miRNA biomarkers for luminal A-like breast cancer. PLoS One 9:e87032. https://doi.org/10.1371/journal.pone.0087032

Mehridehnavi A, Ziaei L (2013) Minimal gene selection for classification and diagnosis prediction based on gene expression profile. Adv Biomed Res 2:26. https://doi.org/10.4103/2277-9175.107999

Nagano T, Fraser P (2011) No-nonsense functions for long noncoding RNAs. Cell 145:178–181. https://doi.org/10.1016/j.cell.2011.03.014

Pass HI, Liu Z, Wali A, Bueno R, Land S, Lott D, Siddiq F, Lonardo F, Carbone M, Draghici S (2004) Gene expression profiles predict survival and progression of pleural mesothelioma. Clin Cancer Res 10:849–859

Petalidis LP, Oulas A, Backlund M, Wayland MT, Liu L, Plant K, Happerfield L, Freeman TC, Poirazi P, Collins VP (2008) Improved grading and survival prediction of human astrocytic brain tumors by artificial neural network analysis of gene expression microarray data. Mol Cancer Ther 7:1013–1024. https://doi.org/10.1158/1535-7163.MCT-07-0177

Quinlan JR (1993) C4.5: programs for machine learning. Morgan Kaufmann Publishers, San Francisco

Rumelhart DE, Hinton GE, William RJ (1996) Learning internal representation by error propagation. Parallel distributed processing: explorations. In: The micro-structure of cognition, Foundations, vol 1. MIT Press, Cambridge, MA

Takahashi M, Hayashi H, Watanabe Y, Sawamura K, Fukui N, Watanabe J, Kitajima T, Yamanouchi Y, Iwata N, Mizukami K, Hori T, Shimoda K, Ujike H, Ozaki N, Iijima K, Takemura K, Aoshima H, Someya T (2010) Diagnostic classification of schizophrenia by neural network analysis of blood-based gene expression signatures. Schizophr Res 119:210–218. https://doi.org/10.1016/j.schres.2009.12.024

Tong DL, Boocock DJ, Dhondalay GK, Lemetre C, Ball GR (2014) Artificial neural network inference (ANNI): a study on gene-gene interaction for biomarkers in childhood sarcomas. PLoS One 9:e102483. https://doi.org/10.1371/journal.pone.0102483

Vapnik V (1995) The nature of statistical learning theory. Springer, New York
Wei JS, Greer BT, Westermann F, Steinberg SM, Son CG, Chen QR, Whiteford CC, Bilke S, Krasnoselsky AL, Cenacchi N, Catchpoole D, Berthold F, Schwab M, Khan J (2004) Prediction of clinical outcome using gene expression profiling and artificial neural networks for patients with neuroblastoma. Cancer Res 64:6883–6891. https://doi.org/10.1158/0008-5472.CAN-04-0695
Witten IH, Frank E (2005) Data mining: practical machine learning tools and techniques. Morgan Kaufmann Publishers, San Francisco
Zou H, Hastie T (2005) Regularization and variable selection via the elastic net. J R Stat Soc Ser B Stat Methodol 67:301–320

3 Soft Computing Approaches to Extract Biologically Significant Gene Network Modules

Swarup Roy, Hazel Nicolette Manners, Monica Jha, Pietro H. Guzzi, and Jugal K. Kalita

Abstract

A group of functionally related genes that take part in similar biological activities constitutes a functional module. Genes collaborating in a common module might induce similar pathological disease and share common genetic origins for the associated disease phenotypes. Computationally isolating such functional modules is useful in unveiling biological and cellular processes or molecular basis of associated diseases. As a result detecting such functional modules is an important and burning issue in the computational biology research.

Various techniques have been proposed for the last few decades to find functional modules or target factor modules in gene regulation networks. Biological modules are overlapping in nature where the same gene may take part in multiple network modules. In addition, data used for inference or detection of modules in silico are noisy in nature. Traditional hard computing methods appear to be ineffective in handling uncertainty, impreciseness, or fuzzy nature in the solutions. The soft computing paradigm is effective in handling such issues. In this work, we discuss a few soft computing methods for detecting regulatory modules and validate the effectiveness of the candidate methods in the light of publicly available expression data with respect to various statistical and topological validation measures.

S. Roy (✉) · H. N. Manners · M. Jha
Department of Computer Applications, Sikkim University, 6th Mile, Samdur, Tadong, Gangtok, Sikkim, India

Department of Information Technology, North-Eastern Hill University, Shillong, Meghalaya, India
e-mail: sroy01@cus.ac.in; swarup@nehu.ac.in

P. H. Guzzi
Department of Surgical and Medical Sciences, University of Catanzaro, Catanzaro, Italy

J. K. Kalita
Department of Computer Science, University of Colorado, Colorado Springs, CO, USA

H. J. Purohit et al. (eds.), *Soft Computing for Biological Systems*,
https://doi.org/10.1007/978-981-10-7455-4_3

Keywords
Gene network · Gene co-expression · Fuzzy membership · Overlapping modules · Clustering

3.1 Introduction

In all living cells, biological processes occur as a result of a complex interplay among macromolecules. Genes and their products such as proteins and noncoding nucleic acids play a prominent role in biological processes (Brown and Botstein 1999). Consequently, the study of the set of interplays (or associations) among them is an important area of research in bioinformatics and systems biology. The exploration of such interactions is based on three main pillars: (1) a set of technologies for data production, (2) a set of methodologies for data representation and storage, and (3) a set of algorithms and models for knowledge extraction from data and subsequent dissemination. The production of raw data is often performed using microarray technologies (Heller. 2002) and more recently using next-generation sequencing (NGS) techniques, such as RNA-Seq (Tu et al. 2005). Such technologies enable the simultaneous investigation of the level of activity of genes and noncoding fragments of the genome. Experiments may compare two different conditions, e.g., healthy vs diseased cells, or a sequence times for the same sample to perform a time series analysis. The result of such experiments is often organized into a data matrix in which each element represents the value of the expression of the ith gene in the jth condition or time. Therefore, data may be analyzed using the clustering-based techniques to infer patterns of co-expression among genes. Genes are co-expressed if they show a similar pattern of behavior under different experimental conditions (i.e., different drug treatments or different time points). The pattern of co-expressions is usually related to similar behaviors; thus, the analysis of such patterns may reveal relations among genes that have a similar role.

Clustering of co-expressed genes is not sufficient to model interplay among genes. Therefore, researchers have introduced formalisms based on graph theory to represent such complex relationships. In co-expression graphs, nodes symbolize genes, whereas edges (directed or undirected) are the symbolic representation of biological association among the genes in terms of co-expression (Pandey et al. 2010) or regulation (i.e., positive or negative control) (Das 2009) (Fig. 3.1).

Many biological problems can be described as graph problems. For instance, an important problem is the individuation of regulatory modules (Ravasz et al. 2002), i.e., sets of genes that act collectively in a genome to perform a distinct biological function. Regulatory modules are co-expressed, co-evolved, and regulated by the same set of transcription factors to respond to different conditions. *Transcription factors* (TFs) are proteins involved in the process of converting, or transcribing, DNA into RNA. Some genes even play multiple roles and become members of more than one module (Kohonen 1993, Zhang et al. 2010). Identifying regulatory modules is vital to understand cellular activities in response to various external or internal stimuli. In turn, it may help to uncover the disease mechanisms in a living organism.

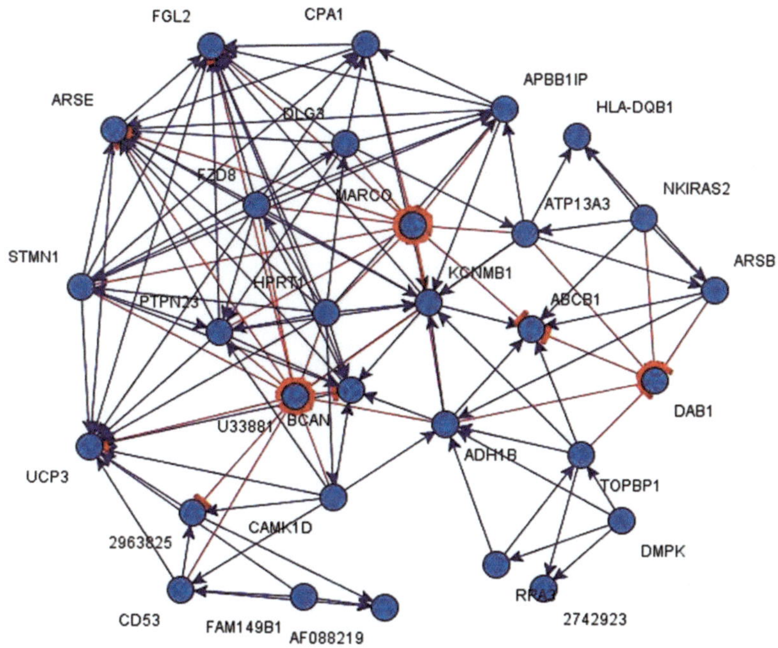

Fig. 3.1 *Homo sapiens* gene regulatory network reconstructed from the GDS825 (NCBI). Labeled nodes are human genes. Directed arrow represents positive regulation, and a nondirected edge (red) represents a negative relationship between the genes. The arrow points from the regulator and their target

Computationally, module finding techniques aim to cluster biological components that have similar functions. *Module* is a biological term for a cluster. Research in this area has been going on for more than a decade. Hence, many approaches have been proposed to identify such modules. Below, we present a brief sketch of available methods. We emphasize mainly on soft computing methods that have inherent ability to handle vagueness and uncertainty. Finally, we validate the merit of the results produced by various candidate methods using several publicly available expression datasets.

3.2 Computational Methods for Detecting Network Modules

Finding network modules from gene expression data basically involves two major independent phases: network construction and application of suitable module detection methods to find compact quasi-clique-like structures in the constructed networks. The problem of network module finding can be framed as follows.

Definition 1 (Network Module) *Given a gene expression data matrix D of order $N \times M$ where N is the number of genes and M is the number of samples or conditions, a module is a group of genes sharing strong correlation or association within a gene regulatory network derived from D.*

A biological module is described as a collection of genes or their products which might be related by means of one or greater genetic or cellular interactions; however, characteristics are separable from those of other modules. These interactions may be co-regulation, co-expression, or membership of a protein complex or of a metabolic or signaling pathway. Genes which are regulated by the same regulators tend to co-express or co-regulate and hence are grouped in the same module or subnetwork. Modules can be understood as separate substructures in a network. Some genes may belong to more than one module and have a fuzzy nature as they are involved in more than one function in the organism forming overlapping modules (Fig. 3.3).

The simplest representation of gene regulatory networks uses an undirected graph (Roy et al. 2014), while more refined models use directed and weighted edges to combine facts about the sort of biochemical association and its course. An undirected graph represents co-expression over a sequence of gene expression measurements. These are often referred to as gene association networks or gene co-expression networks (Fig. 3.2). The directed edges in GRNs correspond to causal influences between a pair of genes. Causal influences include regulation of transcription through transcription elements (Fig. 3.1).

A majority of the detection methods use classical clustering techniques and their variations (Mahanta et al. 2012). Some are based on hierarchical clustering (Immermann and Huang, 2003), k-means clustering, and self-organizing maps (Jobson 2012). A few algorithms based on graph theory or network techniques to

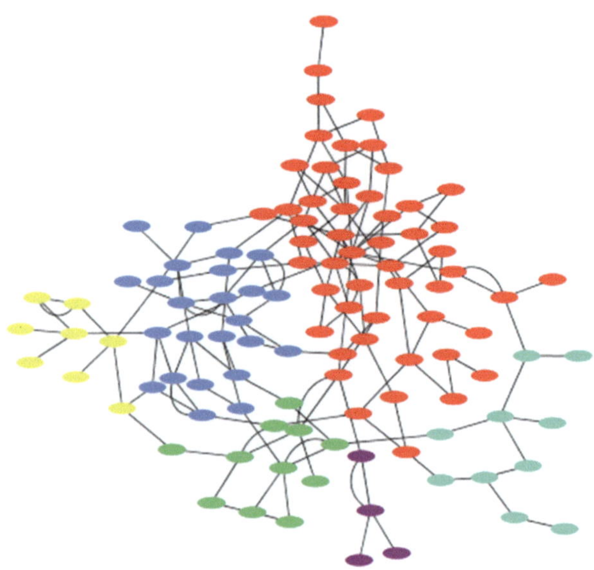

Fig. 3.2 A co-expression network without any directed edges. Nodes with the same color constitute a functionally similar group of genes having high interconnectivity among them. A co-expression network is usually silent about transcription factor (TF) target relationships

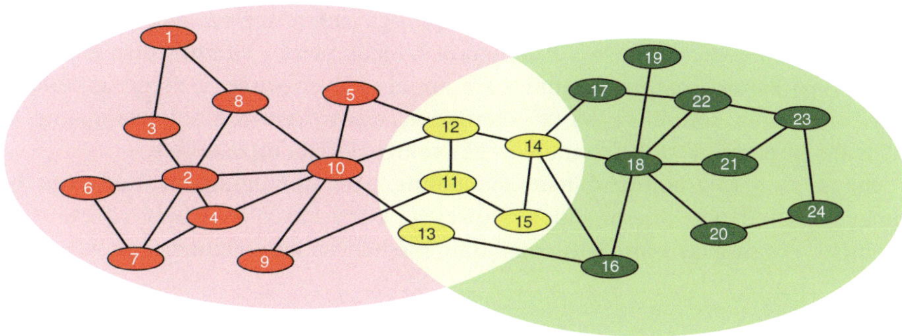

Fig. 3.3 A diagrammatic representation of network modules extracted from a co-expression network. Two distinct sets of genes grouped into two different modules are shown in red and green colors. Yellow color genes or nodes are the overlapping members participating in two other modules. Biologically such genes play both functions along with the other members of the two modules

form modules have been proposed such as CLICK or based on minimum spanning trees (MST) (Chanthaphan et al. 2009, Manners et al. 2016). However, module detection methods integrating biological information such as gene ontology (GO) along with gene expression data tend to perform better than traditional clustering approaches (Sharan et al. 2003). Recent advances in biological research revealed that some genes or proteins play multiple functional roles in a cell depending on experimental conditions. For example, it has been observed that yeast gene CMR1/YDL156W participates in many of the DNA metabolism processes such as replication, repair, and transcription (Newman and Cooper 2010). Out of 1628 proteins in the hand-curated yeast complex dataset, 207 proteins are participating in more than one complex (Huang et al. 2009). The genes responsible for such proteins participate in different functional modules or complexes. They exhibit distinct overlapping structures. Conventional clustering techniques and variants are not suitable for detecting overlapping modules. A good module detection algorithm should also be capable of extracting modules in presence of background noise using as little prior knowledge (Manners et al. 2016).

The soft computing paradigm is an effective alternative for handling uncertainty or vagueness in module detection. Soft computing methods may be explored to achieve effective outcomes.

3.3 Soft Computing Methods for Network Module Extraction

The soft computing paradigm uses a mix of various methods derived from mathematics, machine learning, and meta-heuristic optimization. It mainly comprises of fuzzy logic, rough set theory, artificial neural networks, and evolutionary approaches. Hybridization is also used by combining merits of the different methods in a mutually compatible way. Soft computing approaches are effective in handling computational problem involving impreciseness, uncertainty, and vagueness. Biological data are usually noisy, incomplete, and imprecise in nature. Traditional hard computing methods are not always effective in deriving biologically significant solutions. As an alternative, researchers use soft computing to generate empirical results from the large biological data repositories.

In addition to traditional hard computing, soft computing methods can be applied to find network modules. However, there have been only a few such attempts. We discuss some soft computing module finding methods below.

3.3.1 Weighted Gene Co-expression Network Analysis (WGCNA)

WGCNA reconstructs co-expression networks using correlation as a measure of similarity between a pair of genes. Unlike hard thresholding which hypothesizes that an edge exists between a pair of genes if the similarity score is above a certain threshold, it creates a weighted network using correlation coefficient as the weight of the edge, also called soft thresholding. It then uses hierarchical clustering (Huang et al. 2009) with a dynamic tree cut method to get the clusters or modules. Traditional clustering creates clusters that are disjoint or exclusive and assigns each gene exclusively to one particular module. WGCNA uses a fuzzy module membership function to allow a single gene to participate in multiple modules. In order to extract biologically significant modules, WGCNA uses a gene significance function that assigns a nonnegative number to each gene. Higher the score of the function, the more biologically relevant the genes are.

3.3.2 Fuzzy Network Module Extraction

An approach known as the fuzzy network module extraction approach for gene expression records (FUMET) is proposed in Mahanta et al. (2014). This technique uses what is called the NMRS similarity measure. Construction of the co-expression network is performed using soft thresholding, where the edges carry a weight equivalent to their NMRS similarity measure, creating a weighted co-expression network. The genes may belong to more than one module and hence the concept of fuzziness is useful. It uses topological overlap measure (TOM) to calculate the overlap score between the genes. FUMET accepts a co-expression network G, a membership threshold, and the number of modules as inputs, and it extracts highly correlated network modules. The initial modules are formed with one of the gene pairs in each of these modules. Then, for each module, the memberships of all the genes which are not in the module are checked with the following membership function. The membership value of gene d_i for class C_i is computed as follows.

$$fm(C_i, d_j) = \frac{\sum_{d_j \varepsilon C_i} Adj_{i,j}}{\min(|C_i|, degree(d_i))} \tag{3.1}$$

where, degree(d_i) is the number of nodes connected to d_j in the co-expression network and $Adj_{i,j}$ is the weight of the edge corresponding to genes d_i and d_j. For a gene, if the membership function produces a value greater than the membership threshold for a module, the gene is included in the class.

All the above methods are general in nature and have been shown to work with any gene expression data. There are some efforts that have used to extract cancer regulatory modules from the data. We discuss two of them below.

3.3.3 GA-RNN Hybrid Approach

Chiang and Chao (Chiang and Chao, 2007) use a genetic algorithm (GA) and a recurrent neural network (RNN) to construct cancer regulatory modules. Both microarray and sequence data of TF (transcription factor) binding sites are used to train the RNN model. They can capture the architecture of real-world gene network modules where the networks have feedback loops from the target genes to TFs. They discover novel feed-forward relations in a regulatory network using modified multilayer RNN architectures. They extract regulatory modules containing known TF genes and their target regulated genes.

3.3.4 Multisource Integrative Framework

Wang et al. (2009) detect regulatory modules from the cancer cells by integrating multiple sources of data combining expression profiles, gene ontology (GO), protein-protein interaction, and protein-DNA interaction data. Gene expression patterns are first grouped into biologically meaningful groups using fuzzy c-mean clustering. The optimal numbers of clusters are determined using information gleaned from GO categories of genes. Network motifs are then assigned to every TF, detected from PPI and PDI data. Next, the connections among TF and gene clusters are inferred by RNN.

Below, we discuss a method which is able to detect overlapping modules from expression data using a minimum spanning tree. Interestingly, it does not use any soft computing method to detect fuzzy clusters.

3.3.5 AutoSOME

AutoSOME detects fuzzy clusters or overlapping gene modules by combining self-organizing maps (SOM) with ideas from graph theory. They use a three-step process to extract the modules. Initially, randomly selected SOM lattice nodes

Table 3.1 A summary of some network-based module finding methods

Sl no.	Algorithm	Type of network	Soft computing approaches	Implementation source	References
1.	WGCNA	Undirected	Fuzzy	R package: WGCNA	Newman and Cooper (2010)
2.	FUMET	Undirected	Fuzzy	–	Liu et al. (2015)
3.	Chang and Chao	Directed	Genetic algorithm, recurrent neural network	–	Chiang and Chao (2007)
4.	Zhang et al.	Directed	Fuzzy c-means, GA, recurrent neural network	–	Warde-Farley et al. (2010)
5.	AutoSOME	Undirected	MST, SOM	http://jimcooperlab.mcdb.ucsb.edu/some	Langfelder and Horvath (2008)

are projected onto the planar SOM surface. Error surface is computed after training. A density equalization method is applied to treat nodes with high errors as high density. On the other hand, it forces the nodes far away from each other that are low errors (within clusters) having low density. Subsequently, rescaled node coordinates are used to construct a minimum spanning tree (MST). The MST graph connects all nodes via edges with minimal total distance and without any loops. p-Values of all the edges in the MST are computing using Monte Carlo sampling. Finally, a fuzzy clustering matrix is constructed where a data item can be a part of more of more than one fractional membership. AutoSOME identifies more than 3400 upregulated genes with pluripotency.

A brief summarization of methods discussed above is presented in Table 3.1.

3.4 Assessment

To measure effectiveness, we use three different module finding techniques, AutoSOME, WGCNA, and FUMET. We use the original implementation of the first two methods and implemented FUMET in Java. Due to unavailability of codes for the other two methods used in cancer TF network extraction, we do not consider them for our experimental evaluation.

3.4.1 Dataset

We use five different expression datasets for evaluating the module finding methods. Fibroblast serum dataset from *Homo sapiens* is one such dataset. We also use subset of a *Rattus norvegicus* (also known as rat) dataset related to the

Table 3.2 Brief description of expression datasets used for assessment

Sl no.	Dataset	No. of genes	No. of time points/conditions	Source
1.	Subset of yeast cell cycle	387	17	http://faculty.washington.edu/kayee/cluster
2.	Yeast sporulation	474	17	http://cmgm.stanford.edu/pbrown/sporulation/index.html
3.	Rat CNS	112	9	http://faculty.washington.edu/kayee/cluster
4.	Human fibroblast serum	517	13	http://www.sciencemag.org/feature/data/984559.shl
5.	Subset of *Mus musculus*	693	12	ncbi.nlm.nih.gov/gds-GDS958

central nervous system (CNS). In addition, a subset of yeast cell cycle, yeast sporulation, and house mouse (*Mus musculus*) expression profiles are also used for experimentation. Table 3.2 contains description of the datasets, used for our assessment.

3.4.2 Validation

The biological networks are divided into modules by various methods where genes belonging to the same module are likely to perform similar tasks and share common properties. However, some genes that are grouped as modules may be noise and may contribute to the instability of the module. These may arise due to tuning of parameters when the network is created. Hence, such spurious genes should be detected as much as possible before validating the correctness of the module structures based on known data or biological networks. We validate the significance of the discovered modules and validate their architectural characteristics based on topological analysis score in addition to looking at functional enrichment and biological validation. Quite a few validation tools are available for evaluating the biological merits of the modules inferred by different computational methods. A few of them are listed in Table 3.3.

3.4.2.1 Functional Enrichment Analysis

For functional enrichment analysis of the extracted modules by different techniques, we use DAVID and report the statistical significance values (p-values) for the gene ontology terms linked with the genes or group of genes in a module (28). p-Values test the statistical hypothesis of the observed genes belonging to a module and measure false-positive rates. Before the test is carried out, a threshold value is selected, known as the significance level of the test, traditionally 5% or 1%. A low p-value indicates that the genes are biologically

Table 3.3 Freely available gene module assessment and analysis tools

Tool	Purpose	Source	Platform	References
DAVID	Gene enrichment, KEGG pathway map, etc.	https://david.ncifcrf.gov/	Web (online)	–
Genemania	Functional enrichment analysis	http://genemania.org/	Web (online)	Berriz et al. (2003)
FuncAssociate 3.0	Functional enrichment analysis	http://llama.mshri.on.ca/funcassociate	Web (online)	Eden et al. (2009)
Clean	Functional enrichment analysis	gorilla.cs.technion.ac.il/	Web (online)	Freudenberg et al. (2009)
TopoGSA	Gene module enrichment analysis	http://Clusteranalysis.org	R (offline)	Glaab et al. (2010)

significant and belong to enriched functional categories. For the gene ontology (GO) of the modules, the p-value is calculated as

$$p = \sum\nolimits_{i=x}^{n} \frac{\binom{A}{i}\binom{N-A}{n-i}}{\binom{N}{n}} \tag{3.2}$$

where n is the total quantity of genes in the module and A is the wide variety of genes with a specific annotation. The p-value is described as the probability of observing at the least x genes in the annotation of a module with n genes. For the gene ontology (GO) of the modules, we measure p-values using the Web application, namely, FuncAssociate 3.0 (Berriz et al. 2003).

3.4.2.2 Topological Validation

We also perform various topological analyses of the subnetworks for the individual modules. Topological validation accesses the topology or structure of the extracted modules based on known cellular pathways and processes. We use a Web application called TopoGSA (Glaab et al. 2010) to compute topological properties for the whole network. The network topologies produced by TopoGSA are as follows.

Degree of a Node The degree of a node is the average wide variety of interactions incident on a particular node or gene. A high number interaction a gene has with other genes in a module usually signifies that it serves as a central node in the module or network. These genes play important roles in the functional composition of the module. A high degree is desired, meaning that a node has a lot of interactions. A node with no interaction has zero degree.

Local Clustering Coefficient Local clustering coefficient evaluates the likelihood that the neighbors of a hub are associated. If this probability is high, it means that the neighboring genes interact with each other forming a strong network module. The value of this coefficient can range from 0 to 1, with 1 being the most desired value.

Shortest Path Length Shortest path length (SPL) is the average for all the minimum number of edges needed to traverse between two nodes of the entire network. The lower this length, the more strongly two nodes or genes interact with each other. The lower the SPL value, the lesser the distance to other nodes and hence the stronger is the interaction.

Node Betweenness Hub betweenness can be registered from the quantity of the briefest ways between two hubs or qualities "a" and "b," experiencing hub "c." A high node betweenness value signifies that the node acts as a bridge that helps the other parts of the module or network to interact with each other. Such a node acts as a central node. The higher the value of node betweenness, the more important is the node.

Eigenvector Centrality The eigenvector centrality rating of a node relies upon on the rating of its neighbors. A node or gene gets a higher rank when it is connected to nodes that are important or have high degree and their functions contribute highly to the network module it is a part of. A score of 1 is the best value when the eigenvector is normalized.

3.4.2.3 Experimental Results

Some of the GO annotations with the lowest p-values exhibited by the methods are selected at random and compared for the five different datasets given in Table 3.2. We present the average enrichment scores of all the modules extracted by each module in Fig. 3.4. Functional enrichment analysis reveals that AutoSOME obtains lower average p-values compared to WGCNA and FUMET for all datasets except dataset 3. FUMET performs better than the other two for dataset 3.

Topological validations from TopoGSA are shown in Table 3.4. Here, the term uploaded gene set indicates the values obtained from the uploaded network, while the row labeled 100 random simulations (mean) shows the range of values that are correct that were retrieved from known pathways or networks that incorporate the list of genes uploaded. From the results, it is evident that FUMET outperforms all the other methods in deriving biologically true subnetworks participating in a network module.

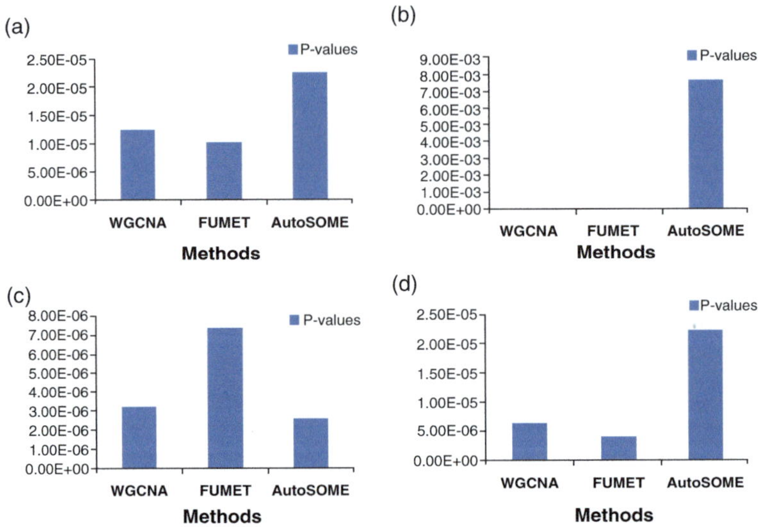

Fig. 3.4 Average p-values for four different datasets (**a–d**)

3.5 Conclusion and Future Scope

Many approaches have been developed to detect modules from gene co-expression networks. Some methods use approaches based on graph theory to partition the graph into subgraphs that are highly interconnected. Other methods require biological information to produce more meaningful modules or use more than one graph clustering algorithm in order to refine the modules. Biological modules are overlapping in nature. A subset of gene may participate in multiple modules and perform different biological functions. Traditional methods based on concepts of hard computing are not effective in detecting overlapping network modules. Conventional clustering techniques and their variants are not suitable for nonexclusive clustering that can detect overlapping structures. One of the important features of clustering algorithm should also be able to detect clusters in the presence of background noise. Module finding algorithm should be able to handle highly connected and highly intersecting structures or even embedded structures that are known to occur in most gene expression data.

Soft computing techniques such as fuzzy learning, rough set, and various optimization techniques are potential alternatives that can handle the issues raised above. However, very little work has been done to obtain biologically relevant solutions to the network modules. Any future endeavor in applying soft computing techniques in module extraction will be of great interest to the system biologists, working toward unveiling cellular mechanism for biological activities including diseases.

Table 3.4 Topological validations of the subnetwork for *yeast* cell cycle

Algorithm		Shortest path length	Node betweenness	Degree	Clustering coefficient	Eigenvector centrality
WGCNA	*Uploaded gene set*	3.336	10343.8	16.994	0.11	0.028
	100 random simulations (mean)	3.514 (0.06)	6961 (4537.4)	11.684 (2.934)	0.022 (0.022)	0.022 (0.012)
FUMET	*Uploaded gene set*	3.45	7236.66	14.31	0.09	0.013
	100 random simulations (mean)	3.5 (0.07)	4826.66 (2461.33)	10.84 (3.46)	0.1 (0.03)	0.02 (0.013)
AutoSOME	*Uploaded gene set*	3.6551	3923.5	11.25	0.08	0.12
	100 random simulations (mean)	3.53 (0.3)	7843.5 (11979)	13.4 (17.985)	0.22 (0.24)	0.02 (0.02)

Acknowledgment SR thanks the Department of Biotechnology, Government of India, for giving him the Overseas Research Associateship (vide sanction BT/NE/20/2011) and the University of Colorado, Colorado Springs, for the support to conduct part of the research at LINC Lab in the Department of Computer Science.

References

Berriz GF, King OD, Bryant B, Sander C, Roth FP (2003) Characterizing gene sets with FuncAssociate. Bioinformatics 19:2502–2504. https://doi.org/10.1093/bioinformatics/btg363

Brown PO, Botstein D (1999) Exploring the new world of the genome with DNA microarrays. Nat Genet 21:33–37

Chanthaphan A, Prom-on S, Meechai A, Chan J (2009) Identifying functional modules using MST-based weighted gene coexpression networks. BIBE'09. Ninth IEEE international conference, pp 192–199. https://doi.org/10.1109/BIBE.2009.35

Chiang JH, Chao SY (2007) Modeling human cancer-related regulatory modules by GA-RNN hybrid algorithms. BMC Bioinf 8:1. https://doi.org/10.1186/1471.-2105-8-91

Das S (2009) Handbook of research on computational methodologies in gene regulatory networks. IGI Global

Eden E, Navon R, Steinfeld I, Lipson D, Yakhini Z (2009) GOrilla: a tool for discovery and visualization of enriched GO terms in ranked gene lists. BMC Bioinf 10(1). https://doi.org/10.1186/1471-2105-10-48

Freudenberg JM, Joshi VK, Hu Z, Medvedovic M (2009) Clean: clustering enrichment analysis. BMC Bioinf 10(1):234

Glaab E, Baudot A, Krasnogor N, Valencia A (2010) TopoGSA: network topological gene set analysis. Bioinformatics 26:1271–1272. https://doi.org/10.1093/bioinformatics/btq131

Heller MJ (2002) DNA microarray technology: devices, systems, and applications. 4:129–153 doi: https://doi.org/10.1146/annurev.bioeng.4.020702.153438

Huang DW, Sherman BT, Lempicki RA (2009) Systematic and integrative analysis of large gene lists using david bioinformatics resources. Nat Protoc 4:44–57. https://doi.org/10.1038/nprot.2008.211

Immermann F, Huang Y (2003) An introduction to cluster analysis. An introduction to Toxicogenomics. In: Burczynski ME (ed), vol 200, CRC Press, Boca Raton, pp 45–78

Jobson J (2012) Applied multivariate data analysis: volume II: categorical and multivariate methods. Springer, New York. https://doi.org/10.1007/978-1-14612-0921-8

Kohonen T (1993) Physiological interpretation of the self-organizing map algorithm. Neural Netw 6:895–905. https://doi.org/10.1016/S0893-6080(09)80001-4

Langfelder P, Horvath S (2008) WGCNA: an R package for weighted correlation network analysis. BMC Bioinf 9:1. https://doi.org/10.1186/1471-2105-9-559

Liu R, Cheng Y, Yu J, Lv QL, Zhou HH (2015) Identification and validation of gene module associated with lung cancer through coexpression network analysis. Gene 563(1):56–62

Mahanta P, Ahmed HA, Bhattacharyya DK, Kalita JK (2012) An effective method for network module extraction from microarray data. BMC Bioinf 13:1. https://doi.org/10.1186/1471-2105-13-S13-S4

Mahanta P, Ahmed HA, Bhattacharyya DK, Ghosh A (2014) FUMET: a fuzzy network module extraction technique for gene expression data. J Biosci 39:351–364. https://doi.org/10.1007/s12038-014-9423-2

Manners HN, Jha M, Guzzi PH, Veltri P, Roy S (2016) Computational methods for detecting functional modules from gene regulatory network. In: Information and Communication Technology for Competitive Strategies, ICTCS Proccedings of second international conference on, 3. ACM. doi:https://doi.org/10.1145/2905055.2905209

Newman AM, Cooper JB (2010) Autosome: a clustering method for identifying gene expression modules without prior knowledge of cluster number. BMC Bioinf 11:1. https://doi.org/10.1186/1471-2105-11-117

Pandey G, Zhang B, Chang AN, Myers CL, Zhu J, Kumar V, Schadt EE (2010) An integrative multi-network and multi-classifier approach to predict genetic interactions. PLoS Comput Biol 6:e1000928. https://doi.org/10.1371/journal.pcbi.1000928

Ravasz E, Somera AL, Mongru DA, Oltvai ZN, Barabási AL (2002) Hierarchical organization of modularity in metabolic networks. Science 297:1551–1555. https://doi.org/10.1126/science.1073374

Roy S, Bhattacharyya DK, Kalita JK (2014) Reconstruction of gene co-expression network from microarray data using local expression patterns. BMC Bioinf 15:1. https://doi.org/10.1186/1471-2105-15-S7-S10

Sharan R, Maron-Katz A, Shamir R (2003) Click and expander: a system for clustering and visualizing gene expression data. Bioinformatics 19:1787–1799. https://doi.org/10.1093/bioinformatics/btg232

Tu K, Yu H, Zhu M (2005) Mego: gene functional module expression based on gene ontology. BioTechniques 38:277–283. https://doi.org/10.2144/05382RR04

Wang Z, Gerstein M, Snyder M (2009) RNA-Seq: a revolutionary tool for transcriptomics. Nat Rev Genet 10:57–63. https://doi.org/10.1038/nrg2484

Warde-Farley D, Donaldson SL, Comes O, Zuberi K, Badrawi R, Chao P, Franz M, Grouios C, Kazi F, Lopes CT, Maitland A (2010) The genemania prediction server: biological network integration for gene prioritization and predicting gene function. Nucleic Acids Res 38:214–220. https://doi.org/10.1093/nar/gkq537

Zhang Y, Xuan J, Benildo G, Clarke R, Ressom HW (2010) Reconstruction of gene regulatory modules in cancer cell cycle by multi-source data integration. PLoS One 5:e10268. https://doi.org/10.1371/journal.pone.0010268

A Hybridization of Artificial Bee Colony with Swarming Approach of Bacterial Foraging Optimization for Multiple Sequence Alignment

4

R. Ranjani Rani and D. Ramyachitra

Abstract

A key and a primary phase for performing additional successive responsibilities in bioinformatics such as critical residue identification, conserved motif findings, phylogenetic analysis, predicting the secondary structure of protein, protein function prediction, the classification of proteins, and much more can be done by aligning multiple protein sequences termed as multiple sequence alignment (MSA). As a result of this, MSA expands as a vast active research field in bioinformatics. Generally, MSA is the process of aligning three or more sequences of DNA/nucleotides simultaneously. Nowadays the number of sequences in databases is increasing expeditiously. Due to this, many methods aroused drastically to implement the MSA problem within few years. Miserably it is an NP-complete problem, still the biologically perfect alignment methods are difficult to find. The objective of this chapter is to shed light on various recent techniques used to solve MSA. And also, a hybridization of artificial bee colony with swarming approach of bacterial foraging optimization (ABC-BFO) algorithm has been proposed to solve the MSA problem with better objective values, which leads to identify the biological features.

Keywords

Multiple sequence alignment · Optimization techniques · Multi objective optimization · Similarity · Gap penalty · BAliBase

R. R. Rani · D. Ramyachitra (✉)
Department of Computer Science, Bharathiar University, Coimbatore, Tamil Nadu, India
e-mail: ranjaniRSR91@gmail.com; jaichitra1@yahoo.co.in

H. J. Purohit et al. (eds.), *Soft Computing for Biological Systems*,
https://doi.org/10.1007/978-981-10-7455-4_4

4.1 Introduction

One among the significant approaches employed in computational biology and bioinformatics is the alignment of biological sequences. Many tasks can be done using the sequence alignment as a primary step. They aim to discover the structure, function, and biological composition of organisms by using mathematics and computer science.

The alignment of more than two DNA, RNA, or protein sequences of related dimension is termed as multiple sequence alignment. It is mainly focused on correlating the similarity among sequences and also recognizes the related homologous residues from the database (Purohit et al. 2003). Frequently it has been employed to assess the 2D and 3D structure of protein, the prediction of functional site, homology of sequence, phylogenetic analyses, protein motif and domain similarity, polymerase chain reaction, etc. Often the unexpected patterns of DNA or protein sequences are suspected due to their biological relevance. Repeating patterns of any such sequences can be inspected in large databases (Raje et al. 2006). It is a problematic computational effort to evolve a precise multiple sequence alignment for diverse protein sequences (Xu and Lei 2010). To solve the MSA problem initially, the exact method known as dynamic programming is an effective solution that can be resolved by separating a problem into overlapping subproblems to detect an optimal alignment by comparing biological sequences. An instance of the dynamic programming method is Needleman–Wunsch algorithm which was utilized to align a pair of sequences. The drawback of this is to align only a small number of amino acids (Feng et al. 1984). Nowadays, the MSA problem grows considerably with their dimensions and quantity of sequences. The progressive and iterative methods were introduced to overcome the drawbacks of the dynamic programming method to solve the problem of aligning multiple sequences.

Although the progressive methods are effective and powerful, they do not assure the optimum alignment because sequences are joined in an irregular direction in the guide tree and cannot be altered (Phillips 2006; Thompson et al. 1999). Thus an iterative method was established (Gotoh 1996) to bury the restrictions of the progressive method to align, and later the preliminary alignment of the progressive assembly of MSA is employed first and the iterative enrichment is accomplished. Until the method does not take any additional improvement of alignments, this process gets repeated. This chapter used the iterative method with stochastic technique for aligning multiple sequences. The remaining sections of this chapter are ordered as follows: Sect. 4.2 illustrates the numerous associated tasks for solving MSA. Section 4.3 describes approaches of MSA; multi-objective-based optimization and steps for the projected hybrid ABC-BFO method for MSA are also explained. Section 4.4 investigates the experiments employed on diverse datasets and compared the outcomes with existing approaches that results in multiple sequence alignment. Section 4.5 displays the implementation and discussion of MSA results. Finally, Sect. 4.6 discusses the conclusion and suggests for an upcoming enrichment.

4.2 Literature Review

Most of the aligners of protein sequence utilize iterative and progressive approaches for alignment. Some of them are ClustalW – graphical user interface, rendition of ClustalW is the Clustal-X, Match-Box, DIALIGN, T-Coffee, MUSCLE, M-Coffee, and Clustal Omega. And the modern formation of MSA integrates the constraint-based approach into progressive methods especially COBALT (Notredame 2002).

Initially, the alignment based on consistency approach will bind the information that is restricted within the sections which constantly aligned between pairwise alignments, and then it built an alignment of multiple sequences (Ebert and Brutlag 2006). Some common examples are MAFFT, ProbCons, Kalign, and Probalign (Notredame 2002). Align-m is used to align conflicting sequences and yield the outcomes in optimal alignment that integrates a nonprogressive local alignment method (Walle et al. 2004).

A probabilistic method of aligning RNA, DNA, and protein sequences is accomplished using PRANK (Loytynoja and Goldman 2005), and it is projected for aligning similar sequences. The GUI form of PRANK method is PRANKSTER. To obtain numerous diverse alignment results of the same set of residues, the consensus approach of alignment is accomplished to discover the optimal solution like M-Coffee and MergeAlign (Collingridge and Kelly 2012). ClonAlign employed a new iterative method (Layeb and Deneche 2007), while PicXAA (Sahraeian and Yoon 2010) employed both consistency- and iterative-based sequence alignment.

The profile Hidden Markov model (HMM) can incorporate the probabilistic pattern of sequence profiles. Thus, rather than using typical profiles in iterative and progressive sequence alignments, the HMM can be employed. Some of the examples are MUMMALS, FSA, and MSAProbs (Notredame 2002). HAlign (Zou et al. 2015) is a fast sequence alignment method based on center star strategy for aligning RNA and DNA sequences. Two software tools are developed which engaged trie trees method to speed up the center star MSA approach. Parallelism was implemented using Hadoop platform to tackle the larger datasets which in result turns to reduce the running time complexity.

Recently, the magnitude and dimension of multiple sequence alignment problem have enlarged in enormous volume. A remarkable proficient approach termed as stochastic optimization is used to handle the above restriction of MSA. The outstanding methods of stochastic optimization are simulated annealing, genetic algorithm, Gibbs sampling, and other evolutionary algorithms that are employed for the purpose of solving the problem of MSA (Bucak and Uslan 2011). Various optimization algorithm concepts for aligning multiple sequences are discussed below.

4.2.1 Genetic Algorithm (GA)

The genetic algorithm is an optimization approach which gets inspired by the biological systems through evolution. Chromosomes is a set of string comprises of a set of elements called genes which is the solution for the given problem which holds the values for optimization variables. SAGA defined as sequence alignment by genetic algorithm describes the approach for aligning multiple sequences using genetic algorithm (Notredame and Higgins 1996). It utilizes 22 various parameters for mutating or joining the alignment among the generations and is controlled by an automatic scheduling system. It attains an optimal solution and performs better than other methods like MSA and ClustalW. Also, it compared the results with known reference sequences of tertiary structure. The method portrayed in SAGA has led to noticeable attention toward the evolutionary computation of MSA.

Recently the more complex biological functions such as non-gaps percentage, structural information, and conserved blocks findings are included as objectives for MSA (Ortuno et al. 2013). The non-dominated sorting genetic algorithm is used along with multi-objective algorithm, which outperforms other approaches such as SAGA, ClustalW, MULTIALIGN, DIALIGN, etc. An innovative genetic algorithm using multi-objective has been used for aligning multiple sequences (Kaya et al. 2016). Within a single iteration, a huge amount of trade-off alignments has been achieved with contrary objective functions such as increasing the similarity value and decreasing the alignment length. Many of the MSA solutions used by genetic algorithm have intricate problem specifications and time-exhausting mutation operators. Thus (Narimani et al. 2013) a novel technique of initialization of population, straightforward mutation operator, and recombination operators are used to align the multiple sequences and proved to achieve good accuracy and also less computational complexity when compared to other existing methods. All the methods using genetic algorithm are refined easily, but the drawing of genetic operators and selection of parameter constraints is intricate which affects the solution.

4.2.2 Particle Swarm Optimization (PSO)

The PSO algorithm was developed by Kennedy and Eberhart which is inspired by the behavior of flock of birds migrating to any unknown destination. It is very famous for its fast convergence of optimal solution and simple implementation with few parameter selections. To undertake the MSA problem, a hybridization of PSO algorithm is used to train Hidden Markov model to align the protein sequences which yields better alignment results (Rasmussen and Krink 2003). Then a new concept of chaotic optimization method is adopted for MSA using PSO algorithm. The chaotic variables take the values between 0 and 1 where various particles are dispersed consistently in the solution space. Also, the chaotic exploration and the variety of population have been improved by using logistic mapping function (Lei et al. 2009).

A novel variant of PSO algorithm called random drift PSO is used along with Hidden Markov model to train and solve the MSA problem (Sun et al. 2014). The overall search capacity of the algorithm and also the performance of algorithm have been increased by including the diversity control method and diversity-guided search. A new fragmented protein sequence alignment is implemented by using two-layer particle swarm optimization technique. This method splits the longer datasets into number of fragments, and two-layer PSO algorithm is implemented to each fragment which increased the diversity of particles and can deal with unconstrained optimization problems (Moustafa et al. 2016).

4.2.3 Artificial Bee Colony (ABC)

The MSA problem is solved by using ABC algorithm which is inspired by an intellectual foraging behavior of honey bees (Lei et al. 2010). Still when using few evolutionary algorithms for MSA, it finds get trapped by local optima. The precision of MSA is measured by using the sum-of-pairs, over other techniques by determining the food source to the neighborhood. An improved method which integrates the artificial bee colony and simulated annealing (ABC-SA) is projected to avoid the local optimal sliding by discovering Metropolis acceptance eligibility into the process of searching food. This made the method to direct toward the global optimal result (Xu and Lei 2010).

A discrete ABC algorithm has been implemented (Aslan and Ozturk 2016) to solve MSA by modifying the onlooker and employed bee phase which ends in a good fine-tuned alignment. A hybridization of multi-objective for MSA using ABC and Kalign2 (Largo et al. 2016) was done. The affine gap penalty, weighted sum-of-pairs, and total column score are used to evaluate the stability of the approach. Due to the running time consumption of alignment process for large sets of biological data, a massively parallel MSA based on ABC algorithm is implemented on supercomputer Blue Gene/P (Borovska et al. 2013) which is highly scalable in nature.

4.2.4 Ant Colony Optimization (ACO)

The ACO was evolved (Dorigo and Blum 2005) by getting motivated by the common behavior of ants which can locate the shortened route between their nest shell and target food source. The ambiguous communication between the ants is performed through a substance called pheromone, which the ants drop whenever they travel from one place to another.

The multiple sequence alignment has also been implemented using ant colony optimization algorithm (Moses and Johnson 2003) which is inspired by the way of ants' organization in search of food. They consume more time for aligning large number of sequences. Thus, the ACO algorithm is integrated with divide and conquer method to align the MSA (Chen et al. 2006). This method bisects the set of sequences vertically in a recursive manner using the ACO technique. This avoids the local optima and reduces the execution time for aligning and achieves a quality alignment. Later the integration of ACO with genetic algorithm is implemented for multiple sequence alignment (Lee et al. 2008) by which the local search has been incorporated using genetic algorithm and alignment is by ACO. So, this provides much diversity of alignment results in avoidance of local optima. Instead of considering an optimization problem of MSA (Guinand and Pigne 2007), a novel idea of constructing and maintaining a structure in a set of biological sequences is done using MSA. The structures are constructed by ant-based algorithm, and the blocks are obtained as conserved motifs. This result is compared with the best-known MSA tool ClustalW, and ant-based model algorithm outperforms the ClustalW.

4.2.5 Bacterial Foraging Optimization (BFO)

The principle behind the bacterial foraging optimization includes the chemotaxis, reproduction, and elimination dispersal, which is used in multiple sequence alignment problems which explores the space and leads to global optimization (Gheraibia and Moussaoui 2011). A BFO algorithm has been integrated with Tabu Search (TS-BFO) for discovering motifs in a sequence. It alters the BFO algorithm by introducing the self-control multilength chemotactic step and Rao metric mechanisms which produces optimal solutions (Shao and Chen 2009). A multi-objective-based BFO method was employed for the MSA problem and detected the conserved regions as a biological outcome with high accuracy (Rani and Ramyachitra 2016).

4.2.6 Bat and Firefly Optimization

The bat algorithm was inspired by echolocation behavior of microbats with fluctuating pulse rates of radiation and intensity. The firefly algorithm is a landscape-stimulated, metaheuristic algorithm that gets its inspiration from the common behavior of fireflies in the sky (Yang 2009, 2010). In previous works, the Hidden Markov model is trained by using a new PSO variant. Now to get optimal solution, the Hidden Markov model is trained by two various optimization algorithms, namely, bat optimization and firefly optimization algorithms. Two techniques, namely, random drift bat optimization and random drift firefly optimization algorithms with the diversity-guided search, have fine adjustments of parameters in this technique and achieved an efficient solution with rate effective method (Priyanka and Sathiyakumari 2015).

4.2.7 Cuckoo Search

The cuckoo search is a new metaheuristic method which gets encouraged by the cuckoo bird by laying eggs in the shells of host birds (Yang and Deb 2009). A multiple sequence alignment can be solved by a novel quantum cuckoo search technique (Kartous et al. 2014). It integrates global and local pairwise alignments in a randomized progressive method and achieves a feasibility solution effectively. Multiple sequence alignment has been achieved in a good score when compared to ClustalW by implementing the hybridization of genetic algorithm and cuckoo search algorithm (Srhan and Daoud 2013). The total column score has been assessed, and this method result outperforms the ClustalW.

4.2.8 Frog Leap Algorithm

A shuffled frog leap algorithm was developed for solving the combinatorial optimization problem (Eusuff et al. 2006). There considers being a cluster of frogs hopping in marshland. With the mass numbers of water lilies present at distinct locations, the frogs interact and jump one by one to seek the attention on discovering the lily with the extreme quantity of nutrition. It was implemented for multi-objective metaheuristic approach for multiple sequence alignment problems. The hybrid multi-objective memetic metaheuristics for multiple sequence alignment algorithm provide an efficient accuracy for low similarity sequences against 16 major MSA tools and proved to be effective (Largo et al. 2015). The motif discovery problem has been addressed as a hybridization of multi-objective method with shuffled frog leap algorithm (Alvarez et al. 2015).

4.2.9 Multiple Sequence Alignment Using Fuzzy Logic

Fuzzy logic is an outline of numerous-valued logic which is a superset of traditional logic that has prolonged to manipulate the theory of incomplete truth. Lotfi Zadeh introduced the phrase "fuzzy logic," and it has been used in many fields from expert systems to artificial intelligence. This logic accepts the uncertainty or errors in subsequence matching. A prototype for a fuzzy assembler has been developed using fuzzy logic for matching the subsequences approximately (Nasser et al. 2007). It is created to work with low similarity sequences.

A fuzzy inference method is used along with quality information for DNA alignment. An exact DNA alignment can be obtained for the lowest similarity sequences by improving the traditional approaches using quality information. Alignment scores are calculated using mapping score parameters which are dynamically tuned by the fuzzy logic system. From the experimental results, it is proved that the alignment using the fuzzy logic method yields better alignment score compared to traditional methods (Kim et al. 2008).

Multiple sequence alignment can be employed using fuzzy logic concept, and the similarities of the sequences are computed depending on the fuzzy parameters. Dynamic programming approach has been used to align sequences. The SinicView tool has been used to judge performance measures. The results revealed that approaches based on fuzzy logic give better matching alignment score (Gill and Singh 2011a, b).

Gill and Singh (2011a, b) compared two different approaches such as Boolean algebra and fuzzy logic for aligning multiple sequences. Final outcomes indicate that the fuzzy logic method outperforms the Boolean method when the number and length of sequences are large. When the dimension of protein sequences is less than or equivalent to 600 bp, the Boolean approach is resourceful in nature.

Many tools contribute partial optimal solutions by concentrating only on one specific biological feature for MSA which is not an effective way to find optimal alignment. Thus, the multi-objective optimization concept evolved. A trade-off or non-dominated solution has been created which encompasses a firm of compromised solutions by communicating among these conflicting objectives. The multi-objective optimization method does not deliver a solitary optimal solution (Chow et al. 2014). Recently, numerous tasks have been executed regarding evolutionary algorithms depending on the multi-objectives (Abbasi et al. 2015; Soto and Becerra 2014). NSGA-II, MO-SAStrE, MSAGMOGA, and MOMSA are some multi-objective-based sequence alignment techniques (Notredame 2002). To attain precise alignments, the integration of protein structural knowledge can be acquired along with the protein sequence. For example, Ortuno et al. (2013) developed MO-SAStrE, and Sullivan et al. (2004) developed 3DCoffee.

MSA is a method to analyze the biological sequences which carry out optimal multiple sequence alignment beneath the performance measures such as sum-of-pairs (SP) and total column score (TC). An assembly of similarity of protein sequence, the penalty of a gap inserted, and the proportion of non-gap in protein sequences are used as the multi-objectives to get a trade-off or non-dominated optimal solution.

4.3 Methodology

The MSA problems are a more common form of distinctive alignment between more than three sequences. Let the input sequence be $seq_1, seq_2, \ldots\ldots\ldots\ldots seq_n$ with at least three sequences in quantity. The alphabet set can be denoted as Σ, and to make equal length of sequences to align, the gap ("$-$") is included. The MSA S is determined as n dimensional character array against the alphabet where $\Sigma' = \Sigma \cup \{-\}$. When an input protein sequence is given, it stores as alignment array S that consists of n rows, respectively, and every row of A_i is the alignment for sequence seq_i (Abbasi et al. 2015).

The objective of MSA is to align sequences of RNA, DNA, or amino acid as a result of attaining optimal alignment outcomes. The similarity of sequences is computed to the corresponding amino acids and penalties that are calculated for the existence of a gap or mismatched amino acids. The substitution matrix score is the universally used scoring scheme to assess the amino acid similarity. A matrix constructed for nucleotides is 4×4 and for amino acids 20×20 which signifies all probable conversions among the DNA and proteins. PAM and BLOSUM are the two frequently used matrices. The weak similarity of protein alignments can be identified by calculating these matrix scores (Henikoff and Henikoff 1992).

4.3.1 Optimizing the Multi-objectives

The fitness score is 1:1 direct mapping of objectives (Gondro and Kinghorn 2007) (the objective function's values are straightly allocated as the fitness values of the candidate alignment). This chapter includes three objectives to detect the optimal solution for MSA, and the objectives are increase sequence similarity (S), decrease penalty of gap (S), and increase percentage of non-gap (S).

4.3.1.1 Sequence Similarity

Primarily, the computation of the position weight matrix for the sequence alignment is produced from the outcome of aligned result. The domination score (dos) of the foremost residues in each column is recognized as follows:

$$\text{dos}(seq) = \text{max}_n\{f(n, seq)\}, seq = 1, 2, 3 \ldots l \tag{4.1}$$

where $f(n, seq)$ is the score of nucleotide or amino acid n on the column seq in the position weight matrix against the existence of gaps, l is the dimension of the alignment of sequence, and dos(seq) is the dominant score of the leading residue on column seq.

The sequence similarity of the alignment S is determined as the moderate of the dominant score of all columns in the position weight matrix. It is also conveyed as follows:

$$\text{Similarity}(S) = \frac{\Sigma_{m=1}^{l}\text{dos}(seq)}{l} \tag{4.2}$$

If the score of similarity is closer to 1, then the highest probability of the candidate alignment S is revealed as the finest alignment (Kaya et al. 2014).

4.3.1.2 Penalty of a Gap

The aligning of related sections of residues into a good alignment can be done using a fake inclusion and exclusion of an empty object into a sequence called gap. A gap inserted in identical columns is nonacceptable which has no sense. Numerous categories of gap penalty values are constant gap penalty, linear gap penalty, convex gap penalty, affine gap penalty, and variable gap penalty. In this chapter, the affine gap and the variable gap penalty are employed for MSA.

Affine Gap Penalty

Affine gap penalty is used to score the insertions and deletions that penalize the gap at one time for opening and then equivalently to its length. Gap opening and gap extension are the two constraints employed in affine gap penalty (Altschul 1998). The formulation of affine gap penalty in the pairwise alignment of m and n rows is well-defined by

$$\text{Gap}_{\text{st}}(a) = \text{Gap}_{\text{open}} + \text{Gap}_{\text{extend}}(\text{len} - 1), \text{where len} > 1 \tag{4.3}$$

Gap_{open} – rate of opening a gap
$\text{Gap}_{\text{extend}}$ – rate of expanding a gap
len – span of gap length

The goal of the affine gap is to cluster the gaps that diminish the penalty scores.

Variable Gap Penalty

The common procedure of an affine gap penalty used for MSA is not suitable. The penalty value of a gap is determinant by using the affine scores as fixed and used correspondingly in all spots. In variable gap penalty, the scores of the gap vary according to the amino acids when employing a new position-specific gap penalty to detect the optimal solution. ClustalW and MAFFT methods applied this kind of penalty for the gap.

The foremost gap penalties are computed depending on the static scores fixed by end users. Mainly, two frequently used penalties of the gap are presented.

The gap opening penalty (GO) is referred as the rate of opening a novel gap at any dimension, and the rate of each single piece in a gap is indicated by gap extension penalty (GE).

The penalty of opening a gap is recomputed depending on the following aspects: (i) based on the matrix weight, (ii) based on the protein sequence similarity, and (iii) based on the dimension of the protein sequences (Thompson et al. 1994; Hung et al. 2008):

$$\text{GO} \rightarrow \{\text{GO} + \log[\min(A, B)]\} * (c) * (d) \tag{4.4}$$

where A and B are the distances of two sequences, c is the residue divergence average score, and d is the scaling feature percent of similar identity.

The penalty of extending a gap is recomputed depending on features such as:

(i) Based on the variance in the distances of the sequences, and they are depicted as

$$\text{GE} \rightarrow \text{GE}^* \ [1.0 \ + \ |\log A/B|] \tag{4.5}$$

where A and B are the distances of the two protein sequences and GE is the score of extending a gap.

(ii) Position-specific gap penalties.
(iii) Penalties of the lowered gap at prevailing gaps, and they are depicted by Eq. (4.6) as

$$\text{GOTable} \rightarrow \text{GOTable}^{*} 0.3^{*} (N_s) \tag{4.6}$$

where N_s is the total quantity of sequences without a gap; GOTable is the table for the penalty of a gap opening which documents the forfeit along the distance of protein sequences i, for every couple of i and j protein sequences; and GETable is the table for the penalty of gap extension.

(iv) Increased penalties of the gap close to standing gaps and they are depicted as

$$\text{GOTable} \rightarrow \text{GOTable}^{*} \{2 + [(8 - \text{radius from the gap}) * 2]/8\} \tag{4.7}$$

(v) Diminished penalties of the gap in hydrophilic radius is depicted as

$$\text{GOTable} \rightarrow \text{GOTable}^{*} 0.5 \tag{4.8}$$

(vi) Penalties for residue-specific is depicted as

$$\text{GOTable} = \text{GOTable}^{*} T[S_x] \tag{4.9}$$

where S_x is the value of amino acid situated on the xth location of sequence S in the amino acid table.

Finally, the GO and GE are calculated based on Eqs. (4.6), (4.7), (4.8), and (4.9):

$$\text{GO}(c, d) = \text{GOTable}(c) + \text{GOTable}(d) \tag{4.10}$$

$$\text{GE}(c, d) = \text{GETable}(d) \tag{4.11}$$

Depending on the sequences given as input, the variable gap penalty is injected by the above aspects.

The combination of an optimization of objective functions is generally utilized to convey the MSA problem. The hybridization of ABC and BFO is proposed to find the MSA alignment and obtain accurate results (Fig. 4.1). Figure 4.5 shows the graphical abstract of this work.

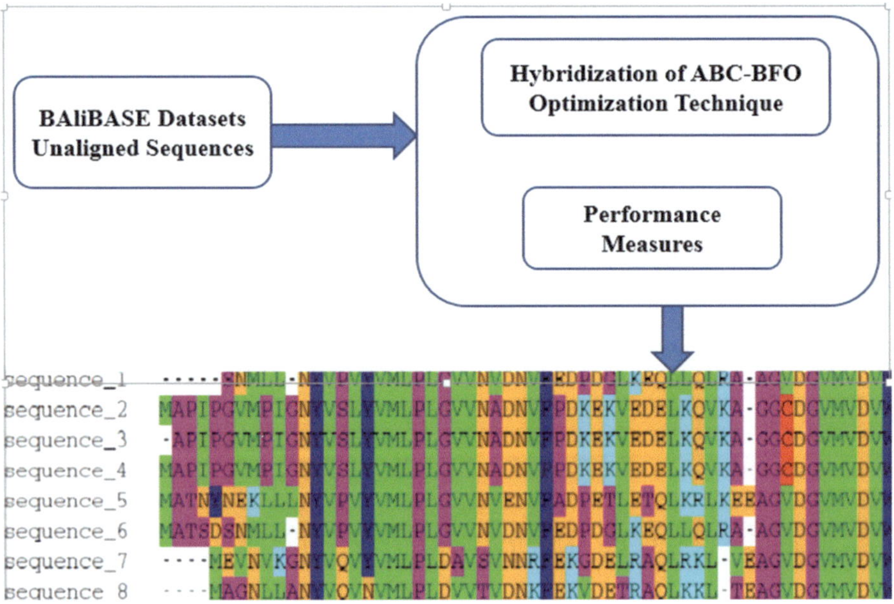

Fig. 4.1 Graphical representation of hybrid of ABC-BFO

4.3.2 Hybrid of ABC-BFO

Artificial bee colony is an efficient swarm optimization algorithm when compared to other swarm intelligence methods such as ACO and PSO. It has very limited control parameters to optimize, and it has a high universal search capacity and an easy execution. Through the exploration of the solutions, simultaneously it is very weak in exploitation of solutions. This happens because of ineffectiveness of the local search of solution space. This in case gives an issue to get optimal solution in few circumstances. The hybridization of ABC algorithm gives better results in all types of problems.

On the divergent case, the bacterial foraging optimization has a slow convergence of the solution, but it has a good capacity to achieve the global optimum solution. The swarming approach of BFO technique makes the bacteria to assemble into a group of clusters and travel in a concentric pattern with high density of bacteria. By this, the bacteria which reached the best route of the food source must let other bacteria to travel against the same direction. Thus, in turn can make other bacteria to reach the destination more quickly and accurately (Li et al. 2015).

To improve the efficiency of the MSA solution, a hybrid of ABC and BFO is implemented. The swarming approach of BFO is employed within the employed bee phase and onlooker bee phase of the artificial bee colony method. The steps for hybrid ABC-BFO algorithm are given (Figs. 4.2 and 4.3).

1. Input the set of unaligned sequences of population of solutions i=1,2,3…S_n.
2. Initialize the variables and randomize positions.
3. while ((Iterations < MaximumCycle))
4. *Employed BeesPhase*
5. FOR(i=1:(Foodelement))
6. create a new food target;
7. compute fitness value of the new food target;
8. Swarming approach; Greedy selection process;
9. end for :
10. Calculate the probability P_i;
11. *Onlooker BeesPhase*
12. FOR(i=1:(Foodelement))
13. Parameter P_i is determined arbitrarily;
14. Onlooker bees locate food sources depending on Pi;
15. create a new food target;
16. assess the fitness value of the new food target;
17. Swarming approach; Greedy selection process;
18. end for
19. *Scout BeePhase*
20. IF(any employed bee becomes scout bee)
21. Parameter P_i is determined arbitrarily;
22. The scout bees locate food sources depending on P_i;
23. end if
24. Terminate the process only if there is a neglected solution and bacterial Chemotaic N_c is employed with population of S_n bacteria in multimodal plane for feasible solution in an iteration and Reproduce N_{re},
25. Reproduce the N_{re}, the bacteria which hold healthy solution and now bacteria will scatter to new environment.
26. estimate the distance d_c as the bacteria will converge to the certain place as its final cluster centers .otherwise, go to step 4
27. Memorize the best solution;
28. Iteration=Iteration+1;
29. end while

Fig. 4.2 Steps for hybrid ABC-BFO algorithm

Fig. 4.3 Steps for BFO swarming approach

1. Initialize variables
2. Let v = 0;
3. While v < Ns
4. IF(the mutant solution is healthier than the current Solution)
5. Revise the solution by the healthier mutant solution;
6. END IF
7. Let v = v + 1;
8. Else,
9. let v = Ns.

4.4 Results

The proposed hybrid algorithm has been experimented to report the difficulties stumble upon when aligning sequences by using benchmark BAliBASE datasets (Bahr et al. 2001) and the PDB database. It has eminent standard sequences to classify the merits and demerits of several alignment techniques to examine the efficiency and accuracy of the algorithm. In addition, the proficiency of the proposed (hybrid ABC-BFO) method has been verified as best by correlating with various MSA methods, namely, GA, ABC, ACO, and PSO, and existing online tools specifically Kalign, MUSCLE, Clustal Omega, and T-COFFEE. Here the multi-objective optimization technique is employed for the MSA problem. The maximization of SP and TCS scores directs to discover the best optimal solution. The objectives, such as sequence similarity, the penalty of a gap, and the proportion of non-gap, are also considered for optimal solution. The datasets used for this chapter are displayed in Table 4.1.

The datasets are collected from the benchmark database named BAliBASE 3.0 and Protein Data Bank (PDB). The BAliBASE database contains manually discriminated protein sequences specially created for the assessment and comparison of numerous MSA methods. The database is separated into five different reference sets. Reference 1 consists of RV11 and RV12 subsets. RV11 has <25% identity of sequences and RV12 has 20–40% identity of sequences. Reference 2 has

Table 4.1 Datasets used for this experimental work

Dataset	Classification name	Total sequence number	Shortest sequence length	Longest sequence length	Alignment length
3ZNG	*Transcription/ DNA*	18	40	268	281
5JA1	*Ligase*	17	104	1295	1360
5AE6	*Hydrolase*	36	56	767	805
4ZHQ	*Structural protein*	4	132	451	474
4N9F	*Viral protein*	15	20	311	327
4V4B	*Ribosome*	26	119	842	884
1M9N	*Transferase hydrolase*	11	464	613	650
2TGT	*Hydrolase zymogen*	8	223	229	240
4X51	*Cytokine*	8	151	162	170
5JTW	*Immune system*	20	56	656	688
2YM9	*Cell invasion*	10	233	346	363
2VK9	*Toxin*	4	153	551	579
5GQH	*Hydrolase inhibitor*	17	99	1090	1144

highly different orphan sequences. Reference 3 has subgroups with less than 25% of the residue identity among the subgroups itself. Reference 4 has N-/C-terminal sequences, and finally Reference 5 has inner insertions (Thompson et al. 2005).

The most generally exploited scoring measures for MSA are sum-of-pairs (SP) and total column score (TCS). SP is represented as the quantity of precisely aligned residues against the total quantity of amino acid pairs in resource alignment.

Study the instance of trial alignment with magnitude P^*Q and a resource alignment of magnitude P^*Q_r, where P is the total quantity of protein sequences and Q and Q_r are the sum of the quantity of columns in the trial and resource alignment (Thompson et al. 1999). Here $A_{\mathrm{j1}}, A_{\mathrm{j2}}.\dots.A_{\mathrm{jX}}$, is the jth column in the alignment; $V_{\mathrm{jab}} = 1$ is defined for each pair of residues $A_{\mathrm{ja}}\ and\ A_{\mathrm{jb}}$ only if $A_{\mathrm{ja}}\ and\ A_{\mathrm{jb}}$ are associated with one another in the resource alignment; else $V_{\mathrm{jab}} = 0$. The value of SP_j for the jth column will be the sum of V_{jab} for entire pairs of amino acids in this column:

$$SP_j = \Sigma_{a=1, a\neq b}^{P} \Sigma_{b=1}^{P} V_{\mathrm{jab}} \tag{4.12}$$

Similarly, SP_{rj} is the score SP_{j} for the jth column in the resource alignment. The SP score of the trial alignment is

$$SP = \Sigma_{j=1}^{Q} SP_j / \Sigma_{j=1}^{\mathrm{Qr}} SP_{\mathrm{rj}} \tag{4.13}$$

TCS is represented as the quantity of precisely associated columns against the overall quantity of columns in the resource alignment.

Study the instance of trial alignment with magnitude $P \times Q$ and a resource alignment of magnitude $P \times Q_r$, where P is the total quantity of sequences and Q and Q_r are the total quantity of columns in the trial and resource alignment. At this time, we describe the value of $\mathrm{Col}_j=1$ if every residue is aligned in the resource alignment; otherwise $\mathrm{Col}_j = 0$ (Thompson et al. 1999).

The TCS value of test alignment is

$$\mathrm{TCS} = \Sigma_{j=1}^{Q} \mathrm{Col}_j / Q \tag{4.14}$$

Universally, the size of the preliminary population used as per literature is 200 individuals. The end point of the algorithm is established, if the best result provided in all generations persists to be similar for 100 following generations of the algorithm or the greatest quantity of generations acquired. Depending on the experimental results, the gap percentage initialization was adjusted. Among several percentage values, it was identified that 5% of the gap offered improved outcomes, and henceforth it was allocated.

In Table 4.2, the average outcomes for 5% of the gap score is shown. Here two sets of empirical results were attained where the initial set is to exhibit the standards of objective: sequence similarity, the penalty of a gap, and the proportion of non-gap for six methods (GA, ACO, ABC, PSO, BFO, and the proposed hybrid

Table 4.2 Comparisons of multi-objectives of BAliBASE 3.0 datasets using various algorithms

Comparisons of multi-objectives using gap penalty

		Similarity		Gap penalty		
Dataset	Algorithms	Affine gap	Variable gap	Affine gap	Variable gap	Non-gap percentage (NGP)
3ZNG	GA	0.6382	0.6721	2681	2691	65
	ACO	0.6472	0.6835	2575	2601	67
	ABC	0.6116	0.6892	2553	2481	62
	PSO	0.6284	0.7183	2532	2476	64
	BFO	0.7426	0.7335	2416	2264	68
	ABC-BFO	0.7618	0.7694	2174	2083	72
5JA1	GA	0.1583	0.1754	11,982	11,679	22
	ACO	0.1762	0.1893	11,539	11,539	31
	ABC	0.2063	0.2317	11,072	11,284	34
	PSO	0.2462	0.2653	11,001	11,174	35
	BFO	0.2964	0.2381	10,976	10,783	40
	ABC-BFO	0.3157	0.3471	10,357	10,289	55
5AE6	GA	0.2042	0.2163	9962	9987	36
	ACO	0.2474	0.2281	9782	9698	36
	ABC	0.2671	0.2759	9709	9629	49
	PSO	0.2532	0.2809	9656	9602	58
	BFO	0.2579	0.2881	9589	9610	69
	ABC-BFO	0.2971	0.3182	9477	9481	75
4ZHQ	GA	0.5952	0.6872	2923	2781	57
	ACO	0.6289	0.6371	2871	2699	56
	ABC	0.6419	0.6532	2791	2518	60
	PSO	0.6587	0.6643	2801	2621	65
	BFO	0.6687	0.6699	2819	2681	79
	ABC-BFO	0.6701	0.6878	2687	2591	72
4N9F	GA	0.6478	0.6592	2567	2481	62
	ACO	0.6681	0.6956	2681	2571	60
	ABC	0.6761	0.6853	2419	2431	66
	PSO	0.6682	0.6911	2408	2418	67
	BFO	0.6812	0.6928	2395	2342	64
	ABC-BFO	0.6807	0.7041	2228	2298	68
4V4B	GA	0.2571	0.2692	10,642	10,541	28
	ACO	0.2781	0.2782	10,521	10,511	26
	ABC	0.2861	0.2817	10,482	10,412	29
	PSO	0.2892	0.2809	10,390	10,365	28
	BFO	0.2961	0.3082	10,310	10,302	36
	ABC-BFO	0.3071	0.3412	10,231	10,210	58

(continued)

Table 4.2 (continued)

Comparisons of multi-objectives using gap penalty						
		Similarity		Gap penalty		
Dataset	Algorithms	Affine gap	Variable gap	Affine gap	Variable gap	Non-gap percentage (NGP)
1M9N	GA	0.7843	0.8271	1073	1067	69
	ACO	0.7968	0.8290	1058	1056	72
	ABC	0.8189	0.8476	1067	1062	71
	PSO	0.8352	0.8481	1052	1038	73
	BFO	0.8482	0.8499	1019	1005	78
	ABC-BFO	0.8742	0.8981	973	968	83
2TGT	GA	0.8952	0.9271	390	376	80
	ACO	0.8992	0.9341	387	365	85
	ABC	0.9052	0.9549	368	352	87
	PSO	0.9282	0.9599	356	342	87
	BFO	0.9371	0.9596	372	349	88
	ABC-BFO	0.9659	0.9745	321	310	90
4X51	GA	0.7952	0.8419	628	621	78
	ACO	0.8173	0.8438	619	598	75
	ABC	0.8379	0.8536	601	629	79
	PSO	0.8472	0.8691	596	623	78
	BFO	0.8567	0.8536	589	578	80
	ABC-BFO	0.8686	0.8974	547	538	85
5JTW	GA	0.3173	0.3372	8963	8165	32
	ACO	0.3912	0.3892	8481	7936	39
	ABC	0.4262	0.4572	7972	7892	43
	PSO	0.4571	0.4601	7765	7698	54
	BFO	0.4973	0.5082	7221	7681	65
	ABC-BFO	0.5281	0.5629	6964	7018	76
2YM9	GA	0.4387	0.4672	2653	2573	72
	ACO	0.5987	0.5193	2590	2382	74
	ABC	0.5482	0.5481	2418	2397	72
	PSO	0.5678	0.5289	2381	2316	75
	BFO	0.5872	0.5732	2235	2214	78
	ABC-BFO	0.6052	0.6391	2183	2093	84
2VK9	GA	0.5273	0.5863	6732	6482	53
	ACO	0.5985	0.6128	6539	6322	58
	ABC	0.6481	0.6372	6318	6281	60
	PSO	0.6598	0.6538	6276	6189	63
	BFO	0.6742	0.6981	6081	6023	67
	ABC-BFO	0.6982	0.7091	5731	5719	74

(continued)

Table 4.2 (continued)

Comparisons of multi-objectives using gap penalty						
		Similarity		Gap penalty		
Dataset	Algorithms	Affine gap	Variable gap	Affine gap	Variable gap	Non-gap percentage (NGP)
5GQH	GA	0.1972	0.2153	10,432	10,328	24
	ACO	0.2371	0.2461	10,382	10,271	25
	ABC	0.2561	0.2567	10,251	10,187	35
	PSO	0.2789	0.2861	10,178	10,091	50
	BFO	0.2918	0.3171	10,981	10,013	57
	ABC-BFO	0.3451	0.3221	10,003	9993	60

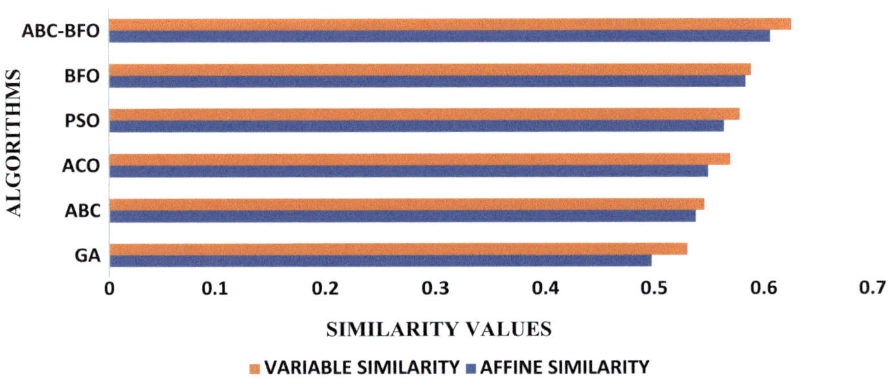

Fig. 4.4 Comparison of average similarity values with respect to gap penalty

ABC-BFO). And the final set is to estimate performance metrics such as SP and TCS scores. The projected method and the remaining existing methods have been implemented for 30 runs, and the average outcomes are demonstrated.

The graph representation of Table 4.2 has been displayed in Figs. 4.4, 4.5, and 4.6. Figure 4.4 shows the average comparison of similarity values with respect to the gap penalty. Figure 4.5 depicts the average comparison of gap penalty values, and finally Fig. 4.6 represents the average non-gap percentage values for the given populations of dataset.

In Table 4.3, performance measures such as SP and TCS for BAliBASE datasets are estimated for different algorithms and various tools. From those results, the confirmation of high performance of hybrid ABC-BFO algorithm when compared to other existing algorithms is displayed. Our algorithm works equally well against other standard existing tools for MSA. The graph representation of the average performance measures of BAliBASE datasets is displayed in Fig. 4.7.

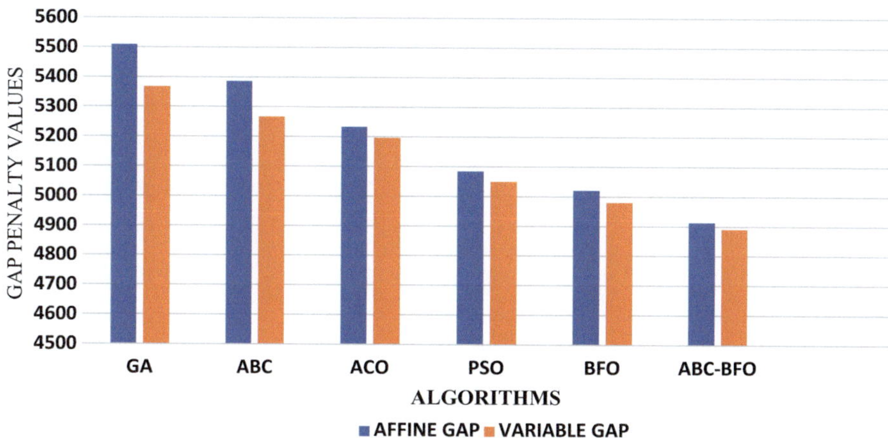

Fig. 4.5 Comparison of average gap penalty values for various algorithms

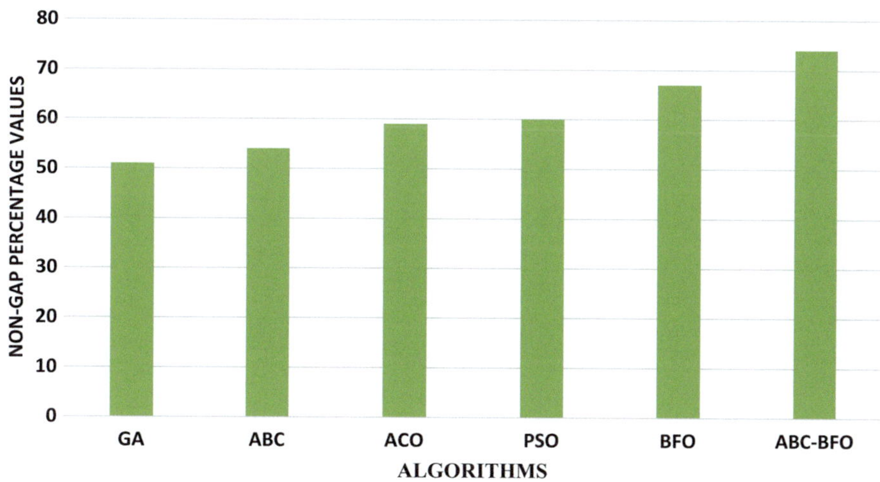

Fig. 4.6 Comparison of various non-gap percentage values for different algorithms

4.4.1 Applications of MSA

Multiple sequence alignment is the primary tool in almost every application of bioinformatics, such as homologous protein primers (Fredslund et al. 2005), structure prediction (Cuff and Barton 2000; Chu et al. 2006), domain identification, conserved region identification (Hertz and Stormo 1999), protein function prediction (Pierri et al. 2010), and phylogenetic tree analysis (Potter 2008). The phylogenetic trees are constructed using subunits of sequences, which is a stimulating observation on the evolution of organisms (Lal et al. 2008, 2015).

Table 4.3 Comparisons of SP and TCS performance measures for various tools and algorithms

Dataset	Performance measure	T-COFFEE	Clustal Ω	MAFFT	Kalign	GA	ACO	ABC	PSO	BFO	BFO-ABC
3ZNG	SP	84	74	76	63	69	72	70	75	78	79
	TCS	64	61	62	54	53	56	59	62	64	65
5JA1	SP	45	40	39	32	34	30	36	37	39	40
	TCS	26	23	21	18	20	19	23	25	24	27
SAE6	SP	59	54	54	53	56	58	60	62	60	61
	TCS	47	43	42	36	38	39	41	45	47	49
4ZHQ	SP	70	65	63	60	62	65	68	69	70	73
	TCS	52	49	50	45	47	51	53	56	54	55
4N9F	SP	77	72	75	70	69	71	74	78	77	78
	TCS	59	57	54	52	53	56	59	62	60	60
4V4B	SP	46	43	41	39	37	39	44	45	46	45
	TCS	30	27	28	23	19	23	24	27	29	29
1M9N	SP	83	81	78	71	76	79	81	83	84	86
	TCS	72	71	68	62	61	63	69	71	72	73
2TGT	SP	96	87	83	81	83	85	87	90	92	94
	TCS	75	71	69	64	68	69	72	75	76	79
4X51	SP	79	76	72	68	74	75	78	77	79	82
	TCS	70	68	64	61	62	65	62	68	70	72
5JTW	SP	36	32	39	26	27	23	28	32	35	38
	TCS	24	21	19	17	17	15	19	23	25	26
2YM9	SP	53	51	49	38	36	39	43	54	56	57
	TCS	39	35	32	25	21	24	32	37	38	41
2VK9	SP	61	59	55	52	50	53	56	63	65	66
	TCS	51	48	45	41	39	43	45	52	52	57
5GQH	SP	35	32	27	25	23	26	29	35	37	38
	TCS	21	16	15	18	16	19	17	23	28	27

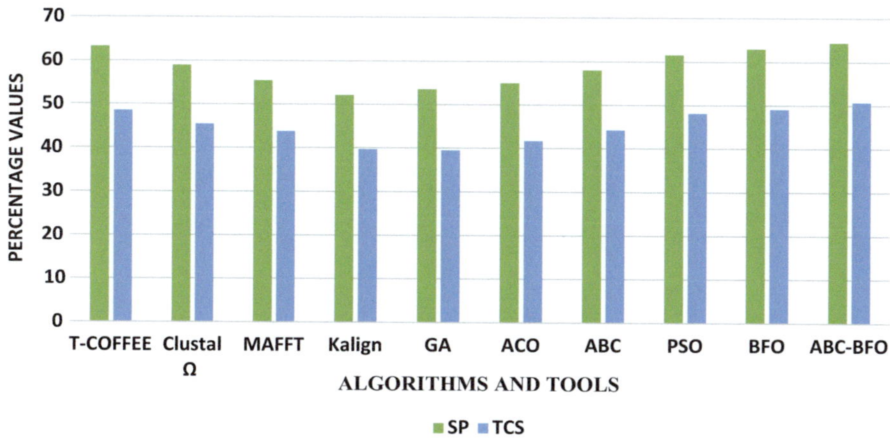

Fig. 4.7 Comparison of average performance measures of BAliBASE 3.0 datasets

4.4.2 Statistical Analysis

The concluding phase produces the statistical significance of the proposed algorithm which is estimated using nonparametric test, namely, Friedman rank test, among every pair of techniques by utilizing a substantial confidence level of 5% (P-score < 0.05) (Largo et al. 2016). Each entry in Table 4.4 consists of P-score allocated to the difference between the pair of approaches using Friedman rank test. The right upper edge of the matrix is attained from SP value, and the left lower edge is attained from TCS value. From Table 4.4, it is detected that ABC-BFO attains statistically noteworthy accuracy enhancement against all other well-known techniques. The P-scores lower than 0.05 are determined as an extremely significant method; contrarily larger than 0.05 are determined as an insignificant method.

4.5 Implementation and Discussion

Even now the computation of MSA with immense accuracy is a challenging task. In this chapter, to solve the MSA problem, multi-objective-based hybridization of ABC-BFO technique is employed. A remarkable enhancement in the alignment accuracy of the ABC-BFO method over the other several well-known methods has been proved. The objectives, namely, sequence similarity, the penalty of a gap, and the proportion of non-gap, are also examined for optimal solution.

To achieve a few invaluable conclusions from the proposed method, a confident level of statistical significance was measured. The application was deployed in 2.00 GHz Intel CPU with 1GB of memory and running in Windows 8.1. While raising the number of iterations of execution, the sequence similarity value rises up, and the penalty value of a gap declines progressively. Correspondingly, the

Table 4.4 Statistical significance of proposed algorithm and existing methods based on BAliBASE 3.0 datasets

	T-COFFEE	CLUSTAL Ω	MAFFT	Kalign	GA	ACO	ABC	PSO	BFO	ABC-BFO
T-Coffee		$<10^{-10}$	$<10^{-10}$	$<10^{-10}$	$<10^{-10}$	$<10^{-10}$	$<10^{-10}$	$<10^{-10}$	$<10^{-10}$	$<10^{-10}$
CLUSTAL Ω	$<10^{-10}$		$<10^{-10}$	0.651	0.982	$<10^{-10}$	0.278	$<10^{-10}$	$<10^{-10}$	$<10^{-10}$
MAFFT	$<10^{-10}$	$<10^{-10}$		$<10^{-10}$	$<10^{-10}$	0.826	$<10^{-10}$	0.768	$<10^{-10}$	$<10^{-10}$
Kalign	$<10^{-10}$	$<10^{-10}$	$<10^{-10}$		0.426	0.326	0.372	$<10^{-10}$	$<10^{-10}$	$<10^{-10}$
GA	$<10^{-10}$	$<10^{-10}$	0.217	0.376		$<10^{-10}$	$<10^{-10}$	$<10^{-10}$	$<10^{-10}$	$<10^{-10}$
ACO	$<10^{-10}$	0.059	0.527	0.158	$<10^{-10}$		$<10^{-10}$	$<10^{-10}$	$<10^{-10}$	$<10^{-10}$
ABC	$<10^{-10}$	0.098	0.232	$<10^{-10}$	0.061	0.351		$<10^{-10}$	$<10^{-10}$	$<10^{-10}$
PSO	$<10^{-10}$	$<10^{-10}$	$<10^{-10}$	$<10^{-10}$	$<10^{-10}$	$<10^{-10}$	$<10^{-10}$		$<10^{-10}$	$<10^{-10}$
BFO	$<10^{-10}$	$<10^{-10}$	$<10^{-10}$	$<10^{-10}$	$<10^{-10}$	$<10^{-10}$	$<10^{-10}$	0.0597		$<10^{-10}$
ABC-BFO	$<10^{-10}$	$<10^{-10}$	$<10^{-10}$	$<10^{-10}$	$<10^{-10}$	$<10^{-10}$	$<10^{-10}$	$<10^{-10}$	$<10^{-10}$	

alignment accuracy declined slowly, when the percentage of a gap rises; concurrently the quantity of gaps also increased.

Besides, the accuracy of multiple sequence alignment is entirely reliant on input sequence characters, and also the performance alignment algorithm is dependent on the features of the sequences to be associated. For instance, 5JA1 and 5GQH data has a huge difference between the smallest and the largest length of the sequence and also the total number of sequences are more, which leads to low similarity and accuracy. Also, 2TGT dataset has less number of protein sequences, and difference between the smallest and longest sequence is very low that stimulates to high sequence similarity and accuracy. The accuracy results of the best prominent methods are matched or fall behind that of the proposed method's accuracy. The algorithm was tested by 500 numbers of generations and observed that after in the average of 370th generation, the value of pareto optimal solution was commenced. In the experiment, the entire performance metrics were oscillated at the time of the first six execution runs, and in later runs, reliability was recognized. The average scores of the first 30 runs of the proposed algorithm were displayed in Tables 4.2 and 4.3. In particular, the algorithm gets eliminated when the best value of pareto solution originated in each generation is persistent for 100 uninterrupted iterations or when the entire number of iterations touched its end. Here the proposed algorithm is the hybridization of the optimization algorithm which leads to the population creation and reproduction. That takes more computation time to display the result of non-dominated optimal solution.

4.6 Conclusion

The alignment of multiple biological sequences derives a resolution for examining the sequence similarity, features, and structure and the protein function of a novel-discovered sequence. There should be an intermittent improvement in alignment techniques as they play a central role in the analysis of huge data contributed by high-throughput experiments and next-generation sequencing. By this, the next stage of protein structure prediction, function prediction, can be accomplished which leads to the drug design which is important. The proposed algorithm delivers a promising outcome that can explore the biological sequence with high performance and efficiency of merging artificial bee colony and the idea of the swarming approach of bacterial foraging optimization algorithm to solve MSA. The multi-objective optimization approach is utilized to determine the MSA problem by increasing the sequence similarity and the proportion of non-gap scores and decreasing the penalty of a gap which directs to the non-dominated optimal solution. Compared to other existing algorithms, the hybridization of ABC-BFO gets better results. To correlate the connotation of the proposed algorithm with other prevailing methods, the statistical measure is also computed. In the future, this method can be prolonged or merged with any alternative evolutionary algorithm to discover the best optimal alignments. To identify the most outstanding outcomes of multiple sequence alignment and also to acquire additional biological insights, many diverse objectives may be announced. Also, by employing this method, 2D and 3D structure of a given protein can also be predicted.

References

Abbasi M, Paquete L, Pereira FB (2015) Local search for multiobjective multiple sequence alignment. Bioinform Biomed Eng 9044:175–182. https://doi.org/10.1007/978-3-319-16480-9_18

Altschul SF (1998) Generalized affine gap rates for protein sequence alignment. Proteins 32:88–96

Alvarez DLG, Rodríguez MAV, Largo AR (2015) A comparative study of different motif occurrence models applied to a hybrid multiobjective shuffle frog leaping algorithm. Comput J 59:384–402. https://doi.org/10.1093/comjnl/bxv055

Aslan S, Ozturk C (2016) Alignment of biological sequences by discrete Artificial Bee Colony algorithm. Conference on Signal Processing and Communications Applications Conference (SIU). doi: https://doi.org/10.1109/SIU.2015.7129916

Bahr A, Thomson JD, Thierry JC, Poch O (2001) BAliBASE (benchmark alignment dataBASE): enhancements for repeats, transmembrane sequences and circular permutations. Nucleic Acids 29:323–326

Borovska P, Gancheva V, Landzhev N (2013) Massively parallel algorithm for multiple biological sequences alignment. IEEE 36th International Conference on Telecommunications and Signal Processing (TSP). doi https://doi.org/10.1109/TSP.2013.6614014

Bucak IM, Uslan V (2011) Sequence alignment from the perspective of stochastic optimization: a survey. Turk J Electr Eng Comput Sci 19:157–173. https://doi.org/10.3906/elk-1002-410

Chen Y, Pan Y, Chen J, Liu W, Chen L (2006) Multiple sequence alignment by ant colony optimization and divide-and-conquer. Comput Sci ICCS:646–653. https://doi.org/10.1007/11758525_88

Chow JF, Savic DA, Fortune D, Kapelan Z, Mebrate N (2014) Using multi-objective optimization to maximize multiple benefits for sustainable drainage design. 11th International Conference on Hydroinformatics HIC 2014, New York City, USA

Chu W, Ghahramani Z, Podtelezhnikov A, Wild DL (2006) Bayesian segmental models with multiple sequence alignment profiles for protein secondary structure and contact map prediction. IEEE/ACM Trans Comput Biol Bioinform 3:99–113

Collingridge PW, Kelly S (2012) MergeAlign: improving multiple sequence alignment performance by dynamic reconstruction of consensus multiple sequence alignments. BMC Bioinform 113:117. https://doi.org/10.1186/1471-2105-13-117

Cuff JA, Barton GJ (2000) Application of multiple sequence alignment profiles to improve protein secondary structure prediction. Proteins 40:502–511

Dorigo M, Blum C (2005) Ant colony optimization theory: a survey. Theor Comput Sci 344:243–278. https://doi.org/10.1016/j.tcs.2005.05.020

Ebert J, Brutlag D (2006) Development and validation of a consistency based multiple structure alignment algorithm. Bioinformatics 22:1080–1087. https://doi.org/10.1093/bioinformatics/btl046

Eusuff M, Lansey K, Pasha F (2006) Shuffled frog-leaping algorithm: a memetic meta-heuristic for discrete optimization. Eng Optim 38:129–154. https://doi.org/10.1080/03052150500384759

Feng DF, Johnson MS, Doolittle RF (1984) Aligning amino acid sequences: comparison of commonly used methods. J Mol Evol 21:112–125

Fredslund J, Schauser L, Madsen LH, Sandal N, Stougaard J (2005) PriFi: using a multiple alignment of related sequences to find primers for amplification of homologs. Nucleic Acids Res 33:W516–W520. https://doi.org/10.1093/nar/gki425

Gheraibia Y, Moussaoui A (2011) Protein multiple sequence alignment using Bacterial Foraging Optimization algorithm. The Second International Conference on Complex Systems (CISC'11)

Gill N, Singh S (2011a) Biological sequence matching using fuzzy logic. Int J Sci Eng Res 2:1–5

Gill N, Singh S (2011b) Multiple sequence alignment using Boolean algebra and fuzzy logic: a comparative study. Int J Comput Technol Appl 2:1145–1152

Gondro C, Kinghorn BP (2007) A simple genetic algorithm for multiple sequence alignment. Genet Mol Res 6:964–982

Gotoh O (1996) Significant improvement in accuracy of multiple protein sequence alignments by iterative refinement as assessed by reference to structural alignments. J Mol Biol 38(264):823. https://doi.org/10.1006/jmbi.1996.0679

Guinand F, Pigne Y (2007) An Ant-based model for multiple sequence alignment. International Conference on Large Scale Scientific Computing, LSSC 2007, pp 553–560. doi: https://doi.org/10.1007/978-3-540-78827-0_63

Henikoff S, Henikoff JG (1992) Amino acid substitution matrices from protein blocks. Proc Natl Acad Sci USA 89:10915–10919

Hertz GZ, Stormo GD (1999) Identifying DNA and protein patterns with statistically significant alignments of multiple sequences. Bioinformatics 15:563–577

Hung CL, Lin CY, Chung YC, Tang CY (2008) Introducing variable gap penalties into three-sequence alignment for protein sequences. IEEE International Conference on Advanced information networking and applications – Workshops, pp 726–731. doi:https://doi.org/10.1109/WAINA.2008.101

Kartous W, Layeb A, Chikhi S (2014) A new quantum cuckoo search algorithm for multiple sequence alignment. J Intell Syst 23:261–275. https://doi.org/10.1515/jisys-2013-0052

Kaya M, Sarhan A, Alhajj R (2014) Multiple sequence alignment with affine gap by using multi-objective genetic algorithm. Comput Methods Prog Biomed 114:38–49. https://doi.org/10.1016/j.cmpb.2014.01.013

Kaya M, Kaya B, Alhajj R (2016) A novel multi-objective genetic algorithm for multiple sequence alignment. Int J Data Min Bioinform 14:139–158. https://doi.org/10.1504/IJDMB.2016.074684

Kim K, Kim M, Woo Y (2008) A DNA sequence alignment algorithm using quality information and a fuzzy inference method. Prog Nat Sci (5):595–602. https://doi.org/10.1016/j.pnsc.2007.12.011

Lal S, Cheema S, Kalia VC (2008) Phylogeny vs genome reshuffling: horizontal gene transfer. Indian J Microbiol 48:228–242. https://doi.org/10.1007/s12088-008-0034-1

Lal S, Raje DV, Cheema S, Kapley A, Purohit HJ, Kalia VC (2015) Investigating the phylogeny of hydrogen metabolism by comparative genomics: horizontal gene transfer. In: Kalia VC (ed) Microbial factories. Springer, New Delhi, pp 317–345. https://doi.org/10.1007/978-81-322-2595-9_20

Largo AR, Rodrıguez MAV, Alvarez DLG (2015) A hybrid multiobjective memetic metaheuristic for multiple sequence alignment. IEEE Trans Evol Comp 20:499–514. https://doi.org/10.1109/TEVC.2015.2469546

Largo AR, RodrIguez MAV, Alvarez DLG (2016) Hybrid multiobjective artificial bee Colony for multiple sequence alignment. Appl Soft Comput 41:157–168. https://doi.org/10.1016/j.asoc.2015.12.034

Layeb A, Deneche AH (2007) Multiple sequence alignment by immune artificial system. IEEE/ACS International Conference on Computer Systems and Applications, pp 336–342. doi: https://doi.org/10.1109/AICCSA.2007.370903

Lee ZJ, Su SF, Chuang CC, Liu KH (2008) Genetic algorithm with ant colony optimization (GA-ACO) for multiple sequence alignment. Appl Soft Comput 8:55–78. https://doi.org/10.1016/j.asoc.2006.10.012

Lei XJ, Sun JJ, Ma QZ (2009) Multiple sequence alignment based on Chaotic PSO. In: Computational intelligence and intelligent systems, pp 351–360. doi: https://doi.org/10.1007/978-3-642-04962-0_40

Lei XJ, Sun JJ, Xu X, Guo L (2010) Artificial Bee Colony algorithm for solving multiple sequence alignment. IEEE Fifth International Conference on Bio-Inspired Computing: Theories and Applications (BIC-TA). doi: https://doi.org/10.1109/BICTA.2010.5645304

Li L, Zhang FF, Liu C, Niu B (2015) A Hybrid Artificial Bee Colony algorithm with bacterial foraging optimization. IEEE International Conference on Cyber Technology in Automation, Control, and Intelligent Systems (CYBER). doi: https://doi.org/10.1109/CYBER.2015.7287922

Loytynoja A, Goldman N (2005) An algorithm for progressive multiple alignment of sequences with insertions. Proc Natl Acad Sci USA 102:10557–10562. https://doi.org/10.1073/pnas.0409137102

Moses J, Johnson CG (2003) An ant colony algorithm for multiple sequence alignment in bioinformatics. In: Artificial Neural Nets and Genetic Algorithms, pp 182–186. doi: https://doi.org/10.1007/978-3-7091-0646-4_33

Moustafa N, Elhosseini M, Taha TH, Salem M (2016) Fragmented protein sequence alignment using two-layer particle swarm optimization (FTLPSO). J King Saud Univ Sci. https://doi.org/10.1016/j.jksus.2016.04.007

Narimani Z, Beigy H, Abolhassani H (2013) A new genetic algorithm for multiple sequence alignment. Int J Comput Intell Appl 4. https://doi.org/10.1142/S146902681250023X

Nasser S, Vert GL, Nicolescu M, Murray A (2007) Multiple sequence alignment using fuzzy logic. IEEE symposium on Computational intelligence and bioinformatics and computational biology, CIBCB'07, pp 304–311

Notredame C (2002) Recent progress in multiple sequence alignment: a survey. Pharmacogenomics 3:131–144

Notredame C, Higgins DG (1996) SAGA: sequence alignment by genetic algorithm. Nucleic Acids Res 24:1515–1524

Ortuno FM, Valenzuela O, Rojas F, Pomares H, Florida JP, Urquiza JM, Rojas I (2013) Optimizing multiple sequence alignments using a genetic algorithm based on three objectives: structural information, non-gaps percentage and totally conserved columns. Bioinformatics 29:2112–2121

Phillips AJ (2006) Homology assessment and molecular sequence alignment. J Biomed Inform 39:18–33. https://doi.org/10.1016/j.jbi.2005.11.005

Pierri CL, Parisi G, Porcelli V (2010) Computational approaches for protein function prediction: a combined strategy from multiple sequence alignment to molecular docking-based virtual screening. Biochim Biophys Acta (BBA) – Proteins and Proteomics 1804:1695–1712. https://doi.org/10.1016/j.bbapap.2010.04.008

Potter RM (2008) Constructing phylogenetic trees using multiple sequence alignment. Thesis submitted to University of Washington

Priyanka A, Sathiyakumari K (2015) A comparative study of hidden Markov models learned by optimization techniques using DNA data for multiple sequence alignment. Int J Sci Eng Res 6:15–19

Purohit HJ, Raje DV, Kapley A (2003) Identification of signature and primers specific to genus pseudomonas using mismatched patterns of 16S rDNA sequences. BMC Bioinform 4:19. https://doi.org/10.1186/1471-2105-4-19

Raje DV, Purohit HJ, Lijnzaad P, Singh RN (2006) Statistical analysis of counts and spacing of consistent repeating patterns in a set of homologous DNA sequences. Curr Sci 91:789–795

Rani RR, Ramyachitra D (2016) Multiple sequence alignment using multi-objective based bacterial foraging optimization algorithm. Biosystems 150:177–189. https://doi.org/10.1016/j.biosystems.2016.10.005

Rasmussen TK, Krink T (2003) Improved hidden Markov model training for multiple sequence alignment by a particle swarm optimization – evolutionary algorithm hybrid. Biosystems 72:5–17. https://doi.org/10.1016/S0303-2647(03)00131-X

Sahraeian SME, Yoon BJ (2010) PicXAA: greedy probabilistic construction of maximum expected accuracy alignment of multiple sequences. Nucleic Acids Res 38:4917–4928. https://doi.org/10.1093/nar/gkq255

Shao L, Chen Y. (2009) Bacterial foraging optimization algorithm integrating Tabu search for Motif discovery. Proceedings of the 2009 I.E. International conference on Bioinformatics and biomedicine, pp 415–418. doi: https://doi.org/10.1109/BIBM.2009.12

Soto W, Becerra D (2014) A multi-objective evolutionary algorithm for improving multiple sequence alignments. In: Advances in bioinformatics and computational biology, vol 8826, pp 73–82. doi: https://doi.org/10.1007/978-3-319-12418-6_10

Srhan AAA, Daoud EA (2013) A hybrid algorithm using a genetic algorithm and cuckoo search algorithm to solve the traveling salesman problem and its application to multiple sequence alignment. Int J Adv Sci Technol 61:29–38. 10.14257/ijast.2013.61.04

Sullivan OO, Suhre K, Abergel C, Higgins DG, Notredame C (2004) 3DCoffee: combining protein sequences and structures within multiple sequence alignments. J Mol Biol 340:385–395. https://doi.org/10.1016/j.jmb.2004.04.058

Sun J, Palade V, Wu X, Fang W (2014) Multiple sequence alignment with hidden Markov models learned by random drift particle swarm optimization. IEEE/ACM Trans Comput Biol Bioinform 11:243–257. https://doi.org/10.1109/TCBB.2013.148

Thompson JD, Higgins DG, Gibson TJ (1994) CLUSTAL W: improving the sensitivity of progressive multiple sequence alignment through sequence weighting, position-specific gap penalties and weight matrix choice. Nucleic Acids Res 22:4673–4680

Thompson JD, Plewniak F, Poch O (1999) A comprehensive comparison of multiple sequence alignment programs. Nucleic Acids Res 27:13

Thompson JD, Koehl P, Ripp R, Poch O (2005) BAliBASE 3.0: latest developments of the multiple sequence alignment benchmark. Proteins: structure, function, and bioinformatics. doi: https://doi.org/10.1002/prot.2052

Walle IV, Lasters I, Wyns L (2004) Align-m: a new algorithm for multiple alignment of highly divergent sequences. Bioinformatics 20:1428–1435. https://doi.org/10.1093/bioinformatics/bth116

Xu X, Lei X (2010) Multiple sequence alignment based on ABC_SA. In: Artificial intelligence and computational intelligence, pp 98–105. doi: https://doi.org/10.1007/978-3-642-16527-6_14

Yang XS (2009) Firefly algorithms for multimodal optimization. In: Stochastic algorithms: foundations and applications. Proceedings of 5th International symposium, SAGA, pp 169–178. doi: https://doi.org/10.1007/978-3-642-04944-6_14

Yang XS (2010) A new metaheuristic Bat-Inspired algorithm. Nature Inspired Cooperative Strategies for Optimization (NICSO 2010), vol 284, pp 65–74. doi: https://doi.org/10.1007/978-3-642-12538-6_6

Yang XS, Deb S (2009) Cuckoo search via Lévy flights. World Congress on Nature & Biologically Inspired Computing (NaBIC, 2009). doi: https://doi.org/10.1109/NABIC.2009.5393690

Zou Q, Hu Q, Guo M, Wang G (2015) HAlign: fast multiple similar DNA/RNA sequence alignment based on the center star strategy. Bioinformatics 31:2475–2481. https://doi.org/10.1093/bioinformatics/btv177

Construction of Gene Networks Using Expression Profiles

5

Harun Pirim

Abstract

Large biological data sets require powerful tools such as co-expression network construction for detailed analysis. Analyzing the gene co-expression data of a species using a clustering method is the crucial step in order to mine the relevant information to identify the key genes or the groups (modules) of key genes. In other words, clustering the expression data helps identify the genes co-expressed significantly in the species of interest. Similarly expressed genes may have a common function; they may be residing in the same pathway, regulatory and signaling mechanisms, while their products form complexes. Clusters of highly interacting genes can be identified by construction and analysis of co-expression networks. Furthermore, each cluster may be summarized using eigengene or a hub gene. Network analysis can relate clusters to each other or to external experiment traits. The network may also be employed in the calculation of cluster membership quality measures. By the application of graph mining algorithms, tight clusters of co-expressed genes might be discovered leading to finding out new gene functions, revealing biomarkers and disease-related genes. The chapter reviews the state-of-the-art gene co-expression network construction studies and discusses the recent applications while explaining the network concepts related to the gene co-expression network analysis.

Keywords

Network Construction · Gene Co-expression networks · Regulatory networks · Weighted Gene Co-expression Network Analysis · Gene clusters

H. Pirim (✉)
Systems Engineering Department, King Fahd University of Petroleum and Minerals, Dhahran, Saudi Arabia
e-mail: harunpirim@gmail.com

H. J. Purohit et al. (eds.), *Soft Computing for Biological Systems*,
https://doi.org/10.1007/978-981-10-7455-4_5

5.1 Introduction

Modeling is a perception of reality. Mathematical models give us insights about the way nature is working. Graph theory has unique tools to capture interesting features of genome-level interactions and associations. Network modeling emerges as a fruitful field and a fundamental tool to represent the interactions between biological molecules/agents/objects. Network representation of biological systems or systems biology enables and guides discoveries for disease mechanisms effecting prognosis, epidemics, cancer therapies, revealing complicated biological processes and diagnosis among many others. In a way, biological network construction triggers insights on how genetic blueprints with environmental factors characterize a living system (Wang and Huang 2014). Network models involve many parameters, and fitting them to data is nontrivial (Filkov 2006). Genetic regulatory networks research employs modeling often at multiple levels, statistical models being at the most basic level for high-throughput data analysis (Bolouri 2014). The analysis may predict the genetic or environmental impacts on cells as well as potential drug targets (Bolouri 2014).

High-throughput data generation and sequencing as well as technological advances (i.e., cloud computing) rapidly change the way computational biology is performed and what biological insights it can deliver (Bolouri 2014). The post-genome era brings the challenge of biological data analysis to help biologists in their real-life experiments (Lee and Tzou 2009).

Since the technological advances in DNA sequencing that occurred at the end of the twentieth century, quantitative skills have become essential to distill meaning from the vast emerging and increasingly diverse data sets. Modern technological developments in high-throughput data-producing technologies such as microarrays and RNA sequencing enable generation of terabytes of data in a quite short time. The type of data generated comprises levels regarding abundance of RNA, quantification of protein-protein interactions (PPI), and many other biological molecular interactions. The generated data is embraced for statistical inference and computational analysis including low-level data processing and high-level algorithmic analysis with computations and machine-learning techniques. Making use of the data is a reverse engineering approach as illustrated in Fig. 5.1, adapted from Lee and Tzou (2009). Gene expression microarrays measure interactive activities of thousands of genes. The nodes of the co-expression matrix are gene products, and the edges of the matrix represent the relationship between the products (usually expressed by correlations). The rows of the matrix are gene products, while the columns are the experiments/samples/tissues. The numeric values of the matrix are the expression values of genes across the experiments. The experiments maybe "control vs. treated" or time course.

To gain insight into both co-regulation of genes on the same pathway and functional similarity between genes, researchers typically resort to clustering of genomic data. Both co-regulation and functional similarity studies aim to reveal

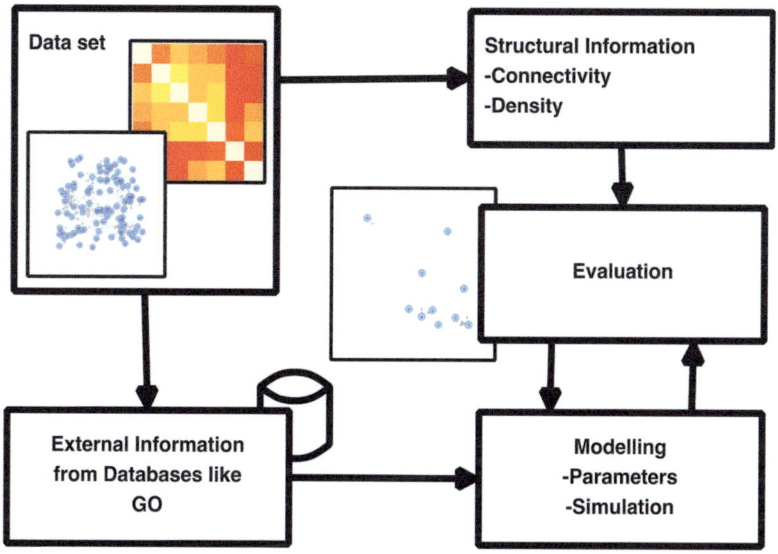

Fig. 5.1 Reverse engineering approach for co-expression network inference

patterns hidden in gene expression data to help uncover putative functional annotations and biological pathways. Many different approaches and methods have been applied for clustering genomic data, an area which is at the focus of diverse disciplines such as bioinformatics, statistics, and computer science.

Availability of the abundant biological data from microarrays, next-generation sequencing (NGS), or similar technologies require efficient modelling and analysis approaches to guide biological experiments and infer about biological processes. Clustering is a high-level data analysis technique employed for grouping biological objects exhibiting similar patterns under some conditions. It is also an important research problem in both bioinformatics and medical research (Liu et al. 2009). Similar objects might reveal important information regarding biological functions, disease markers, and drug targets. In healthcare, clustering may help diagnosis, prognosis, and treatment decisions (Liu et al. 2009). The way similarity between objects is defined and the way the similarity is used result in a large number of clustering algorithms developed and applied for clustering biological data. Most of them are general purpose clustering algorithms such as k-means or hierarchical clustering. These algorithms are biased toward a certain criterion. So, it is important to figure out the circumstances that an algorithm outperforms the other one (D'haeseleer 2005). Model-based statistic applications as in Wang and Zhu's (2008) work are assuming that biological data fit a statistical distribution. Self-organizing maps (SOM) and meta-heuristics such as genetic algorithms are two different categories of clustering algorithms developed by computer science community among others. These clustering approaches lack customization for specific features of biological data on hand.

The best model for a biological network cannot be found due to the NP complete nature of the problem (Janjic and Przulj 2012). Based on the fact, clustering models used in biological networks are general such as clique finding (see Balasundaram et al. 2005).

Genes are co-expressed under certain conditions, while they behave independently under some conditions (Dharan and Nair 2009). Bi-clustering clusters rows and columns of data sets, simultaneously grouping genes co-expressed under certain conditions. Hence, bi-clustering deserves a special focus while developing clustering approaches and heuristics. Developing clustering quality metrics is essential as there is no best cluster validation index (Almeida et al. 2011).

5.2 Genetic Regulatory Networks

Genetic regulatory networks (GRNs) are sets of interactions between genes and their product RNAs and proteins that determine the rate of RNA expression (Bolouri 2014). Gene regulation has sequential processes involving transcription and translation that control the gene expression. Expression triggers specific protein production (Filkov 2006). Gene regularity networks hold biological interactions at different molecular levels. Biomolecules such as RNA and transcription factors (TF) may induce or suppress gene expression. The interactions may be physical attachments directly (i.e., TF binding on a local sequence) or indirect effects of genes on each other including some intermediates (Wang and Huang 2014). GRNs can exhibit nonlinear behaviors, and they are composed of many components and interactions requiring mathematical and computational approaches supported by high-throughput technologies; integrative analysis of complementary data, making use of public data sets; and computational modeling and analysis (Bolouri 2014). Chai et al. (2014) present a review on computational approaches for gene regulatory network construction including Boolean, Bayesian neural networks, and ordinary differential equations. The direct and indirect relations of genes can be validated either by biological experiments such as gene knockouts or computational analysis, the latter one being cost-efficient. Computational analysis of co-expression networks employs graph-theoretic approaches such as network decomposition (graph partition), network statistics (connectivity, degrees distribution, clustering coefficient, etc.), special network motif finding, and topological network inference (i.e., scale-free networks). The way of network construction is chosen based on the biological question and hypothesis, the available data attributes, and computational background of the analyst. The co-expression network construction procedure is shown in Fig. 5.2, adapted from Serin et al. (2016). The gene co-expression networks are built using microarray data unless otherwise stated. RNA-seq data-based co-expression network construction is not at its maturity level yet (Ballouz et al. 2015). Ballouz et al. (2015) find a major difference between RNA-seq and microarray co-expression network topologies in terms of overlaps between hub-like genes from each network. They discuss the advantages and disadvantages of using RNA-seq over microarrays. Microarrays are widely used since they are relatively

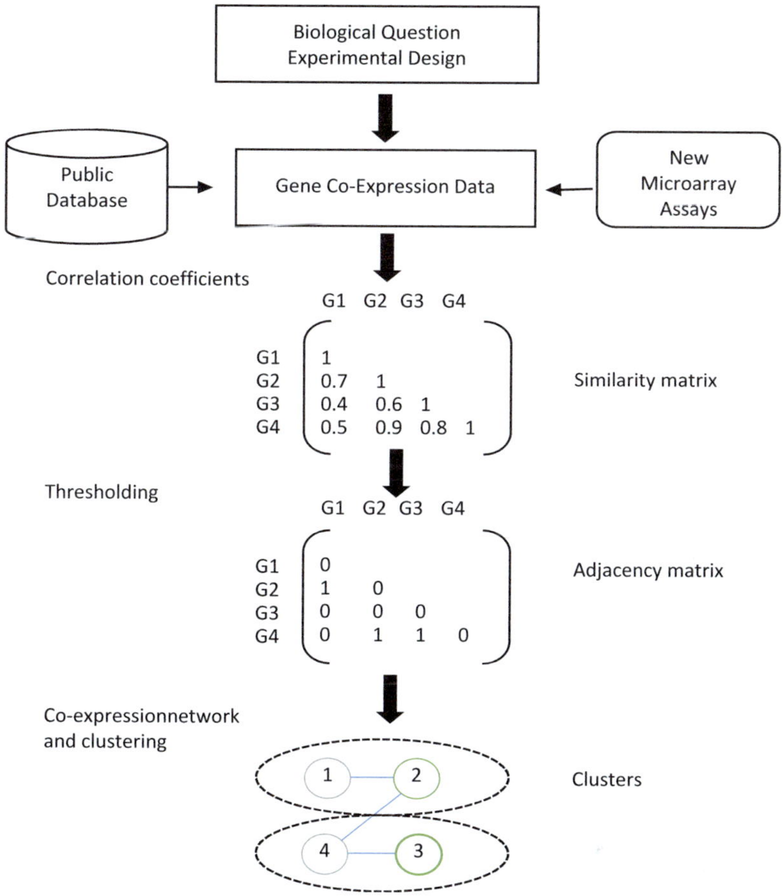

Fig. 5.2 Co-expression network construction

cheap, and their analysis is highly standardized (Serin et al. 2016). Some of the public microarray databases are GEO, ArrayExpress, and Genevestigator. Interestingly, approximately one in four studies uses public data to address a biological problem instead of generating new data (Rung and Brazma 2013). Co-expression cannot predict the outcomes of perturbations since the relationships are not directed (Yeunga et al. 2011). Other types of data are required to be able to define direction of the relations. Time regulations data may provide such an information. A typical microarray data is illustrated in Table 5.1.

In order to see the big picture of co-expression network construction and analysis, a few common processes can be summarized (Bolouri 2014):

1. Obtaining a filtered data
2. Making use of network inference or guilt by association as in clustering

Table 5.1 A sample microarray data (Iyer et al. 1999)

Gene name	OHR	15MIN	30MIN	1HR	2HR	4HR	6HR
ESTW95908	1	0.72	0.1	0.57	1.08	0.66	0.39
SID487537 ESTAA045003	1	1.58	1.05	1.15	1.22	0.54	0.73
SID486735	1	1.1	0.97	1	0.9	0.67	0.81
Genes	–	–	–	–	–	–	–
	Expression values						
	–	–	–	–	–	–	–
MAP kinase phosphatase-1	1	2.09	3.37	5.52	4.89	3.05	3.27
MAP kinase phosphatase-1	1	1.52	4.39	7.03	5.45	2.93	3.91
MAP kinase phosphatase-1	1	2.25	4.67	7.94	5.94	3.76	4.46

3. Enrichment analysis to see biological relevance of computational outputs
4. Extension of the model(s) integrating multiple data types such as mRNA, miRNA data from RNA-req, TF DNA-binding data from ChIP-seq, and protein interaction data from mass spectrometry

Ideally, network decomposition results in tight clusters/modules with dense intracluster and sparse intercluster connections. Tight clusters are supposed to include biologically relevant genes in terms of functions or residing in the same pathway.

The networks may be static or dynamic. Static networks rely on a similarity measure between genes. Dynamic networks change over time giving a more realistic picture of genetic changes. They forecast future states of a living organism. As a matter of fact, dynamic networks require more samples making biological experiments more expensive. Once the similarity (or distance) is defined, the co-expression network is constructed using guilt by association logic meaning genes with similar expression levels under different treatments have similar functions, and they are potentially co-regulated (Wang and Huang 2014). In construction of static networks, correlation values are used. Then the network is called correlation-based co-expression network. The correlation may be Pearson, rank-based ones, or some other distance measures such as Euclidean. Pearson measure is a stronger indicator than rank-based correlations; however, it is less resistant to outliers in the data. Kumari et al. (2012) compare different distance measures on both simulated and real data sets. Mutual information (MI) is another way of measuring of relationship for construction of co-expression networks. MI is able to capture nonlinear relations in practice (Wang and Huang 2014). MI is more expensive computationally. Partial correlation measure gives a way for conditional correlations such that all genes may not have pairwise relationships. The networks constructed with abovementioned measures are undirected. Bayesian networks are used to construct casual, directed relationships between genes. Some other types of data are required besides co-expression data to be able to define the directed relationship. High-throughput data integration remains a challenging and necessary task.

5.3 Co-expression Networks

Networks constructed based on gene expression similarity are called gene co-expression networks (Serin et al. 2016). They can be named association, correlation, and influence networks (Fuller et al. 2011). Co-expression network analysis requires selection of a similarity measure between genes and a clustering algorithm to decompose the network into functional clusters/modules after a meaningful experiment design. However, there exists clustering algorithms that do not require a distance matrix as an input; rather they require the network itself (e.g., some community structure finding algorithms). Then, the modules require biological inference. Co-expression analysis is used for many species including yeast, humans, plants, flies, mice, and worms. A lot of studies report that gene co-expression networks exhibit a scale-free feature. The degree distribution of the nodes and the hierarchical organization of an expression network are different from other biological networks (Ruan et al. 2010). However, Ruan et al. (2010) report that once the co-expression networks are constructed based on rank-based methods compared to the value-based ones, they tend to have common topological properties with other biological networks.

5.3.1 Identifying Genes with Key Roles

The genomic resources culminated in the release of the whole-genome sequence database (Yin et al. 2008). For example, poplar is an ideal model plant system to investigate both the spatial and the temporal arrangements of local and systemic resistances against herbivores due to its size and longevity (Consortium 2007). One computational way to identify the genes with key roles is microarray analysis of poplar genome. Microarray data from some poplar experiments may be "control vs. treatment" or "time course." Gene expression data is extracted from microarrays.

Analyzing the gene expression data of poplar using a clustering method is the crucial step in order to mine the relevant information to identify key (hub) genes or groups (modules) of key genes. In other words, clustering the expression data helps identify the genes co-expressed significantly in poplar.

Co-expressed genes are likely to have similar functions. Hence, in case one of the genes with an unknown function is grouped with genes with a known function using a clustering method, it is probable that the gene has a similar function with the other genes in the group. However, it is likely that a gene with an unknown function is grouped with the genes whose functions are unknown as well. At this point, the sequence of the gene is searched through a relevant database (e.g., JGI, AspenDB, TIGR, TAIR) holding sequence information of poplar and similar plants (*Arabidopsis thaliana* for poplar). The sequence of the gene whose function is unknown is searched in the database of the most sequence-similar species with an idea that there may be homolog of the gene for which the database is queried. Using a BLAST algorithm is one of the most common ways of searching a gene sequence through many databases.

As mentioned above, clustering is a reasonable start point to identify the genes having key roles in the poplar genome. It can help focus the research on the genes that have a high probability of importance (Yin et al. 2008). A researcher is usually interested in finding the hub genes (the genes having many connections to other genes) since the hub genes are thought to be the crucial genes. Usually, clustering algorithms are employed to detect the hub genes.

For example, the floral induction is essential for poplar to have early-flowering genotype. This is a start point for identifying the key genes in poplar (there may be many different hub genes based on the biological experiment; here flowering process is given as an example) since flowering is an inherent feature of a poplar. Conducting exhaustive research on genotypes that are known to flower early is the first computational step. Using mining tools like Chilibot (http://www.chilibot.net) that searches through PubMed abstract database to find specific relationship between proteins, genes, and keywords regarding the flowering of poplar is a computational way. A list of genes obtained as an output of a clustering method for poplar genome can be used as an input for Chilibot as well. The list of genes is searched with different relationship criteria through PubMed abstracts. The pulled results may give insights for the functions of the genes in the list, and one of these genes may be the focus for various follow-up treatments (chemical, physical, cultural) shown to be successful in other woody angiosperms as mentioned in the *Populus* genome science plan 2004–2009 (http://web.ornl.gov/sci/ipgc/).

The second computational step to identify the key genes in poplar is identifying the gene modules using an appropriate clustering approach. An appropriate clustering approach is bi-clustering which groups similarly expressed genes over some of the samples. It clusters both rows (genes of poplar) and columns (conditions or samples from various treatments on poplar genes) of a data matrix (gene expression matrix of poplar) simultaneously (Mitra et al. 2009). Some of the justifications to use bi-clustering for poplar microarray data are:

1. Poplar has thousands of genes which may not be relevant to the analysis a researcher is interested, and even these genes may hide the contribution of the relevant genes in the expression data of poplar.
2. Co-expressed poplar genes mostly behave independently under certain conditions (Dharan and Nair 2009).

Based on the treatment used in the experiment for poplar, for instance time-course samples collected and embedded to microarray chips, bi-clustering results are evaluated focusing on the module in which a hypothetical key gene for poplar's flowering exists. This evaluation is realized in multiple ways: by comparing the genes with homologs as mentioned above, designing further experiments for the genes of the focused module from bi-clustering of poplar expression data, and analyzing the network topology (microarray data can be inferred using network formalism) of expression data to infer about hub genes using the measures from graph theory such as connectivity, number of neighbors, etc.

In conclusion, a typical work flow for identifying key genes in polar is:

1. Use Chilibot to have an idea about key role genes.
2. Bi-cluster microarray dataset from a poplar experiment.
3. Focus on the modules of clusters to identify the key role genes using databases such as JGI, AspenDB, and GO, and use Chilibot for querying the genes of the same module.
4. Use network topology to infer about hub genes which are thought to be the key genes of poplar.

5.3.2 Construction of Large-Scale Regulatory Networks

Constructing large-scale regulatory networks to understand how a complete system (i.e., human system) functions and interacts with environmental factors requires unveiling regulatory mechanisms of each type of molecular elements (e.g., genes, transcripts, TFs, microRNAs) and integration of all these mechanisms to be able to see the big picture at system level as well as linking the dynamic regulatory mechanisms of molecular elements with the environmental factors or stimuli they are triggered.

Some of the environmental factors triggering gene regulation are temperature, pressure, oxygen concentration in air, drinking water, nitrates, burns, and allergic stress. Building a model to represent the complex nature of large-scale regulatory system of human functions and interactions is the first step to construct the large-scale regulatory network. This model should be able to integrate high-dimensional data of different formats (expression data, real numbers; relevance data, binary; qualitative data, high, low, average) to be able to make the best use of available experimental biology information available. For example, Huttenhower et al. (2009) state that a part of a complex regulatory system can be modeled as a combination of regulatory modules including co-regulated genes, co-regulation conditions, and regulatory motifs at sequence level. Linear methods are useful in obtaining information about large-scale networks, whereas focusing on smaller networks for functional information about genes, the probabilistic nature of biological processes, should be taken into account making use of Bayesian statistics.

In unicellular systems, regulation often occurs based on TF binding sites and activation or repression of transcription (Huttenhower et al. 2009). By this assumption microarray data are clustered, promoter sequences of genes in each cluster are tested for enriched motifs, and the consensus sequences are matched with known TF binding sites (Huttenhower et al. 2009). However, in multicellular organisms such as humans, predicting regulatory modules from microarray gene expression data is a very difficult problem where a clustering method and de novo motif discovery from DNA sequences need to be combined (Huttenhower et al. 2009).

In human genome, regulatory motifs on DNA sequences may be short, degenerate, and present without an indication of function. In addition to the need of combining clustering and motif discovery tasks, the way they are combined is also an important step toward achieving a system-wide understanding of human functions. While most existing approaches to regulatory module discovery are sequential in the sense that motif finding follows clustering or they are approached as separate tasks, it is reasonable to perform these tasks simultaneously as is (Huttenhower et al. 2009). In other words, it is an integrative approach to cluster genes regarding both co-expression and enrichment of regulatory motifs. Clustering algorithms scalable to high-dimensional data sets and flexible to use data of different formats are necessary for understanding human functions through constructing regulatory networks. Regulation data, alternate splicing data, noncoding regulatory element data, any kinds of experimental data (Huttenhower et al. 2009), metabolic network data, signal transduction network data, protein-protein interaction data, DNA-binding regulator data, temporal gene expression data, cellular population harvested from different individuals (Margolin et al. 2006), etc. would be necessary to accomplish constructing large-scale regulatory networks to understand how a human system functions and interacts with environmental factors.

There are many approaches for forming regulatory networks such as Boolean circuits and complicated nonlinear spatial models (Gustafsson et al. 2005). A network formation based on just mRNA data would not represent the complete picture of a network involving metabolites, proteins, etc. (Gustafsson et al. 2005). For example, transcriptional data is used to construct a gene-to-gene regulatory network where many physical molecular connections and some intermediate products in regulatory cascades are hidden (Gustafsson et al. 2005).

Available approaches for constructing regulatory networks involve some drawbacks: overfitting, reliance on nonrealistic network models, computational complexity, and critical dependency on supporting data which is just available for unicellular organisms (Margolin et al. 2006). Large gene expression data derived from perturbations to unicellular organisms are not easily obtained for humans; however, it is suggested that an equivalent dynamic content fullness could be obtained using natural and experimentally generated phenotypic variations of a given cell type like B cell (Margolin et al. 2006).

In conclusion, one of the work flows for constructing large-scale regulatory network of humans is:

1. Obtain genome-wide data sets for different stress conditions from humans in which B cells are used.
2. Assemble data sets for B cell expression profile.
3. Use complementary data (protein, metabolites, etc.) for a clustering algorithm (GRAM (Joseph et al. 2003)).
4. Construct regulatory network based on the results of clustering with simplification on edge weights (converting to binary network) using a measure of association.

5.4 Weighted Gene Co-expression Network Analysis (WGCNA)

WGCNA is a powerful framework/tool supported in R computing environment to find groups of genes with similar expression profiles, summarize clusters using cluster eigengene and intramodular hub gene, relating clusters to each other and to external sample traits (by network eigengene), and calculate cluster membership metrics (Fuller et al. 2011). WGCNA can be summarized in the following steps:

1. Construction of a co-expression network
2. Computing connectivity for each gene
3. Computing gene significance (please refer to the "network concepts" section)
4. Identifying clusters and cluster eigengenes
5. Finding cluster intramodular connectivity and genetic hubs
6. Relating cluster concepts to each other to identify disease-related genes and clusters (Fuller et al. 2011)

Fuller et al. (2011) define differential expression and connectivity based analysis (DWGCNA) that identifies genes and pathways based on both expression and connectivity. They additionally define IWGCNA that uses correlation to compare gene expression profiles, clinical traits, and genetic markers. Imprialou (2012) aims to extend WGCNA to include RNA-seq data.

Tejera et al. (2013) analyze co-expression network for prioritization of genes in preeclampsia. They use R packages to process microarrays. Affymetrix platform raw data are preprocessed using *mas5* and *log2* transformation. Illumina platform raw data are subject to batch correction, normalization, and log2 transformation. The experimental data are obtained from Gene Expression Omnibus (GEO) including normal (N) and preeclamptic pregnancies (PRE). Statistically different expressed genes between N and PRE groups are considered for construction of co-expression networks. The authors report 1146 such genes. The weighted gene co-expression network analysis (WGCNA) package (Langfelder and Horvath 2008) is used for the network construction. Song and Zhang (2015) developed a framework for co-expression network analysis. The framework comprises controlling the quality of co-expression similarities, construction of the network, and a clustering approach. They criticize WGCNA enforcing the connectivity of network nodes in a way to make the underlying network scale-free. They also mention drawbacks of some other type of co-expression network analysis. While WGCNA uses soft thresholds, the unweighted networks contain false-positive interactions since hard thresholds are applied for the network construction. K-nearest neighbor networks require a subjective criterion called connectedness. Some other methods such as partial correlation-based co-expression network analysis are computationally expensive. Also, the clustering methods within the mentioned network analysis approaches make the process more problematical. For example, *k-means* algorithm asks for the number of clusters as an input argument. Many graph mining approaches lack multi-levels of cluster combinations existing together within a single network (Song and Zhang 2015). Modularity maximization-clustering algorithms fail to detect network structures that involve both course-grained and compact clusters. The authors use planar maximally filtered graph (PMFG) to retrieve information from similarity matrices. They argue that the PMFG is an ideal platform to construct co-expression network for the following reasons:

1. Hierarchy preservation using subgraphs generated by minimum spanning trees
2. A coherent cluster and a connected subgraph correspondence
3. Exhibition of clustering structures such as cliques
4. Presenting different network features in embedded networks such as scale-freedom, small-world characteristic

The new framework is shown to be superior compared to Infomap, Walkstrap, leading eigenvector clustering, and WGCNA. The framework is available as an R package.

Yang et al. (2014) examine prognostic gene characteristics in four different cancer co-expression networks. They report that the prognostic mRNA genes tend not to be hub genes. However, they are enriched in modules. They construct the network using Agilent microarray data employing WGCNA. Cogill and Wang (2014) use WGCNA for co-expression network analysis of human IncRNAs and cancer genes. They visualize the networks using VisANT software (Hu et al. 2013). Liao et al. (2011) employ WGCNA to construct coding-noncoding co-expression network. The experiments include 30–40 data sets with 9 or more samples. Each data set is processed by expression variance ranks of genes to which 75 percentile is retained. WGCNA package is used to implement Fisher's asymptotic test. Gene pairs with *p-value* less than or equal to 0.01 and Pearson correlation value ranked at the top or the bottom 0.05 percentile are accepted as co-expressed. Jiang et al. (2016) apply the WGCNA to construct the network in *Mycobacterium tuberculosis*. They determine the soft thresholding parameter to be five, and the algorithm converts the correlation coefficients between genes into the adjacency coefficients. Topological overlap matrix of dissimilarity is generated using the adjacency coefficients. Then agglomerative hierarchical clustering is applied. Some other assumptions made are average connectivity usage between different classes, a gene module size being at least ten. The R Bioconductor limma package is used to process the raw data. They summarize a typical workflow for the WGCNA as follows:

1. Calculate the similarity matrix (S) values (S_{ij}) indicating relations between gene pairs (S_{ij}) using Pearson's correlation coefficient (cor(i,j)). The following relationship holds

$$S_{ij} = |cor(i,j)|$$

2. Define exponential weighted value β such that the adjacency is

$$a_{ij} = |S_{ij}|^{\beta}$$

3. Convert the adjacency matrix $A = [a_{ij}]$ into the topological overlap matrix $\Omega = [\omega_{ij}]$

$$\omega_{ij} = \frac{l_{ij} + a_{ij}}{min\{k_i, k_j\} + 1 - a_{ij}}$$

$$l_{ij} = \sum_{\mu} a_{i\mu} a_{\mu j} \; k_i = \sum_{\mu} a_{i\mu} \; k_j = \sum_{\mu} a_{j\mu}$$

4. Build the hierarchical clustering tree using dissimilarity measure $d_{ij}^{\omega} = 1 - \omega_{ij}$. Once the tree is cut at a level, the branches of the tree will represent the clusters (modules).
5. Construct the co-expression network.

Ferrari et al. (2016) apply WGCNA on 101 individual non-neurodegenerative disease microarray data sets to investigate co-expression profiles in the frontal and temporal cortices for 12 genes. They suggest a shift in the study of a disease from gene to pathology to gene to networks to pathways strategy. Similarly, Bettencourt et al. (2015) use 101 neuropathologically normal individuals' gene expression data to analyze using WGCNA. Medina and Lubovac-Pilav (2016) apply WGCNA for determining clusters and pathways enriched in functions regarding type 1 diabetes (T1D). Public microarray data sets of 43 T1D and 24 control data sets are used. R affy package is used for normalization of the raw data. They employ module preservation statistics in the package. The statistic tests in a given module in the healthy network (control data) are present in the disease network. They also make use of betweenness centrality metric as a topological analysis. The betweenness centrality values of the genes are computed after obtaining the modules from the WGCNA. The measure reflects influence over the information transfer between genes. The measure is calculated for both the disease and the healthy network. Rodius et al. (2016) employ WGCNA and clustering with overlapping neighborhood expansion (Nepusz et al. 2012) to analyze the dynamic heart regeneration co-expression network in the zebrafish. Pre-processed microarray data are filtered by an FDR method and variance. They indicate that the groups of transcriptionally coordinated genes and the hub genes of the network mediate the regeneration steps. Maschietto et al. (2015) apply WGCNA using 29 schizophrenia (SZP) and 30 control (CTS) brain tissue microarray data sets. They detect important clusters of co-expressed genes. They apply the Wilcoxon rank sum test to identify differentially expressed genes between SZP and CTS considering *p-values* less than 0.01. Guo and Xing (2016) construct weighted gene co-expression networks using

WGCNA. They apply RMA normalization correction in R affy package. Topological characteristics including clustering coefficient and average shortest paths of specific gene modules are calculated. The 12 modules with large clustering coefficients and small average shortest paths possess small-world feature.

5.5 Other Gene Co-expression Network Construction Applications

Zhao et al. (2016) analyze co-expression network of Down syndrome from microarray data sets. The data sets are acquired from GEO database and pre-processed by R Bioconductor oligo package. Microarray data sets are processed through background correction, quantile normalization, and probe summarization. R samr package is used to monitor differentially expressed genes. Zhang et al. (2012) apply Quasi-Clique Merger (QCM) (Ou and Zhang 2007), a graph mining algorithm to determine co-expression networks which are tightly connected from microarray data sets. Their work flow includes computing Pearson correlation coefficients for gene pairs of cancer data sets and normal tissue data sets. Then, frequency tables using cancer and normal data sets for regarding each gene are constructed. The QCM algorithm is used for mining and merging cancer and normal networks. Lehtinen et al. (2015) utilize yeast co-expression networks before and after being subject to stress to model the stress effect on mutational robustness. Robustness is defined as the ability to maintain biological function when a perturbation happens. The study suggests that the stress increased tolerance to loss of function mutations and future perturbations based on the same or different stress types. Interesting results of the study are that damage distribution is different in scale-free and Erdos-Renyi (ER) graphs, essential gene removal causes greater efficiency loss than removal of a nonessential gene, changes from stress are not explained by degree distribution changes, and type of damage distribution has functional consequences. The network generation and network operations are executed using NetworkX package for Python. Ruan et al. (2010) present a robust rank-based network construction method. They propose a parameter-free clustering algorithm and a new reference network-based measure for validating the quality of the partition. A Matlab implementation of their algorithm is available. Mao et al. (2009) construct Arabidopsis gene co-expression network making use of 1094 microarrays. They cluster the network into modules using a graph mining algorithm. They find out 382 hub genes forming a clique. Wei et al. (2015) utilize microarrays to screen gene expression profiles in mice hearts, and they confirmed their results with qPCR analysis. To construct network, they calculate Pearson correlations. Network analysis includes computing degree centrality and detecting k-cores subgraphs. Knott et al. (2010) introduce two algorithms to construct gene networks: one is based on the sensitivity analysis that was a systematic perturbation of nonlinear neural networks; the other one is a heuristic search approach based on gene set stochastic sampling. Leal et al. (2014) show complex plant immune responses based on construction and comparison of gene co-expression networks.

They employ a multivariate approach based on principal component analysis. Jing et al. (2010) present a hybrid approach for gene expression and gene ontology to construct the gene network. They experimentally show that their method was faster than Bayesian networks. Liang et al. (2014) construct gene network using Pearson correlation coefficient gene expression similarity, visualizing expression data by Cytoscape and identifying modular structures using a community structure finding algorithm, Qcut (Ruan and Zhang 2008).

5.6 Determining the Thresholds and Clusters for Co-expression Networks

There isn't any natural way of determining a biologically relevant threshold (Rubinov et al. 2009). The best threshold for converting a complete biological weighted network to an unweighted binary one would emerge by reflecting the relations among the genes on the threshold as much as possible. There are some common ways used for threshold determining. One of them is using a threshold that retains a percentage of strongest (highest weighted) edges on the network (Rubinov et al. 2009; Schwarz et al. 2009). Another way of determining a threshold is based on nodes' connectivity (Bartolomei et al. 2006). The threshold is determined such that nodes have a fixed number of connectivity or connectivity greater than a fixed number. One other way is using a threshold corresponding to a significance level based on Fisher's correlation test (Yip and Horvath 2007).

Rubinov et al. (2009) apply thresholds to maintain from 10% to 30% of the strongest edges. Removing too many edges may result in a disconnected graph. Disconnected graph imposes limitations on calculating the path lengths (Rubinov et al. 2009). Schwarz et al. (2009) threshold networks so that 2% of the strongest edges retain in the unweighted network. Retaining more edges results in dense node connections and loss of the topological uniqueness. On the other hand, retaining fewer edges disconnects the network and suppresses the topological information (Schwarz et al. 2009). They also report that they ran their clustering algorithm using different threshold values for network edges. Values retaining the strongest 1–10% edges resulted in similar cluster memberships independent from the exact value of the threshold. For lower or higher fractions of edges, clusters split and merge, respectively. Bartolomei et al. (2006) choose a threshold such that for an individual graph, they obtain ten connections per node. They want to have a connected network with "small-world" feature. Yip and Horvath (2007) use a threshold to approximate a scale-free topology, corresponding to a significance level of Fisher's correlation test.

It is reasonable to determine a threshold based on the network topology we want. If we have a prior information about the structure of a network, we can determine the threshold value accordingly. For example, Bartolomei et al. (2006) mention that networks of functional connectivity based upon recordings in animals show small-world characteristics. Yip and Horvath (2007) use the threshold for the co-expression network construction of yeast genes. Hence, using a threshold for fixed connectivity may be reasonable for obtaining "small-world" feature while it may not be suitable for obtaining a scale-free network. In other words it may work for brain network; however, it may not be suitable for yeast co-expression network. In a similar manner, using a threshold based on a significance value from a statistical test may be suitable in some biological networks while it may not give good results for the others. Using a threshold to retain a percentage of edges may result in disconnected networks or densely connected networks which are the two undesirable extremes. However, applying a range of edge fractions may help finding the threshold suitable for the application on hand.

Some of the similarity matrix thresholding techniques are ad hoc methods, permutation testing, linear regression, rank-based methods, homogeneity test, spectral graph theory, information theoretic approaches, topological methods, and machine-learning approaches (Gibson et al. 2013). Random matrix theory is one of the thresholding techniques. The technique may be employed in testing the robustness of the co-expression network subject to perturbations in the input. Considering the mentioned aspects of determining threshold values above may improve the quality of a clustering result and decrease the possibility of effecting a clustering algorithm badly.

A gene co-expression network is very complex in terms of nodes and edges, since nodes represent thousands of genes and edges refer to ten thousands of relationships. A typical research question addresses the effect of a treatment at genomic level or time-course genomic changes. The result of the treatment or the time-course changes leads the researcher to focus on the significantly co-expressed genes. The fact that the classes the genes belong to is usually unknown, requires an unsupervised learning technique. Clustering, as an unsupervised learning technique, groups similar objects based on a cluster definition or a similarity metric.

A clustering approach aids the following arguments. One of the common understanding is that a small subset of genes are involved in the cellular process of interest, and a cellular process happens only in subset of samples (Jiang et al. 2004). Another argument is that the genes of the same pathway may be expressed or suppressed simultaneously or sequentially upon receiving stimuli (Zhu et al. 2008). That is, expressed genes have a closer relationship, and they are likely to appear in the same cluster. Genes with similar expression profiles are more likely to regulate each other or to be regulated by a parent gene (Mitra et al. 2009); that is an evidence that clustering may help finding the genes of the same pathway as well. A powerful clustering approach may reveal the relationships that exist in the expression data (Ma and Chan 2009).

We cluster expression data to group the genes with similar expression patterns (different clustering algorithms detect different patterns). Clusters are group of genes supposed to have similar functions or be located in the same pathway or interact with each other. It means that there is a similarity (similarity matrices are constructed upon similarity; one of them is topological overlap matrix (Yip and Horvath 2007) defining the similarity of two genes as the number of common neighbors) or relationship between two genes of the same cluster.

5.7 Network Concepts Useful in Co-expression Network Construction

The concepts mentioned here are detailed in (Horvath 2011). Given an undirected and unweighted network, connectivity (degree) of a node *i* (k_i) is defined as the number of nodes directly connected to node *i*. The connectivity is defined as the summation of the connection weights between node *i* and the others in a weighted network. It is formulated as:

$$k_i = \sum_{j \neq i} A_{ij}$$

where A_{ij} is the adjacency between nodes *i* and *j*.

Scaled connectivity (K_i) is:

$$K_i = \frac{k_i}{k_{max}}$$

k_{max} can be at most $n - 1$ where n is number of nodes.

Biological networks are thought to have scale-free topology. The scale-free topology assumes frequency distribution of node degrees which follows a power law. In other words, a fraction *P(k)* of nodes with *k* connections with others in the network is approximately

$$P(k)^\sim k^{-\gamma}$$

where γ is a positive real number.

Network heterogeneity is defined as variance of the connectivity. It is formulated as

$$\frac{\sqrt{var(k)}}{mean(k)}$$

Maximum adjacency ratio of node *i* is

$$MAR_i = \frac{\sum_{j \neq i} (A_{ij})^2}{\sum_{j \neq i} A_{ij}}$$

The ratio is between 0 and 1. 1 implies deviance from a neutral relationship.

Network density is related with mean connectivity. It is defined as the mean off-diagonal adjacency formulated as

$$\frac{mean(k)}{n-1}$$

The density shows overall affection among the nodes.

Network (degree) centralization is formulated as

$$\frac{max(k)}{n} - density$$

Centralization value 1 gives a star topology, while the value 0 gives each node the same connectivity.

Clustering coefficient is a measure for cliquishness. Clustering coefficient for node *i* in an unweighted network is the proportion of triangles having node *i* among all triangles:

$$\frac{\sum_{j\neq i}\sum_{k\neq i,j} A_{ij}A_{jk}A_{ki}}{\left(\sum_{j\neq i} A_{ij}\right)^2 - \sum_{j\neq i}(A_{ij})^2}$$

Mean clustering coefficient is an indicator for a module structure in a network.

Hub node significance is an association between connectivity and node significance. A node significance (GS_i) for node *i* based on *p* value can be defined as –log ($p\ value_i$). Then, the hub node significance is

$$\frac{\sum_i GS_i K_i}{\sum_i (K_i)^2}$$

Hub genes have been shown to be essential for survival in yeast, while they are not always critical in higher organisms (Fuller et al. 2011). Hubs may refer to genes with a significant biological role or regulation activity (Mitra et al. 2013). Network significance is defined as the average node significance of the nodes.

Topological overlap measure is a normalization of shared neighbors between nodes *i* and *j*:

$$\begin{cases} \dfrac{\sum_{l\neq i,j} A_{il}A_{jl} + A_{ij}}{min\left\{\sum_{l\neq i} A_{il} - A_{ij}, \sum_{l\neq j} A_{jl} - A_{ij}\right\} + 1} & if\ i \neq j \\ 1\ if\ i = j. \end{cases}$$

Some of the network data analysis tasks include:

1. Description of direct or indirect relationships between genes. Adjacency matrix and shared neighbors are used.
2. Network statistics.
3. Module detection.
4. Measurement of different connectivity patterns between data sets.
5. Finding hub nodes. Hub nodes represent the modules.
6. Annotation of genes inside a module.

Horvath (2011) defines similar network concepts for intramodular analysis. He also defines special concepts for a network where its nodes are modules and formulations for comparison of two networks.

5.8 Conclusion

Biochemical methods of placing genes in pathways are not feasible to be applied on candidate genes for which enormous amount of genetic information is generated (Yeunga et al. 2011). The fact makes the network construction and inference necessary for co-expression analysis. Abstract models involving less biological detail are easier to be implemented on large-sized networks compared to concrete models describing networks in more details and closer to biological facts (Lee and Tzou 2009).

Bolouri (2014) lists some of the important available tools for GNRs analysis: Bioconductor (http://bioconductor.org), Cytoscape (http://cytoscape.org/), Galaxy (http://galaxyproject.org/), GenePattern (http://www.broadinstitute.org/cancer/software/genepattern), and GenomeSpace (http://www.genomespace.org/). For reproducible research, data, code, and work flow, Synapse (https://www.synapse.org) offers sharing facilities. GebeMania (http://genemania.org/) integrates different data types such as co-expression and pathway data from distinct resources. For biological inference there are many software and methods available. Some of them are TargetMine (http://targetmine.nibio.go.jp/), GeneTrail (http://genetrail.bioinf.uni- sb.de/), and David (http://david.abcc.ncifcrf.gov/). Serin et al. 2016 overview available resources for co-expression network analysis.

As mentioned in Bolouri (2014), although high-throughput data sets are publicly available, a variety of computational codes and software for biological data analysis exist; the way and expertise to employ any combination of the available resources are still ambiguous. Current gene co-expression models are too simplistic needing updates. Integration of diverse computational biology aspects on different types of data sets is essential.

Acknowledgments Harun Pirim acknowledges the support of KFUPM.

References

Almeida H, Guedes D, Meira W Jr, Zaki MJ (2011) Is there a best quality metric for graph clusters? Mach Learn Knowl Disc Databases, Lect Notes Comput Sci 6911:44–59. https://doi.org/10.1007/978-3-642-23780-5_13

Balasundaram B, Butenko S, Trukhanov S (2005) Novel approaches for analyzing biological networks. J Comb Optim 10:23–39. https://doi.org/10.1007/s10878-005-1857-x

Ballouz S, Verleyen W, Gillis J (2015) Guidance for RNA-seq co-expression network construction and analysis: safety in numbers. Bioinformatics 31:2123–2130. https://doi.org/10.1093/bioinformatics/btv118

Bartolomei F, Bosma I, Klein M, Baayen JC, Reijneveld JC, Postma TJ, Heimans JJ, van Dijk BW, de Munck JC, de Jongh A, Cover KS, Stam CJ (2006) Disturbed functional connectivity in brain tumour patients: evaluation by graph analysis of synchronization matrices. Clin Neurophysiol 117:2039–2049. https://doi.org/10.1016/j.clinph.2006.05.018

Bettencourt C, Forabosco P, Wiethoff S, Heidari M, Johnstone DM, Botía JA, Collingwood JF, Hardy J, for the UK Brain Expression Consortium (UKBEC) 2: , Milward E A, Ryten M, Houlden H (2015) Gene co-expression networks shed light into diseases of brain iron accumulation. Neurobiol Dis 87:59–68. doi:https://doi.org/10.1016/j.nbd.2015.12.004

Bolouri H (2014) Modeling genomic regulatory networks with big data. Trends Genet 30. https://doi.org/10.1016/j.tig.2014.02.005

Chai LE, Loh SK, Low ST, Mohamad MS, Deris S, Zakaria Z (2014) A review on the computational approaches for gene regulatory network construction. Comput Biol Med 48:55–65. https://doi.org/10.1016/j.compbiomed.2014.02.011

Cogill SB, Wang L (2014) Co-expression network analysis of human lncrnas and cancer genes. Cancer Informat 13:49–59. https://doi.org/10.4137/CIN.S14070

Consortium (2007) A physical map of the highly heterozygous Populus genome: integration with the genome sequence and genetic map and analysis of haplotype variation. Plant J 50:1063–1078. https://doi.org/10.1111/j.1365-313X.2007.03112.x

D'haeseleer P (2005) How does gene expression clustering work? Nat Biotechnol 23:1499–1501. https://doi.org/10.1038/nbt1205-1499

Dharan S, Nair AS (2009) Biclustering of gene expression data using reactive greedy randomized adaptive search procedure. BMC Bioinfor 10:S27. https://doi.org/10.1186/1471-2105-10-S1-S27

Ferrari R, Forabosco P, Vandrovcova J, Botía JA, Guelfi S, Warren JD, UK Brain Expression Consortium (UKBEC), Momen P, Weale ME, Ryten M, Hardy J (2016) Frontotemporal dementia: insights into the biological underpinnings of disease through gene co-expression network analysis. Mol Neurodegener 11:21. https://doi.org/10.1186/s13024-016-0085-4

Filkov V (2006) Identifying gene regulatory networks from gene expression data. Handb of Comp Mol Biol, Chapman & Hall/CRC 27-1-27-29. ISBN:1-58488-406-1

Fuller T, Langfelder P, Presson A, Horvath S (2011) Review of weighted gene coexpression network analysis. Springer Handb Comput Stat. https://doi.org/10.1007/978-3-642-16345-6-18

Gibson SM, Ficklin SP, Isaacson S, Luo F, Feltus FA, Smith MC (2013) Massive-scale gene co-expression network construction and robustness testing using random matrix theory. PLoS One 8:e55871. https://doi.org/10.1371/journal.pone.0055871

Guo Y, Xing Y (2016) Weighted gene co-expression network analysis of pneumocytes under exposure to a carcinogenic dose of chloroprene. Life Sci 151:339–347. https://doi.org/10.1016/j.lfs.2016.02.074

Gustafsson M, Hornquist M, Lombardi A (2005) Constructing and analyzing a large-scale gene-to-gene regulatory network—lasso-constrained inference and biological validation. IEEE/ACM Trans Comput Biol Bioinform 2:254–261. https://doi.org/10.1109/TCBB.2005.35

Horvath S (2011) Weighted network analysis. Appl Genet Sys Biol. https://doi.org/10.1007/978-1-4419-8819-5

Hu Z, Chang YC, Wang Y, Huang CL, Liu Y, Tian F, Granger B, Delisi C (2013) VisANT 4.0: integrative network platform to connect genes, drugs, diseases and therapies. Nucleic Acids Res 41:225–231. https://doi.org/10.1093/nar/gkt401

Huttenhower C, Mutungu KT, Indik N, Yang W, Schroeder M, Forman JJ, Troyanskaya OG, Coller HA (2009) Detailing regulatory networks through large scale data integration. Bioinformatics 25:3267–3274. https://doi.org/10.1093/bioinformatics/btp588

Imprialou M (2012) Gene co-expression network design from RNA-seq data in Arabidopsis Thaliana.url:https://www.stats.ox.ac.uk/__data/assets/pdf_file/0010/8398/arabidopsis_rnaseq.pdf

Iyer VR, Eisen MB, Ross DT, Schuler G, Moore T, Lee JC, Trent JM, Staudt LM, Hudson J, Boguski MS, Lashkari D, Shalon D, Botstein D, Brown PO (1999) The transcriptional program in the response of human fibroblasts to serum. Science 283:83–87. https://doi.org/10.1126/science.283.5398.83

Janjic V, Przulj N (2012) Biological function through network topology: a survey of the human diseasome. Brief Funct Genet 11:522–532. https://doi.org/10.1093/bfgp/els037

Jiang D, Tang C, Zhang A (2004) Cluster analysis for gene expression data: a survey. IEEE Trans Knowl Data Eng 16. https://doi.org/10.1109/TKDE.2004.68

Jiang J, Sun X, Wu W, Li L, Wu H, Zhang L, Yu G, Li Y (2016) Construction and application of a co-expression network in *Mycobacterium tuberculosis*. Sci Rep 6:28422. https://doi.org/10.1038/srep28422

Jing L, Ng MK, Liu Y (2010) Construction of gene networks with hybrid approach from expression profile and gene ontology. IEEE Trans Inf Tech Biomed 14. https://doi.org/10.1109/TITB.2009.2033056

Joseph ZB, Gerber GK, Lee TI, Rinaldi NJ, Yoo JY, Robert F, Gordon DB, Fraenkel E, Jaakkola TS, Young RA, Gifford DK (2003) Computational discovery of gene modules and regulatory networks. Nat Biotechnol 21:1337–1342. https://doi.org/10.1038/nbt890

Knott S, Mostafavi S, Mousavi P (2010) A neural network based modeling and validation approach for identifying gene regulatory networks. Neurocomputing 73:2419–2429. https://doi.org/10.1016/j.neucom.2010.04.018

Kumari S, Nie J, Chen HS, Ma H, Stewart R, Li X, Lu MZ, Taylor WM, Wei H (2012) Evaluation of gene association methods for coexperession network construction and biological knowledge discovery. PLoS ONE 7:e50411

Langfelder P, Horvath S (2008) WGCNA: an R package for weighted correlation network analysis. BMC Bioinformatics 9:559. https://doi.org/10.1186/1471-2105-9-559

Leal LG, Lo ́pez C, Kleine LL (2014) Construction and comparison of gene co-expression networks shows complex plant immune responses. Peer J 2:e610. https://doi.org/10.7717/peerj.610

Lee W, Tzou W (2009) Computational methods for discovering gene networks from expression data. Brief Bioinform 104:408–423. https://doi.org/10.1093/bib/bbp028

Lehtinen S, Bähler J, Orengo C (2015) Co-expression network models suggest that stress increases tolerance to mutations. Sci Rep 5:16726. https://doi.org/10.1038/srep16726

Liang Y, Cai B, Chen F, Wang G, Wang M, Zhong Y, Cheng Z (2014) Construction and validation of a gene co-expression network in grapevine (Vitis vinifera. L.) Hortic Res 1:14040. https://doi.org/10.1038/hortres.2014.40

Liao Q, Liu C, Yuan X, Kang S, Miao R, Xiao H, Zhao G, Luo H, Bu D, Zhao H, Skogerbø G, Wu Z, Zhao Y (2011) Large-scale prediction of long non-coding RNA functions in a coding–non-coding gene co-expression network. Nucleic Acid Res 39. https://doi.org/10.1093/nar/gkq1348

Liu J, Li Z, Hu X, Chen Y (2009) Biclustering of microarray data with MOSPO based on crowding distance. BMC Bioinformatics 10:S9. https://doi.org/10.1186/1471-2105-10-S4-S9

Ma PCH, Chan KCC (2009) A novel approach for discovering overlapping clusters in gene expression data. IEEE Trans Biomed Eng 56:1803–1808. https://doi.org/10.1109/TBME.2009.2015055

Mao L, Hemert JLV, Dash S, Dickerson JA (2009) *Arabidopsis* gene co-expression network and its functional modules. BMC Bioinformatics 10:346. https://doi.org/10.1186/1471-2105-10-346

Margolin AA, Nemenman I, Basso K, Wiggins C, Stolovitzky G, Favera RD, Califano A (2006) ARACNE: an algorithm for the reconstruction of gene regulatory networks in a mammalian cellular context. BMC Bioinformatics 7:S7. https://doi.org/10.1186/1471-2105-7-S1-S7

Maschietto M, Tahira AC, Puga R, Lima L, Mariani D, Paulsen B d S, Abreu PB, Vieira H, Krepischi ACV, Carraro DM, Palha JA, Rehen S, Brentani H (2015) Co-expression network of neural-differentiation genes shows specific pattern in schizophrenia. BMC Med Gen 8:23. https://doi.org/10.1186/s12920-015-0098-9

Medina IR, Lubovac-Pilav Z (2016) Gene co-expression network analysis for identifying modules and functionally enriched pathways in type 1 diabetes. PLoS One. https://doi.org/10.1371/journal.pone.0156006

Mitra S, Das R, Banka H, Mukhopadhyay S (2009) Gene interaction: an evolutionary biclustering approach. Inf Fusion 10:242–249. https://doi.org/10.1016/j.inffus.2008.11.006

Mitra K, Carvunis A-R, Ramesh SK, Ideker T (2013) Integrative approaches for finding modular structure in biological networks. Nat Rev Genet 14:719–732. https://doi.org/10.1038/nrg3552

Nepusz T, Yu H, Paccanaro A (2012) Detecting overlapping protein complexes in protein-protein interaction networks. Nat Methods 9:471–472. https://doi.org/10.1038/nmeth.1938

Ou Y, Zhang C-Q (2007) A new multimembership clustering method. J Ind Manag Optim 3:619–624. https://doi.org/10.3934/jimo.2007.3.619

Rodius S, Androsova G, Götz L, Liechti R, Crespo I, Merz S, Nazarov PV, de Klein N, Jeanty C, González-Rosa JM, Muller A, Bernardin F, Niclou SP, Vallar L, Mercader N, Ibberson M, Xenarios I, Azuaje F (2016) Analysis of the dynamic co-expression network of heart regeneration in the zebrafish. Sci Rep 6:26822. https://doi.org/10.1038/srep26822

Ruan J, Zhang W (2008) Identifying network communities with a high resolution. Phys Rev E 77:016104. https://doi.org/10.1103/PhysRevE.77.016104

Ruan J, Dean AK, Zhang W (2010) A general co-expression network-based approach to gene expression analysis: comparison and applications. BMC Syst Biol 4:8. http://www.biomedcentral.com/1752-0509/4/8

Rubinov M, Knock SA, Stam CJ, Micheloyannis S, Harris AWF, Williams LM, Breakspear M (2009) Small-World properties of nonlinear brain activity in schizophrenia. Hum Brain Mapp 30:403–416. https://doi.org/10.1002/hbm.20517

Rung J, Brazma A (2013) Reuse of public genome-wide gene expression data. Nat Rev Genet 14:89–99. https://doi.org/10.1038/nrg3394

Schwarz AJ, Gozzi A, Bifone A (2009) Community structure in networks of functional connectivity: resolving functional organization in the rat brain with pharmacological MRI. Neuroimage 47:302–311. https://doi.org/10.1016/j.neuroimage.2009.03.064

Serin EAR, Nijveen H, Hilhorst HWM, Ligterink W (2016) Learning from co-expression networks: possibilities and challenges. Front Plant Sci:7–444. https://doi.org/10.3389/fpls.2016.00444

Song WM, Zhang B (2015) Multiscale embedded gene co-expression network analysis. PLoS Comput Biol 1:e1004574. https://doi.org/10.1371/journal.pcbi.1004574

Tejera E, Bernardes J, Rebelo I (2013) Co-expression network analysis and genetic algorithms for gene prioritization in preeclampsia. BMC Med Genet 6:51

Wang YXR, Huang H (2014) Review on statistical methods for gene network reconstruction using expression data. J Theor Biol 362:53–61. https://doi.org/10.1016/j.jtbi.2014.03.040

Wang S, Zhu J (2008) Variable selection for model-based high-dimensional clustering and its application to microarray data. Biometrics 64:440–448. https://doi.org/10.1111/j.1541-0420.2007.00922.x

Wei S, Zhao WJ, Zeng XJ, Kang YM, Du J, Li H (2015) Microarray and co-expression network analysis of genes associated with acute doxorubicin cardiomyopathy in mice. Cardiovasc Toxicol 15:377–393. https://doi.org/10.1007/s12012-014-9306-7

Yang Y, Han L, Yuan Y, Li J, Hei N, Liang H (2014) Gene co-expression network analysis reveals common system-level properties of prognostic genes across cancer types. Nat Commun 5:3231. https://doi.org/10.1038/ncomms4231

Yeunga KY, Dombek KM, Loa K, Mittlera JE, Zhuc J, Schadtd EE, Bumgarnera RE, Rafterye AE (2011) Construction of regulatory networks using expression time-series data of a genotyped population. PNAS 108(48). https://doi.org/10.1073/pnas.1116442108

Yin T, DiFazio SP, Gunter LE, Zhang X, Sewell MM, Woolbright SA, Allan GJ, Kelleher CT, Douglas CJ, Wang M, Tuskan GA (2008) Genome structure and emerging evidence of an incipient sex chromosome in populous. Genet Res 18:422–430. https://doi.org/10.1101/gr.7076308

Yip AM, Horvath S (2007) Gene network interconnectedness and the generalized topological overlap measure. BMC Bioinformatics 8:22. https://doi.org/10.1186/1471-2105-8-22

Zhang J, Lu K, Xiang Y, Islam M, Kotian S, Kais Z, Lee C, Arora M, Liu H, Parvin JD, Huang K (2012) Weighted frequent gene co-expression network mining to identify genes involved in genome stability. PLoS Comput Biol 8:e1002656. https://doi.org/10.1371/journal.pcbi.1002656

Zhao J, Zhang Z, Ren S, Zong Y, Kong X (2016) Co-expression network analysis of Down's syndrome based on microarray data. Exp Ther Med 12:1503–1508. https://doi.org/10.3892/etm.2016.3462

Zhu D, Dequeeant ML, Li H (2008) Comparative analysis of clustering methods for microarray data. Anal Microar Dat:27–50. https://doi.org/10.1002/9783527622818.ch2

Bioinformatics Tools for Shotgun Metagenomic Data Analysis

6

Rajesh Ramavadh Pal, Ravi Prabhakar More, and Hemant J. Purohit

Abstract

The tremendous progress in next-generation sequencing (NGS) technology has brought an avalanche of sequence-based data. This huge volume of data has resulted in novel challenges for existing bioinformatics tools in terms of data handling and subsequent analyses. Additionally, complexity of such data makes the task of analysis of metagenomic datasets more complicated for available bioinformatics pipelines. Here we are dealing with various bioinformatics tools, available online for analysis of WGS-based metagenome datasets, and simultaneously comparing their analysis pipelines. In the last one decade, over a dozen of such online tools/servers have been developed which are accessible via public domain. IMG/M and MG-RAST are two of the most popular tools as per the number of citations they received in peer-reviewed scientific journals till December 2016. This chapter discusses and compares 11 online bioinformatics tools detailing their sequence data handling, pipelines for annotation, sequence clustering methods, user-friendly attributes, and feasibility of data repository.

Keywords

Shotgun sequencing · Metagenomics · Bioinformatics tools · IMG/M · MG-RAST

R. R. Pal
Nagarjuna Fertilizers and Chemicals Limited, Hyderabad, Telangana, India

R. P. More
ADBS, National Centre for Biological Sciences (NCBS), Bangalore, Karnataka, India

H. J. Purohit (✉)
Environmental Biotechnology and Genomics Division, CSIR-National Environmental Engineering Research Institute (NEERI), Nagpur, Maharashtra, India
e-mail: hj_purohit@neeri.res.in; hemantdrd@hotmail.com

H. J. Purohit et al. (eds.), *Soft Computing for Biological Systems*,
https://doi.org/10.1007/978-981-10-7455-4_6

6.1 Introduction

First introduced in 1998, the concept of metagenomics has become an inevitable tool in microbial ecology studies (Handelsman et al. 1998). Additionally cited as environmental genomics, it involves the study of microbial community DNA obtained from ecological niche. Thus, it differs with conventional genomics wherein genome sequence data from pure culture or mono isolate is analysed to explore characteristic attributes of respective microbe (Pal et al. 2015; Tikariha et al. 2016). During initial phase, metagenomic studies mainly involved cloning of small DNA sequences retrieved from the environment followed by microbial diversity studies or functional expression screening (Handelsman et al. 1998). The other approach involved extraction of community DNA followed by community profiling with respect to taxonomic marker (16S rRNA gene) or any functional gene (responsible for some biological activity) based on PCR-mediated methods such as ARDRA, RFLP, etc. (Purohit et al. 2003; Dubey and Padmanabhan 2003; Moharikar et al. 2003; Dafale et al. 2010; Sharma et al. 2012). Microbial population dynamics would be studied by monitoring change in 16S rRNA gene profiles in any environment (Moharikar et al. 2005). However, it cannot depict the complete biodiversity and also may omit large number of microbial communities. Thus, 16S rRNA gene profile-based data alone cannot be relied upon to unravel the microbial diversity. The emergence of next-generation sequencing (NGS) technologies has enabled availability of huge sequence-based data conferring great coverage with respect to taxonomic as well as functional community profiles in any metagenomic analyses (Pandit et al. 2016; Kapley et al. 2015). High-throughput sequence data at markedly decreased cost can be obtained by using NGS platforms which have opened new avenues into sequencing data-dependent metagenomic analyses in order to elucidate complete biodiversity (Thomas et al. 2012).

Generally there are two different metagenomic methods: (a) marker-gene metagenomics and (b) whole-genome shotgun (WGS) metagenomics (Dudhagara et al. 2015a). Figure 6.1 illustrates bioinformatics tools being developed in recent times with respect to the above two mentioned metagenomic approaches. The first approach refers to the use of marker genes [taxonomic markers involving 16S rRNA (Ghelani et al. 2015), 18S rRNA, and ITS (Dudhagara et al. 2015b) and an amplicon library data] to explore the microbial community profile for an environmental sample (Carlos et al. 2012; Oulas et al. 2015; Gulhane et al. 2017). However, the second approach has random nature and involves shotgun sequencing which enables sufficient data coverage to determine the complete biodiversity (Yadav et al. 2015; Puranik et al. 2016; Jadeja et al. 2014). This approach has yielded new arsenal for exploring unique structural and functional features from untapped microbial world. Presently, myriad of projects have already been completed based on marker genes and shotgun metagenomic data, rendering deposition of innumerable sequencing data to public domain. Subsequent to metagenome sequencing, the preliminary aim is to process the huge sequence data for annotation of structural and functional features of microbial community present in the sample (Ounit et al. 2015). However, major technical issues still

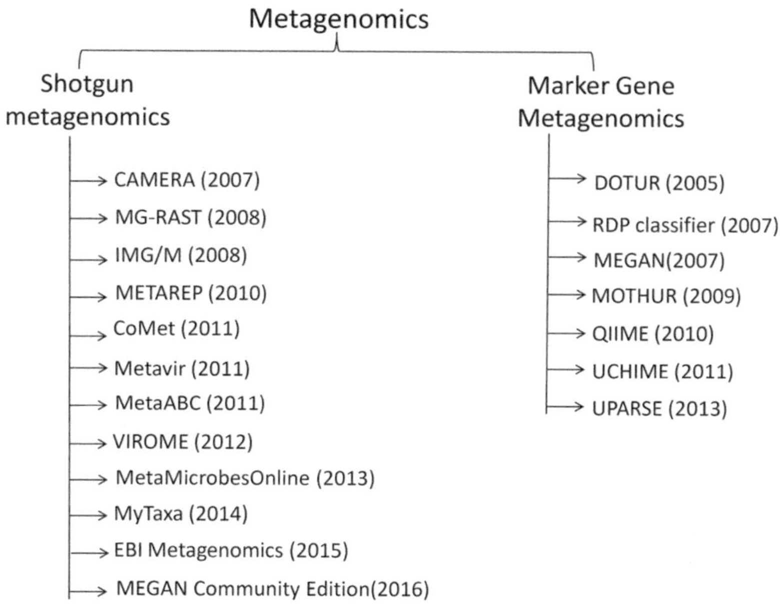

Fig. 6.1 The metagenomic tools available for sequence data analysis

remain in metagenomic sequence analysis approaches corresponding to assembly of raw metagenomic reads followed by annotation of the operational taxonomic units (OTUs) (Behnam and Smith 2014). Additionally, owing to complex nature of sequence data, it is difficult to perform statistical analyses of such metagenomes. Besides, several offline and online tools/softwares are present for annotation of metagenomic sequences employing homology with reference databases (Dudhagara et al. 2015b).

6.2 Shotgun Metagenomics

Shotgun metagenomic sequencing allows researchers to comprehensively sample all genes in all organisms present in a given environmental sample. In shotgun sequencing, DNA is fragmented randomly into several small segments, which subsequently are sequenced by employing the chain termination strategy to yield small reads (Staden 1979; Sharma and Vakhlu 2014). Multiple cycles of this fragmentation and sequencing steps are conducted to achieve numerous overlapping reads for the target DNA. Eventually, by using computer programs, the overlapping ends of various reads are conjoined into an uninterrupted contig. With the advent of NGS, this approach enabled microbiologists to evaluate bacterial diversity and determine the abundance of microbes in various environments such as marine sediment, activated sludge, hot water spring sediment, saline desert, etc. in less time and with reduced cost (Mason et al. 2014; Chao et al. 2013;

Mangrola et al. 2015; Pandit et al. 2014). Employing NGS techniques such as 454 pyrosequencing, Illumina systems, and ion torrent, shotgun metagenomic approach has been exploited in studying microbial dynamics with respect to population as well as functions (Kröber et al. 2009; Hasan et al. 2014). Keeping similar pace of advancement, analytical bioinformatics tools have also advanced with sequencing technologies which generate tremendous amount of sequencing data.

In this chapter, we discuss the bioinformatics tools being developed in the last decade and available online for WGS-based metagenome data analyses, detailing in short for analysis pipeline of each tool (Fig. 6.1). Table 6.1 enlists these tools being developed for shotgun metagenomic datasets, with some relevant information indicating their impact on metagenomic field. This significance was accounted by considering number of scientific citations they have obtained in course of time since their inception. Figure 6.2 depicts the characteristic attributes of these online tools. In order to evaluate and compare each of these tools, they have been discussed in brief in chronological order.

6.2.1 CAMERA

Being introduced in 2007, the Community Cyber infrastructure for Advanced Marine Microbial Ecology Research and Analysis (CAMERA) was one of the first online bioinformatics tools for metagenome analyses. Basically it was meant for studying microbial diversity of the ocean and its response to different environmental habitats (Seshadri et al. 2007). After 2010, CAMERA elaborated its objective to allow analysis of all metagenomic datasets acquired from various environments by discarding "marine" term from its previous abbreviation (Sun et al. 2010). Moreover, the original webpage (http://camera.calit2.net) of CAMERA Data Distribution Center (DDC) has also been moved to different destination (http://data.imicrobe.us) which is governed by the iMicrobe Project. CAMERA was the first online bioinformatics tool being developed as an eminent large-scale database which analyses, shares, and collects metagenome sequences. It offered a repository for huge sequence data obtained from Global Ocean Sampling (GOS) expedition, supplemented with analytical pipelines to integrate metadata information with sequence data to derive correlations between deciphered ecology. In order to submit the data, users must register at iPlant cyberinfrastructure via Discovery Environment (DE) web interface. Normally 100 GB virtual space is allotted to users which on request can be extended to an additional 1 TB. The uploading of huge datasets (>1.9 GB) can be accomplished by either of the two options: (1) Cyberduck for users having Mac and Windows platforms or (2) iDrop Desktop (iCommands) for users running LINUX. However, a simple URL can be used for upload of small data files within the DE. CAMERA's principle objective is to manage a huge, unique data repository and essential bioinformatics tools in interest of overcoming the novel hurdles of metagenome data analyses including complexity, heterogeneity, and truncated data. CAMERA database is comprised of environmental genomic

Table 6.1 Major online bioinformatics tools for shotgun metagenomic data analysis

Tools	Weblink	Inbuilt reference databases and annotation pipeline	Data storage
CAMERA (2007 -Now obsolete)	http://camera.calit2.net/	FragGeneScan, KEGG, COG,, TIGRfam, GO, Pfam, MetaGene	128 projects and 2660 samples
MG-RAST (2008)	http://metagenomics.anl.gov/	SEED subsystem, KEGG, KO, NOG, COG, eggNOG, M5RNA, TrEMBL, SEED,, SwissProt, GenBank, RefSeq, PATRIC	273,897 metagenomes (127.09 Tbp), 21,886 registered users,
IMG/M (2008)	https://img.jgi.doe.gov/cgi-bin/m/main.cgi	KOG, KEGG, COG, KO, TIGR, MetaCyc, GO, Pfam, TIGRfam,	62,994 datasets (Genomes + Metagenomes)
METAREP (2010)	http://jcvi.org/metarep/	GO, NCBI Taxonomy	NA
CoMet (2011)	http://comet.gobics.de/	Pfam, GO	NA
METAVIR (2011)	http://metavir-meb.univ-bpclermont.fr/	Pfam, RefSeq virus database	170 viral metagenomic dataset and 335 projects
MetaABC (2011)	http://metaabc.iis.sinica.edu.tw/	Genome database from NCBI	52 datasets
VIROME (2012)	http://virome.dbi.udel.edu/	SEED, COG, ACLAME, GO, UniRef 100, KEGG, MGOL	50 project, 258 libraries, total proteins – 44,895,778
metaMicrobesOnline (2013)	http://meta.microbesonline.org/	TIGRfam, COG, Pfam	155 metagenomes, 3527 microbial genomes
MyTaxa (2014)	http://enve-omics.ce.gatech.edu/mytaxa/	Reference genomes from NCBI	NA
EBI metagenomics (2015)	https://www.ebi.ac.uk/metagenomics/	InterPro protein signature database, Greengenes database, RDP	784 projects, 72,945 datasets
MEGAN Community Edition (CE) (2016)	https://github.com/danielhuson/megan-ce	NCBI BLAST	NA

NA stands for Not Applicable (Ref: Dudhagara et al. 2015a updated with data till December 2016)

and metagenomic sequence data, corresponding to environmental features, processed search results, and bioinformatics tools to enable efficient cross analysis among different datasets. It was meant to harbour a huge metagenome datasets from

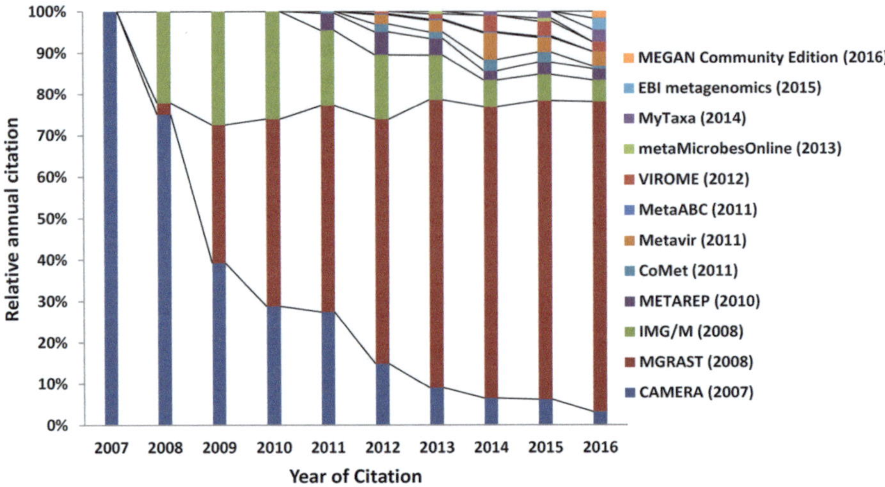

Fig. 6.2 Comparative citation index of the metagenomic tools (for shotgun sequencing dataset) retrieved from research papers published in peer-reviewed articles in December 2016. Year of release of each tool is depicted in the bracket in the legend box. Citation for each tool was pursued from Google Scholar

around the globe; but after 2011, it has been overlooked regularly and citations dropped owing to technical difficulty in data uploading process and advent of more user-friendly online tools like MG-RAST and IMG/M. Furthermore, it has been integrated with QIIME so as to achieve quick online cloud computing which would attract more number of users and confer orderliness to relevant analytical processes.

6.2.2 MG-RAST

Metagenomics Rapid Annotation using Subsystem Technology (MG-RAST) is an easy-to-use open-source server for metagenomic data analyses (Meyer et al. 2008; Overbeek et al. 2014; More et al. 2014). Introduced in 2007, it has been one of the primary online bioinformatics tools which is still being used widely by researchers (Meyer et al. 2008). A new version (version 4.0) is presently functional, which unlike its preliminary versions is not completely dependent on SEED platform, rather employs SEED subsystem (being favoured reference database) for taxonomic and functional annotation of metagenome dataset. As compared to previous versions, the current online platform allows much better scalability and rapid computation attributes. Presently, MG-RAST harbours 273,897 metagenomes comprising 127.09 Tbp data, which have been accessed by 21,886 registered users (As of Dec-2016). Apart from accessing the publicly available data on server, registered users can upload personal metagenome sequence dataset in either of the formats (FASTA, FASTQ, and SFF) supplemented with structured metadata information. A multistep workflow processes the uploaded data including quality

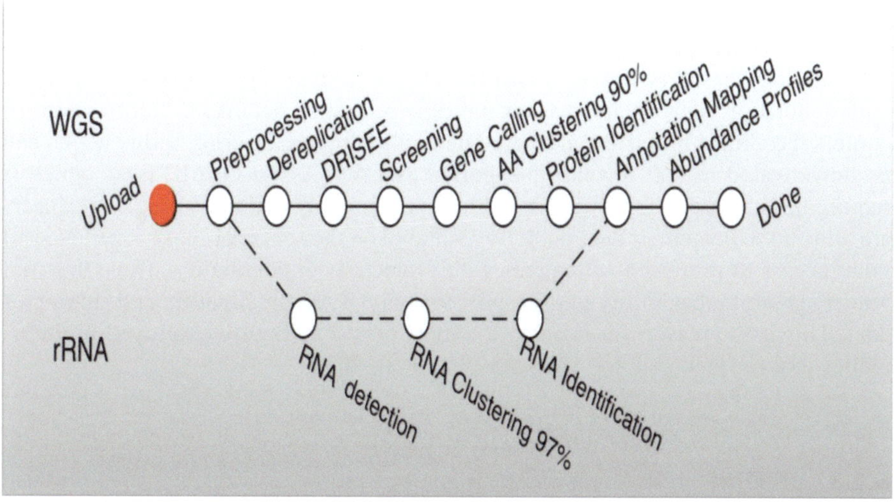

Fig. 6.3 Metagenome data analysis workflow in MG-RAST (Ref – MG-RAST manual Weblink – http://metagenomics.anl.gov/)

control of sequences, automated annotation, and analysis. Figure 6.3 depicts schematically workflow of data analysis in MG-RAST.

Briefly, after uploading data is preprocessed by employing SolexaQA (Cox et al. 2010) which curates low-quality regions from raw sequences. Subsequently, dereplication, that is, removal of artificial duplicate reads (ADRs) from filtered data, is performed by employing duplicate read inferred sequencing error estimation (DRISEE) (Keegan et al. 2012). Further sequence data is screened against model organism's genome sequence data, comprising fly, mouse, cow, and human in order to ensure homogeneity of microbial metagenomic data. The next step involves annotation based on machine learning approach being employed by FragGeneScan tool (Rho et al. 2010). In order to decrease the load on computational system for comparison of complete short read datasets, prior protein clustering is performed at 90% homology threshold employing UCLUST command from QIIME, simultaneously retaining the relative abundances (Edgar 2010). Subsequently, one representative sequence (longest read) from every cluster is processed for homology search by using BLAT algorithm followed by reconstructions of the species content of the sample based on the similarity results.

For homology search MG-RAST utilizes a protein database referred to as M5NR representing nonredundant integration of multiple databases: GenBank, SEED, IMG, UniProt, KEGG, and eggNOGs (Wilke et al. 2012). At the same time, the user is allowed to use either one of these listed reference databases individually. After annotation abundance profiles can be retrieved from MG-RAST's user interface and subsequently subjected to various statistical analyses. Various analyses can be performed on the server involving phylogenetic, metabolic, functional, and comparative analyses of two or multiple metagenome datasets which can be visualized via various diagrammatic illustrations such as pie chart, bar chart, heat maps, etc. Additionally, MG-RAST allows the user to choose a reference database for homology-based analyses. It enables the user to perform comparative analyses with annotation data obtained from multiple reference databases. Furthermore, it confers the choice to retrieve analysed data in multiple clustering forms which can be downloaded in text format or exported as FASTA and QIIME files, which if required can be used directly as input files in other analytical tools. Registered users are allowed to keep their data publicly available on the server or make it private and grant access to preferred colleagues with protected confidentiality. Thus, this tool renders greater adaptability with respect to analysis, confidentiality, and sharing of data. This is the reason for it being the most cited and highly employed bioinformatics tool for metagenome analyses (Sun et al. 2015).

6.2.3 IMG/M

Integrated Microbial Genomes and Metagenomes (IMG/M) is another important online server for repository and analysis of genome and metagenome sequence data (Markowitz et al. 2008). Joint Genome Institute (JGI) of US Department of Energy (DOE) maintains this project. The system store annotated datasets of (1) bacterial, archaeal, viral, and eukaryotic genomes corresponding to cultivated microbes; (2) single cell genomes (SCG) and genomes from metagenomes (GFM) pertinent to uncultivable archaea, bacteria, and viruses; and (3) metagenomes from environmental, host-associated, and engineered artificial microbiome samples. The database of IMG/M is comprised of sequence data being created by DOE's Joint Genome Institute (JGI), uploaded by independent researchers, or fetched from public domain. IMG/M allows users to perform annotation, integration, and comparative analyses of sequence data corresponding to genomic and metagenomic studies. Moreover, it also permits for combined analyses for query sequence data against sequence data (genomes and metagenomes) available in public domain (Chen et al. 2016; Markowitz et al. 2012a). Sequence data and bioinformatics tools are regularly upgraded at the server. As of July 2016, there are 11,004 (among them 5735 public) metagenome datasets from 544 (250 public) metagenome studies with over 45.7 billion (35.9 billion public) protein-coding genes in IMG. The use of IMG/M datasets and analytical tools is allowed for registered users. After registration at the server, personal sequence data can be uploaded online and be kept private for up to 2 years after which data is transferred to the public domain. Additionally, downloading and analysis of sequence data is also allowed via JGI's portal for registered users.

After submission of new sequence data, JGI's metagenome annotation system first performs quality control, and then annotation is accomplished. The annotation is performed with reference to multiple datasets at three levels: (1) phylogenetic composition, (2) functional or metabolic potential of individual microbiomes, and

(3) comparisons among microbiomes. The server supports these bioinformatics analyses by employing consolidated datasets comprised of metagenomes and microbial genomes from the IMG system (Markowitz et al. 2014). The output data is provided in user-friendly multiple-cluster forms, which accelerate its application for different metagenomic studies. Additionally, IMG/ M-HMP (Human Microbiome Project; http://www.hmpdacc.org) have been established by IMG/M, which renders different bioinformatics tools for the annotation of metagenome datasets generated through HMP. The annotation is accomplished with reference to public sequence database available through IMG (Markowitz et al. 2012b). IMG has been proven extremely useful in a variety of research studies requiring high-quality metagenomic assembly, such as the identification and genomic reconstruction of novel phylogenetic lineages, discovery of novel biosynthetic gene clusters, identification of alternative genetic codes, uncovering gaps in amplicon-based detection of microbial diversity, and discovery of novel viruses (Markowitz et al. 2012c). Owing to its comprehensive sequence data analyses and mining via exploration of dataset, IMG/M has been attracting increasing number of researchers.

6.2.4 METAREP

Metagenomics Reports (METAREP) is an online tool for high-performance comparative metagenomic analysis. This bioinformatics tool is commonly used for short sequence read assemblies, metagenome annotation, and comparative metagenome analyses (Goll et al. 2010). In order to upload metagenome sequence data, one has to be acquainted with fundamentals of computer programming and some expertise to explore METAREP as an online software tool. In order to perform comparative studies, multiple datasets can be analysed concurrently at different functional and taxonomic levels of annotation. Furthermore, data download option is also available wherein annotation files can be exported in form of tab-delimited files for subsequent statistical analyses. METAREP also enables data exploration via integrated taxonomic and functional levels in order to examine it from different viewpoints and thus to handle big datasets conveniently. To perform annotations of enormous metagenomic sequence datasets, the software incorporates data-intensive algorithms by employing Solr/Lucene, R, and CAKEPHP. It has been optimized to be user friendly and fast. This bioinformatics tool is being constructed and perpetuated by the J.C. Venter Institute (JCVI). For improvement of the tool, the users are allowed to recommend new analytical attributes to be amended for particular objectives of their research interests. After last update in METAREP showing enhanced performance with respect to data upload, clustering, and annotation, it was shown to be effective in terms of comparative metagenomics. Goll et al. employed the shotgun sequence data from HMP (http://www.hmpdacc.org/) to demonstrate improvement in METAREP tool (Goll et al. 2012).

6.2.5 CoMet

CoMet is another bioinformatics freeware available for comparative metagenome analyses. Here, even user registration is not compulsory for sequence data upload. It possesses good analysis workflow which enables quick and efficient analysis of short sequences from metagenome sequence dataset (Lingner et al. 2011). Metagenomic data generated from NGS platforms such as 454 and Ion Torrent can be processed effectively by CoMet. In order to perform comparative metagenomics, up to 20 metagenome datasets can be uploaded in FASTA format in multiple files; alternatively multiple files can be zipped into single file which enables faster data uploading. Sequence files ranging around 500 MB can be submitted altogether for data processing through the convenient submission portal of CoMet. This tool employs Pfam and the corresponding Gene Ontology (GO) database for functional annotations of various metagenomes. It collates ORF-finding and protein annotations to different Pfam domains with a comparative statistical analysis, enabling an immediate outline of putative functional variations among a group of metagenome datasets. The comparative analysis against reference database of CoMet comprising beyond 1,000 precomputed profiles allows a quick determination of similar datasets and the access to respective metagenome projects. The output includes downloadable text files, result in the form of pictures, and a matrix of the statistical results.

6.2.6 METAVIR

METAVIR is a web server developed for annotation of raw or assembled viral metagenomic sequences (Roux et al. 2011). METAVIR provides users a collection of bioinformatics tools through proprietary platform for analysis of viral metagenome data. After registering at server, the user can upload metagenome sequence data as FASTA files. Moreover, big sequence files in compressed forms such as zip, gzip, or tar.gz can also be uploaded in order to manage large datasets. Subsequently, tool assesses the Virome composition by employing the GAAS tool (Angly et al. 2009). The Virome sequences are analysed against RefSeq database of viral genomes, followed by normalization of annotation results by genome length which yields estimation of total viral particles for each viral clade in metagenomic dataset. Newly an updated version, METAVIR 2, has been released that can be used for extensive viral metagenome analyses (Roux et al. 2014).

This bioinformatics tool is worthwhile for mining viral ecology from metagenome sequence data and performs comparative analyses with reference to predetermined phylogenetic database of viral-specific genes. Diversity data can be obtained with respect to rarefaction curves and multivariate analyses being generated on k-mer signatures and BLAST-based comparisons. The analytical results can be illustrated through various correlative and effective modes, including tables, phylogenetic trees, recruitment plots, and maps. The assembled viral genomes constituting thousands of larger contigs can be handled by METAVIR 2.

Till October 2015, tool's server possessed 335 Virome projects containing more than 64 million sequences from different ecological niches. METAVIR presents convenient data upload portal and allows comprehensive Virome data analyses aided with data privacy, which are important attributes making it highly cited Virome analysis pipeline.

6.2.7 MetaABC

MetaABC is a complete package for metagenomic data management, binning, and clustering. It consolidates two methods of artefact removal ((1) 454 replicate filter and (2) cdhit-454), five methods for taxonomic binning ((1) BLAST, (2) PhymmBL, (3) MEGAN, (4) SOrt-ITEMS, and (5) DiScRIBinATE), and one approach each to re-analyse unassigned reads and to control sampling biases via gene adjacency and genome-length normalization, respectively (Su et al. 2011). MetaABC was developed in order to handle sequencing data generated using different technologies. The database of this tool comprises around 50 analysed metagenomes. In order to analyse personal dataset, 2–20 SFF, FASTA, or FASTQ files can be deposited simultaneously. After sequence data submission, the results can be generated in various forms involving tables, pie charts, and bar charts of abundance profiles. Furthermore, a hierarchical clustering algorithm is incorporated in this tool for comparative metagenomics. However, on the other hand, MetaABC also possesses drawbacks with respect to inability to perform functional and metabolic profiling of metagenomes. Furthermore, the options for data security are a limiting factor followed by the absence of data-sharing attributes. Moreover, the tool cannot handle big (>2 MB) sequence files. In order to analyse larger dataset, MetaABC allows the use of a stand-alone platform.

6.2.8 VIROME

Viral Informatics Resource for Metagenome Exploration (VIROME) is an online tool exclusively created for viral metagenome data analysis corresponding to various environmental habitats. The annotation and classification of metagenomic viral reads is performed via homology search in reference to both known and environmental sequences (Wommack et al. 2012). A web-application interface supports the sequence data upload to VIROME. After online form submission, the user is notified via Email describing subsequent instructions related to upload of sequence data. The tool allows various formats of sequence data files, comprising FASTA, QUAL, FASTQ, and SFF. The analysis pipeline includes homology search of metagenome sequence data with reference to MetagenomesOnline database (MgOl), followed by graphical illustration and integration with different protein metadata allowing interpretation of the output. The analysis pipeline of VIROME comprises two back-to-back steps: (1) quality check of sequence data and analysis in reference to UniVec database and (2) three concurrent steps including

ORF-assignment by MetaGeneAnnotator, identification of known proteins, and annotation of environmental protein against databases comprising MGOL and UniRef 100. The quality check of sequences is accomplished by Ergatis which is a management system for analyses pipeline. The output of sequence data curation and BLAST homology search steps are saved in an in-built database called MySQL. Subsequently, the interface between the MySQL database and the VIROME web page is mediated by one server called Adobe ColdFusion. The output from VIROME can be downloaded for subsequent use as input data for multivariate statistical analyses using different analytical tools. However, VIROME also holds one main drawback which is the length of time required to perform analysis. It takes usually several weeks to months for even a small-sized metagenome dataset.

6.2.9 metaMicrobesOnline

metaMicrobesOnline database is another free online bioinformatics tool and can be accessed at http://meta.MicrobesOnline.org. It allows phylogenetic analysis of gene sequence data from microbial genomes and metagenomes (Chivian et al. 2013). Notably, metaMicrobesOnline execute neither assembly of contigs nor gene annotations, emphasizing rather on phylogeny-based tree analysis for gene sequences. Thus, users have the options to choose a suitable method for read assembly and gene annotations congruous to private data. Majority of metagenomes being deposited to metaMicrobesOnline are the ones which have been annotated from other online tools such as IMG/M or MG-RAST. However, any user can load their sequence data files which should be compatible with the tool (e.g. sequence file in FASTA format for contigs and gene coordinates of respective contigs in tab-delimited format). In order to achieve trustworthy position in gene phylogenetic trees, the tool specifically offers analysis of only those public metagenome datasets having longer contigs (typically above 500 bp) possessing higher chances of comprising complete genes. The analysis starts with loading of sequence data and corresponding annotations into the metaMicrobesOnline analysis pipeline. Subsequently, genes are translated into protein sequences and scanned by employing HMMER3 (Eddy 2011) in reference to canonical gene and protein families such as COG (Tatusov et al. 2001), Pfam (Finn et al. 2013), and TIGRFAMs (Selengut et al. 2007). The alignment outputs obtained by HMMER3 search are employed to assign the metagenomic reads to multiple sequence alignment corresponding to all gene families. Subsequently the curated multiple sequence alignments are subjected to construct phylogenetic trees for each gene/domain family using FastTree-2 (Price et al. 2010). The analysis outcomes are presented via bioinformatics tools such as genome browser based on phylogenetic tree and a similar tree-based domain browser. Thus, one can analyse the multiple genomes and browse genes associated with different functions. With metaMicrobesOnline users are allowed to choose most suitable analysis pipeline for their dataset, including read assembly, gene annotations, and phylogeny of genes. metaMicrobesOnline comprises a protected database of 155 metagenomes encompassing 123 datasets from different environmental niches and another 32 corresponding to different hosts. After July 2010, it was merged with the MicrobesOnline web application (http://www.microbesonline.org), which houses thousands of genome datasets corresponding to all the three domains of life (Dehal et al. 2010). However, this tool is not being used and cited regularly owing to inconsistent updates and the limitation with requirement of longer contig reads leading to limited submission of new metagenomes.

6.2.10 MyTaxa

MyTaxa is a recent bioinformatics tool based on homology search for classification of metagenomic and genomic sequences with high precision (Luo et al. 2014). It utilizes each gene corresponding to every unidentified sequence as classifiers, measuring all genes depending on its recomputed classifying potential for described taxonomic level and prevalence of HGT (horizontal gene transfer). This tool is especially relevant for classification of unidentified genomic and metagenomic datasets. An indexed database is the key component of MyTaxa which is available for free and can be downloaded from MyTaxa's webpage. This database includes guidelines for the gene clusters being exploited during online analysis for taxonomic classification. In order to attain high precision and specificity for taxonomic annotations, the database is frequently updated by integrating extra reference genes being accessible from newly identified genomes or ongoing sequencing projects. MyTaxa is flexible for analysis of bigger sequence data additionally harmonious with newly developed efficient algorithms in addition to homology search. For online analysis with MyTaxa, pipeline starts with upload of two files: (1) a regular GFF file comprising of annotated genes from query sequences and (2) an output file in tabular format corresponding to homology search of the annotated gene sequences in reference to the ones that are being employed to establish the index of gene weights. MyTaxa employs a distinctive classification approach which is dependent on the genome-aggregate average amino acid identity (AAI) phenomenon to ascertain the confidence in uniqueness of sequences depicting the unclassified taxa. This classification scheme supports MyTaxa for being an appropriate tool to find out the level of uniqueness of uncharacterized metagenomic reads. Nonetheless, since there is lack of reference sequences of distant species from public database, the annotation of species (the last taxa in hierarchy) is a difficult work for the tool. Furthermore, the efficiency is hindered due to incompetency to undertake the functional or metabolic profiling of query datasets. Additionally, a main hurdle for new users to operate the tool is the necessity of metagenomic protein file obtained after BLASTx translation. However, if restricted to classification task, this tool has been shown to perform annotation of at least 5% more sequences than any other tool. When tested with simulated and genuine metagenome data of heterogeneous read length ranging from 100–2000 bp, it revealed that ~10% of the assembled read data corresponding to human gut metagenomes constitute unique species with no sequenced representatives. Thus, MyTaxa can be a useful tool in microbial diversity studies and offer further penetration through highly intricate microbial world.

6.2.11 EBI Metagenomics

Introduced in 2011 by EMBL-EBI, EBI Metagenomics is the first platform in Europe for taxonomic, functional, and comparative metagenomic analyses (Hunter et al. 2014). It encompasses QIIME originated algorithms for data analysis, storage, and sharing of metagenomes. To start the analysis, users must register on web server to submit raw sequence data that can be kept private till 2 years, and the sequences are deposited with distinctive accession numbers in the European Nucleotide Archive (ENA). Deposited datasets and corresponding metadata abide by the Genomic Standards Consortium (GSC) and Minimum Information about any (X) Sequence (MIxS) for verification, making it available for the scientific community. Furthermore, by virtue of ENA's Webin tool or ISAcreator, this online tool is compatible with raw sequences yielded from any NGS platforms. The analysis pipeline includes preprocessing of sequence data involving trimming and quality checks to eliminate sequencing artefacts and organize the data. Subsequently, the processed data is analysed by rRNAselector and QIIME for taxonomic identification and functional annotation via consolidated protein database called InterProScan. EBI Metagenomics is an efficient tool for the processing and analysis of both shotgun and marker-gene metagenome datasets. Although basically it is centred on shotgun metagenomes, rRNA data can be retrieved and analysed from shotgun metagenomic datasets via rRNASelector, making this tool suitable for marker-gene metagenomics. The analysis outputs are simply available from the EBI Metagenomics web interface and thus can be downloaded in various formats suitable with subsequent analyses employing online or stand-alone tools. EBI Metagenomics also allows a quick comparative taxonomic and functional profiling of metadata allowing exploration of novel features among different metagenomes. By virtue of numerous merits presented by EBI Metagenomics, in recent years it has been a favourite bioinformatics platform among new researchers. EBI Metagenomics regularly updates the data processing and analysis pipeline complying with emergence of new analysis and visualization tools. Presently, it ventures to develop a common platform for NGS computational pipelines.

6.2.12 MEGAN Community Edition

Community Edition (CE) is another freeware bioinformatics tool and an extension of widely used microbial diversity analysis tool MEGAN for accelerating the taxonomic and functional analyses of huge microbiome datasets (Huson et al. 2016). The input file for MEGAN CE is called RMA (a compressed, indexed file format) including sequence reads, alignments, and annotations (taxonomic and functional both) for a given sample. Such file can be generated interactively using MEGAN CE, as well as by employing command line tool termed blast2rma, yielding RMA file on server for subsequent analyses. Alternatively, if DIAMOND (new alignment tool) is exerted for computing alignments, another command line tool termed Meganizer is utilized to accomplish taxonomic and functional binning of the reads (Buchfink et al. 2015). The output is then adjoined with the file along with supplementary indices needed for coherent sequence accessibility through taxonomic or functional annotations. These meganized diamond files can directly be analysed via MEGAN CE instead of additional data processing. As compared to previous versions of MEGAN (Huson et al. 2011) necessitating presence of files (undergoing analysis) on the computer, MEGAN CE offers flexibility by harbouring new program called MeganServer which can store and allow access to RMA and meganized diamond files for analysis via online mode. Thus, MEGAN CE presents an efficacious pipeline for the analysis of shotgun metagenome dataset by integrating with DIAMOND (a new high-throughput DNA-to-protein alignment tool) and by offering a new platform MeganServer that allows access to analysed metagenome files on the server. This system can facilitate the analysis of huge data comprising of hundreds of metagenomes and billions of sequences through single server in quick time.

Any computational analysis of metagenomic data has two main goals of determining the taxonomy and functional capacity of microbial community at respective environmental niche. Like previous versions, MEGAN CE also implements binning at taxonomic level by attributing reads to nodes from NCBI taxonomy by employing the LCA algorithm (Huson et al. 2007). For the binning of reads with respect to functions, MEGAN CE utilizes various classification systems – (1) *InterPro2GO* analyser which classifies InterPro families (Mitchell et al. 2014) by using a metagenome GO-slim (Hunter et al. 2014), (2) SEED analyser exploiting the phenomenon of subsystems and functional roles (Overbeek et al. 2014), and (3) *eggNOG* viewer on eggNOG platform which is the extension of COGs (Powell et al. 2012). Additionally, MEGAN CE also employs legendary KEGG viewer, by retrieving the files from KEGG in 2011 (Kanehisa and Goto 2000). After annotations, user can perform principal coordinate analysis (PCoA) and cluster analysis by using various diversity indices in addition to standard alpha diversity index. This tool offers a gene-centric method to read assembly data. The reads assigned to any given taxonomic or functional level can be assembled and retrieved as contigs, without any substantial calculations or supplementary tools.

6.3 Conclusion

With advanced NGS platforms, structural and functional microbial dynamics is a common practice in any ecological studies. Moreover, the comparative analyses of multiple metagenomic datasets rapidly expanded the area of microbial diversity studies. For extensive exploration of ecosystems, unique bioinformatics tools, pipelines, and postulates are needed for analysis, data repository, output illustrations, and sharing of massive dataset. Therefore, any single bioinformatics tool would not be sufficient for performing full-fledged metagenome analyses. Development in sequencing platforms yielding longer-read length, new bioinformatics tools for precise assembly, and annotation of bigger datasets are the expected advancement in the forthcoming metagenomic research projects in the coming years. This book chapter has illustrated a précis of some of the existing metagenome analysis tools for data obtained from whole-genome shotgun sequencing approach. The way some of the leading online software tools have evolved after their inception, their newer versions with more advanced analysis pipelines and user-friendly platforms are also anticipated in the coming years. This chapter may confer an understanding to budding research scholars to select the most suitable bioinformatics tools available in public domain.

References

Angly FE, Willner D, Prieto-Davó A, Edwards RA, Schmieder R, Vega-Thurber R, Antonopoulos DA, Barott K, Cottrell MT, Desnues C, Dinsdale EA (2009) The GAAS metagenomic tool and its estimations of viral and microbial average genome size in four major biomes. PLoS Comput Biol 5(12):e1000593. https://doi.org/10.1371/journal.pcbi.1000593

Behnam E, Smith AD (2014) The Amordad database engine for metagenomics. Bioinformatics 30:2949–2955. https://doi.org/10.1093/bioinformatics/btu405

Buchfink B, Xie C, Huson DH (2015) Fast and sensitive protein alignment using DIAMOND. Nat Methods 12(1):59–60. https://doi.org/10.1038/nmeth.3176

Carlos N, Tang YW, Pei Z (2012) Pearls and pitfalls of genomics-based microbiome analysis. Emerg Microbes Infect 1:e45. https://doi.org/10.1038/emi.2012.41

Chao Y, Ma L, Yang Y, Ju F, Zhang XX, Wu WM, Zhang T (2013) Metagenomic analysis reveals significant changes of microbial compositions and protective functions during drinking water treatment. Sci Rep 19:3. https://doi.org/10.1038/srep03550

Chen IM, Markowitz VM, Chu K, Palaniappan K, Szeto E, Pillay M, Ratner A, Huang J, Andersen E, Huntemann M, Varghese N (2016) IMG/M: integrated genome and metagenome comparative data analysis system. Nucleic Acids Res. https://doi.org/10.1093/nar/gkw929

Chivian D, Dehal PS, Keller K, Arkin AP (2013) MetaMicrobesOnline: phylogenomic analysis of microbial communities. Nucleic Acids Res 41:D648–D654. https://doi.org/10.1093/nar/gks1202

Cox MP, Peterson DA, Biggs PJ (2010) SolexaQA: at-a-glance quality assessment of Illumina second-generation sequencing data. BMC Bioinformatics 11(1):485. https://doi.org/10.1186/1471-2105-11-485

Dafale N, Agrawal L, Kapley A, Meshram S, Purohit H, Wate S (2010) Selection of indicator bacteria based on screening of 16S rDNA metagenomic library from a two-stage anoxic–oxic bioreactor system degrading azo dyes. Bioresour Technol 101(2):476–484. https://doi.org/10.1016/j.biortech

Dehal PS, Joachimiak MP, Price MN, Bates JT, Baumohl JK, Chivian D, Friedland GD, Huang KH, Keller K, Novichkov PS, Dubchak IL (2010) MicrobesOnline: an integrated portal for comparative and functional genomics. Nucleic Acids Res 38(suppl 1):D396–D400. https://doi.org/10.1093/nar/gkp919

Dubey SK, Padmanabhan P (2003) Tracking of methanotrophs and their diversity in paddy soil: a molecular. Curr Sci 85(1):93

Dudhagara P, Bhavsar S, Bhagat C, Ghelani A, Bhatt S, Patel R (2015a) Web resources for metagenomics studies. Genomics Proteomics Bioinformatics 13(5):296–303. https://doi.org/10.1016/j.gpb.2015.10.003

Dudhagara P, Ghelani A, Bhavsar S, Bhatt S (2015b) Metagenomic data of fungal internal transcribed Spacer and 18S rRNA gene sequences from Lonar lake sediment, India. Data Brief 4:266–268. https://doi.org/10.1016/j.dib.2015.06.001

Eddy SR (2011) Accelerated profile HMM searches. PLoS Comput Biol 7(10):e1002195. https://doi.org/10.1371/journal.pcbi.1002195

Edgar RC (2010) Search and clustering orders of magnitude faster than BLAST. Bioinformatics 26 (19):2460–2461. https://doi.org/10.1093/bioinformatics/btq461

Finn RD, Bateman A, Clements J, Coggill P, Eberhardt RY, Eddy SR, Heger A, Hetherington K, Holm L, Mistry J, Sonnhammer EL (2013) Pfam: the protein families database. Nucleic Acids Res. https://doi.org/10.1093/nar/gkt1223

Ghelani A, Patel R, Mangrola A, Dudhagara P (2015) Cultivation independent comprehensive survey of bacterial diversity in Tulsi Shyam Hot Springs, India. Genom Data 4:54–56. https://doi.org/10.1016/j.gdata.2015.03.003

Goll J, Rusch DB, Tanenbaum DM, Thiagarajan M, Li K, Methé BA, Yooseph S (2010) METAREP: JCVI metagenomics reports—an open source tool for high-performance comparative metagenomics. Bioinformatics 26(20):2631–2632. https://doi.org/10.1093/bioinformatics/btq455

Goll J, Thiagarajan M, Abubucker S, Huttenhower C, Yooseph S, Methé BA (2012) A case study for large-scale human microbiome analysis using JCVI's metagenomics reports (METAREP). PLoS One 7:e29044. https://doi.org/10.1371/journal.pone.0029044

Gulhane M, Pandit P, Khardenavis A, Singh D, Purohit H (2017) Study of microbial community plasticity for anaerobic digestion of vegetable waste in Anaerobic Baffled Reactor. Renew Energy 101:59–66. https://doi.org/10.1016/j.renene.2016.08.021

Handelsman J, Rondon MR, Brady SF, Clardy J, Goodman RM (1998) Molecular biological access to the chemistry of unknown soil microbes: a new frontier for natural products. Chem Biol 5:R245–R249. https://doi.org/10.1016/S1074-5521(98)90108-9

Hasan NA, Young BA, Minard-Smith AT, Saeed K, Li H, Heizer EM, McMillan NJ, Isom R, Abdullah AS, Bornman DM, Faith SA (2014) Microbial community profiling of human saliva using shotgun metagenomic sequencing. PLoS One 9(5):e97699. https://doi.org/10.1371/journal.pone.0097699

Hunter S, Corbett M, Denise H, Fraser M, Gonzalez-Beltran A, Hunter C, Jones P, Leinonen R, McAnulla C, Maguire E, Maslen J (2014) EBI metagenomics—a new resource for the analysis and archiving of metagenomic data. Nucleic Acids Res 42(D1):D600–D606. https://doi.org/10.1093/nar/gkt961

Huson DH, Auch AF, Qi J, Schuster SC (2007) MEGAN analysis of metagenomic data. Genome Res 17(3):377–386. https://doi.org/10.1101/gr.5969107

Huson DH, Mitra S, Ruscheweyh HJ, Weber N, Schuster SC (2011) Integrative analysis of environmental sequences using MEGAN4. Genome Res 21(9):1552–1560. https://doi.org/10.1101/gr.120618.111

Huson DH, Beier S, Flade I, Górska A, El-Hadidi M, Mitra S, Ruscheweyh HJ, Tappu R (2016) MEGAN community edition-interactive exploration and analysis of large-scale microbiome sequencing data. PLoS Comput Biol 12(6):e1004957. https://doi.org/10.1371/journal.pcbi.1004957

Jadeja NB, More RP, Purohit HJ, Kapley A (2014) Metagenomic analysis of oxygenases from activated sludge. Bioresour Technol 165:250–256. https://doi.org/10.1016/j.biortech.2014.02.045

Kanehisa M, Goto S (2000) KEGG: kyoto encyclopedia of genes and genomes. Nucleic Acids Res 28(1):27–30. https://doi.org/10.1093/nar/28.1.27

Kapley A, Liu R, Jadeja NB, Zhang Y, Yang M, Purohit HJ (2015) Shifts in microbial community and its correlation with degradative efficiency in a wastewater treatment plant. Appl Biochem Biotechnol 176(8):2131–2143. https://doi.org/10.1007/s12010-015-1703-2

Keegan KP, Trimble WL, Wilkening J, Wilke A, Harrison T, D'Souza M, Meyer F (2012) A platform-independent method for detecting errors in metagenomic sequencing data: DRISEE. PLoS Comput Biol 8(6):e1002541. https://doi.org/10.1371/journal.pcbi.1002541

Kröber M, Bekel T, Diaz NN, Goesmann A, Jaenicke S, Krause L, Miller D, Runte KJ, Viehöver P, Pühler A, Schlüter A (2009) Phylogenetic characterization of a biogas plant microbial community integrating clone library 16S-rDNA sequences and metagenome sequence data

obtained by 454-pyrosequencing. J Biotechnol 142(1):38–49. https://doi.org/10.1016/j.jbiotec.2009.02.010
Lingner T, Asshauer KP, Schreiber F, Meinicke P (2011) CoMet – a web server for comparative functional profiling of metagenomes. Nucleic Acids Res 39:W518–W523. https://doi.org/10.1093/nar/gkr388
Luo C, Rodriguez-R LM, Konstantinidis KT (2014) MyTaxa: an advanced taxonomic classifier for genomic and metagenomic sequences. Nucleic Acids Res 42:e73. https://doi.org/10.1093/nar/gku169
Mangrola A, Dudhagara P, Koringa P, Joshi CG, Parmar M, Patel R (2015) Deciphering the microbiota of Tuwa hot spring, India using shotgun metagenomic sequencing approach. Genom Data 4:153–155. https://doi.org/10.1016/j.gdata.2015.04.014
Markowitz VM, Ivanova NN, Szeto E, Palaniappan K, Chu K, Dalevi D, Chen IM, Grechkin Y, Dubchak I, Anderson I, Lykidis A (2008) IMG/M: a data management and analysis system for metagenomes. Nucleic Acids Res 36(suppl 1):D534–D538. https://doi.org/10.1093/nar/gkm869
Markowitz VM, Chen IM, Chu K, Szeto E, Palaniappan K, Grechkin Y, Ratner A, Jacob B, Pati A, Huntemann M, Liolios K (2012a) IMG/M: the integrated metagenome data management and comparative analysis system. Nucleic Acids Res 40(D1):D123–D129. https://doi.org/10.1093/nar/gkr975
Markowitz VM, Chen IM, Chu K, Szeto E, Palaniappan K, Jacob B, Ratner A, Liolios K, Pagani I, Huntemann M, Mavromatis K (2012b) IMG/M-HMP: a metagenome comparative analysis system for the Human Microbiome Project. PLoS One 7(7):e40151. https://doi.org/10.1371/journal.pone.0040151
Markowitz VM, Chen IM, Palaniappan K, Chu K, Szeto E, Grechkin Y, Ratner A, Jacob B, Huang J, Williams P, Huntemann M (2012c) IMG: the integrated microbial genomes database and comparative analysis system. Nucleic Acids Res 40(D1):D115–D122. https://doi.org/10.1093/nar/gkr1044
Markowitz VM, Chen IM, Chu K, Szeto E, Palaniappan K, Pillay M, Ratner A, Huang J, Pagani I, Tringe S, Huntemann M (2014) IMG/M 4 version of the integrated metagenome comparative analysis system. Nucleic Acids Res 42(D1):D568–D573. https://doi.org/10.1093/nar/gkt919
Mason OU, Scott NM, Gonzalez A, Robbins-Pianka A, Bælum J, Kimbrel J, Bouskill NJ, Prestat E, Borglin S, Joyner DC, Fortney JL (2014) Metagenomics reveals sediment microbial community response to Deepwater Horizon oil spill. ISME J 8(7):1464–1475. https://doi.org/10.1038/ismej.2013.254
Meyer F, Paarmann D, D'Souza M, Olson R, Glass EM, Kubal M, Paczian T, Rodriguez A, Stevens R, Wilke A, Wilkening J (2008) The metagenomics RAST server–a public resource for the automatic phylogenetic and functional analysis of metagenomes. BMC Bioinformatics 9(1):386. https://doi.org/10.1186/1471-2105-9-386
Mitchell A, Chang HY, Daugherty L, Fraser M, Hunter S, Lopez R, McAnulla C, McMenamin C, Nuka G, Pesseat S, Sangrador-Vegas A (2014) The InterPro protein families database: the classification resource after 15 years. Nucleic Acids Res. https://doi.org/10.1093/nar/gku1243
Moharikar A, Purohit HJ, Kumar R (2005) Microbial population dynamics at effluent treatment plants. J Environ Monit 7(6):552–558. https://doi.org/10.1039/B406576J
Moharikar A, Kapley A, Purohit HJ (2003) Detection of dioxygenase genes present in various activated sludge. Environ Sci Pollut Res 10(6):373–378. https://doi.org/10.1065/espr2003.07.164
More RP, Mitra S, Raju SC, Kapley A, Purohit HJ (2014) Mining and assessment of catabolic pathways in the metagenome of a common effluent treatment plant to induce the degradative capacity of biomass. Bioresour Technol 153:137–146. https://doi.org/10.1016/j.biortech.2013.11.065
Oulas A, Pavloudi C, Polymenakou P, Pavlopoulos GA, Papanikolaou N, Kotoulas G, Arvanitidis C, Iliopoulos I (2015) Metagenomics: tools and insights for analyzing next-

generation sequencing data derived from biodiversity studies. Bioinf Biol Insights 9:75. https://doi.org/10.4137/BBI.S12462

Ounit R, Wanamaker S, Close TJ, Lonardi S (2015) CLARK: fast and accurate classification of metagenomic and genomic sequences using discriminative k-mers. BMC Genomics 16:236. https://doi.org/10.1186/s12864-015-1419-2

Overbeek R, Olson R, Pusch GD, Olsen GJ, Davis JJ, Disz T, Edwards RA, Gerdes S, Parrello B, Shukla M, Vonstein V (2014) The SEED and the Rapid Annotation of microbial genomes using Subsystems Technology (RAST). Nucleic Acids Res 42(D1):D206–D214. https://doi.org/10.1093/nar/gkt1226

Pal RR, Khardenavis AA, Purohit HJ (2015) Identification and monitoring of nitrification and denitrification genes in Klebsiella pneumoniae EGD-HP19-C for its ability to perform heterotrophic nitrification and aerobic denitrification. Funct Integr Genomics 15(1):63–76. https://doi.org/10.1007/s10142-014-0406-z

Pandit AS, Joshi MN, Bhargava P, Ayachit GN, Shaikh IM, Saiyed ZM, Saxena AK, Bagatharia SB (2014) Metagenomes from the saline desert of Kutch. Genome Announc 2(3):e00439-14. https://doi.org/10.1128/genomeA.00439-14

Pandit PD, Gulhane MK, Khardenavis AA, Purohit HJ (2016) Mining of hemicellulose and lignin degrading genes from differentially enriched methane producing microbial community. Bioresour Technol 216:923–930. https://doi.org/10.1016/j.biortech.2016.06.021

Powell S, Szklarczyk D, Trachana K, Roth A, Kuhn M, Muller J, Arnold R, Rattei T, Letunic I, Doerks T, Jensen LJ (2012) eggNOG v3. 0: orthologous groups covering 1133 organisms at 41 different taxonomic ranges. Nucleic Acids Res 40(D1):D284–D289. https://doi.org/10.1093/nar/gkr1060

Price MN, Dehal PS, Arkin AP (2010) FastTree 2–approximately maximum-likelihood trees for large alignments. PLoS One 5(3):e9490. https://doi.org/10.1371/journal.pone.0009490

Puranik S, Pal RR, More RP, Purohit HJ (2016) Metagenomic approach to characterize soil microbial diversity of Phumdi at Loktak Lake. Water Sci Technol 74(9):2075–2086. https://doi.org/10.2166/wst.2016.370

Purohit HJ, Kapley A, Moharikar AA, Narde G (2003) A novel approach for extraction of PCR-compatible DNA from activated sludge samples collected from different biological effluent treatment plants. J Microbiol Methods 52(3):315–323. https://doi.org/10.1016/S0167-7012(02)00185-9

Rho M, Tang H, Ye Y (2010) FragGeneScan: predicting genes in short and error-prone reads. Nucleic Acids Res 38(20):e191. https://doi.org/10.1093/nar/gkq747

Roux S, Faubladier M, Mahul A, Paulhe N, Bernard A, Debroas D, Enault F (2011) Metavir: a web server dedicated to virome analysis. Bioinformatics 27(21):3074–3075. https://doi.org/10.1093/bioinformatics/btr519

Roux S, Tournayre J, Mahul A, Debroas D, Enault F (2014) Metavir 2: new tools for viral metagenome comparison and assembled virome analysis. BMC Bioinformatics 15(1):1. https://doi.org/10.1186/1471-2105-15-76

Selengut JD, Haft DH, Davidsen T, Ganapathy A, Gwinn-Giglio M, Nelson WC, Richter AR, White O (2007) TIGRFAMs and Genome Properties: tools for the assignment of molecular function and biological process in prokaryotic genomes. Nucleic Acids Res 35(suppl 1):D260–D264. https://doi.org/10.1093/nar/gkl1043

Seshadri R, Kravitz SA, Smarr L, Gilna P, Frazier M (2007) CAMERA: a community resource for metagenomics. PLoS Biol 5:e75. https://doi.org/10.1371/journal.pbio.0050075

Sharma S, Vakhlu J (2014) Metagenomics as advanced screening methods for novel microbial metabolite. Microb Biotechnol Prog Trends 7:43–62. https://doi.org/10.1201/b17587-4

Sharma N, Tanksale H, Kapley A, Purohit HJ (2012) Mining the metagenome of activated biomass of an industrial wastewater treatment plant by a novel method. Indian J Microbiol 52(4):538–543. https://doi.org/10.1007/s12088-012-0263-1

Staden R (1979) A strategy of DNA sequencing employing computer programs. Nucleic Acids Res 6(7):2601–2610. https://doi.org/10.1093/nar/6.7.2601

Su CH, Hsu MT, Chiang S, Cheng JH, Weng FC, Wang D, Tsai HK (2011) MetaABC—an integrated metagenomics platform for data adjustment, binning and clustering. Bioinformatics 27(16):2298–2299. https://doi.org/10.1093/bioinformatics/btr376

Sun S, Chen J, Li W, Altinatas I, Lin A, Peltier S, Stocks K, Allen EE, Ellisman M, Grethe J, Wooley J (2010) Community cyberinfrastructure for advanced microbial ecology research and analysis: the CAMERA resource. Nucleic Acids Res. https://doi.org/10.1093/nar/gkq1102

Sun Q, Liu L, Wu L, Li W, Liu Q, Zhang J, Liu D, Ma J (2015) Web resources for microbial data. Genomics Proteomics Bioinformatics 13(1):69–72. https://doi.org/10.1016/j.gpb.2015.01.008

Tatusov RL, Natale DA, Garkavtsev IV, Tatusova TA, Shankavaram UT, Rao BS, Kiryutin B, Galperin MY, Fedorova ND, Koonin EV (2001) The COG database: new developments in phylogenetic classification of proteins from complete genomes. Nucleic Acids Res 29 (1):22–28. https://doi.org/10.1093/nar/29.1.22

Thomas T, Gilbert J, Meyer F (2012) Metagenomics – a guide from sampling to data analysis. Microb Inform Exp 2:3. https://doi.org/10.1186/2042-5783-2-3

Tikariha H, Pal RR, Qureshi A, Kapley A, Purohit HJ (2016) In silico analysis for prediction of degradative capacity of Pseudomonas putida SF1. Gene 591(2):382–392. https://doi.org/10.1016/j.gene.2016.06.028

Wilke A, Harrison T, Wilkening J, Field D, Glass EM, Kyrpides N, Mavrommatis K, Meyer F (2012) The M5nr: a novel non-redundant database containing protein sequences and annotations from multiple sources and associated tools. BMC Bioinformatics 13(1):141. https://doi.org/10.1186/1471-2105-13-141

Wommack KE, Bhavsar J, Polson SW, Chen J, Dumas M, Srinivasiah S, Furman M, Jamindar S, Nasko DJ (2012) VIROME: a standard operating procedure for analysis of viral metagenome sequences. Stand Genomic Sci 6(3):421. https://doi.org/10.4056/sigs.2945050

Yadav TC, Pal RR, Shastry S, Jadeja NB, Kapley A (2015) Comparative metagenomics demonstrating different degradative capacity of activated biomass treating hydrocarbon contaminated wastewater. Bioresour Technol 188:24–32. https://doi.org/10.1016/j.biortech.2015.01.141

Protein-protein Interactions: Basics, Characteristics, and Predictions

7

Angshuman Bagchi

Abstract

Most of the cellular processes involve protein-protein interactions (PPIs). It therefore necessitates obtaining the detailed information about the amino acid residues involved in PPIs. Available are the different PPI determining experimental techniques. These experimental methods, though very accurate, are time consuming, labor intensive, and very expensive. To solve the aforementioned problems, different labs developed different bioinformatic protocols to build different number of bioinformatic software tools to predict PPIs. The bioinformatic algorithms are used for prediction of three-dimensional structures of proteins as well as protein complexes. Nowadays, different machine learning algorithms are employed for the purpose of prediction of PPIs. The computational structure prediction methods involve homology modeling, threading, and ab initio modeling. These methods have nearly 75%–80% overall accuracies. The other most widely used method is molecular docking which is used to generate the three-dimensional conformations of protein complexes. The docking methods can broadly be categorized as rigid body docking and flexible docking. In this chapter, the different aspects of computational modeling and docking strategies will be covered. The basic terminologies will be revisited. The chapter will aim at providing a firsthand guide on protein interaction prediction methods.

Keywords

PPI · PPI interface · PPI formation · Machine learning · Docking

A. Bagchi (✉)
Department of Biochemistry and Biophysics, University of Kalyani, Kalyani, Nadia, India
e-mail: angshumanb@gmail.com

H. J. Purohit et al. (eds.), *Soft Computing for Biological Systems*,
https://doi.org/10.1007/978-981-10-7455-4_7

7.1 Introduction

In the biological systems, almost all of the biochemical reactions are the outcomes of different forms of protein-protein interactions (PPIs). Proteins bind to themselves as well as with other biomolecules like nucleic acids, organic or inorganic cofactors, and so on (Creighton 1992; Branden and Tooze 2008; Whiteford 2005; Park and Cochran 2009; Kessel and Ben-Tal 2010; Lesk 2010; Tropp 2011; Kurian et al. 2012; Nelson and Cox 2012; Walsh 2002). It is also a well-established fact that PPI dysfunctions may lead to different disease situations (Erickson 1978; Cox et al. 2006; Greene and McEvanely 2009; Bourin et al. 2012; Meyer and Jaspers 2015; Twigg et al. 2015). Thus, the ideas about PPIs are becoming important day by day, and it has therefore become essential to biologists to have a good understanding of PPIs. There are numerous PPI detecting experimental as well as computational approaches developed by different laboratories. The experimental tools to study PPIs include X-ray crystallography, nuclear magnetic resonance imaging, electron microscopy, microarray analysis, co-immunoprecipitation techniques, etc. The aforementioned experimental tools would produce accurate results, but the main problem with them is the time. These experimental techniques are lengthy processes. Besides that, these techniques are labor intensive and very costly. In order to solve these problems, a number of computational algorithms have been developed. The computational PPI prediction techniques can be classified as:

(a) Techniques to build the three-dimensional structures of protein and protein complexes
(b) Techniques to build protein-docking methodologies

Nowadays, different machine learning tools are constantly being used for development of protein-docking algorithms. The basic principle behind such docking techniques is to build a training model on the basis of a gold-standard training dataset which contains a list of positive and negative examples. The machine learning algorithms would build a model on the basis of which new examples may be classified as PPI or non-PPI. The computational PPI identification technologies have various degrees of accuracies. These computational tools though not as accurate as the experimental tools come up with fairly good predictive models of PPIs (Erickson 1978; Ausubel 1987; Phizicky and Fields 1995; Bollag et al. 1996; Rigaut et al. 1999; Puig et al. 2001; Golemis 2002; Piehler 2005; Kerppola 2008; Braun and Gingras 2012; Rao et al. 2014). The computational approaches may, therefore, be considered to be the start point of PPI prediction methodologies (Sims and Wander 2002; Bader et al. 2003; Hermjakob et al. 2004; Peri et al. 2004; Puente and López-Otín 2004; Woessner 2004; Chatr-Aryamontri et al. 2007; Shoemaker and Panchenko 2007; Breitkreutz et al. 2008; Skrabanek et al. 2008; van der Hoorn 2008; Tuncbag et al. 2009; Li et al. 2010; Theofilatos et al. 2011; Kohei 2012). The present chapter is aimed to give firsthand knowledge of different computational PPI prediction methodologies. However, before going into the technical details of the PPI prediction methods, the basic definitions need to be revisited.

7.2 Basic Definitions

PPI Interface PPI interface is the area between the two protein chains. If the amino acid compositions of the two protein chains are the same, the interface is called homomeric interface; otherwise, it is termed as heteromeric interface. The PPI interface has the following characteristics:

- Surfacc area of interface: For heterodimeric proteins, the surface area is generally around 600Å^2. For a homodimers it is even larger than that.
- Shape of the PPI interface: It is known that the PPI interface is nearly flat. A PPI interface has two separate zones, viz., the core which is buried in the interface and the rim which is solvent accessible.
- Composition of amino acids at the PPI interface: It is known that the PPI interface has an abundance of aromatic amino acid residues and Arg. However, Cys is not generally found at the PPI interfaces.
- Secondary structural distribution at the PPI interface: It is known that a PPI interface is made up of beta sheet regions (Bogan and Thorn 1998; Faisal et al. 1999; Sheinerman et al. 2000; Schreiber 2002; Nooren and Thornton 2003; Nooren 2003; Ofran and Rost 2003a; Bahadur et al. 2004; Keskin et al. 2005; Shenoy and Jayaram 2010).

7.2.1 Classification of PPI Interface

The PPIs can broadly be classified into several different classes based on the nature of the interacting partners, the stability of the PPI complexes, the life-span of the interactions between the protein partners, and the nature of the PPI interface between the proteins.

- Nature of interacting protein partners: If the interacting protein partners have the same amino acid compositions, they form homo-oligomers, with structural symmetry. On the other hand, nonidentical protein partners form hetero-oligomers. Hemoglobin is a homo-tetramer and a protease-anti-protease complex is a heteromer.
- Stability of interacting protein complexes: If the individual protein partners forming the PPI complex cannot exist in free state and are stable only in multi-meric association, they are called obligate oligomers (homo-obligomers and/or hetero-obligomers), like the Arc repressor dimer where dimerization is essential for DNA binding. On the other hand, when the protein partners can exist in free states on their own, they are called non-obligate partners like antigen-antibody complex.
- Lifetime of PPI: When an association between the protein partners is highly stable and needs external agencies to break them, they are called permanent

complexes. Hetero-trimeric G protein (Gα, Gβγ, and GDP) forms this type of PPI. In contrast, the interacting partners of sperm lysin, a homodimer, exist in a dynamic equilibrium consisting of association and dissociation of oligomeric forms. This type of PPI is named as transient complex.

- Nature of the interaction interface: When the individual protein partners in a PPI use the same interacting interface to join each other, they are called isologous complexes. On the other hand, in heterologous assembly, the individual protein partners in a PPI complex use different interfaces to form PPI without any closed symmetry (Bogan and Thorn 1998; Faisal et al. 1999; Sheinerman et al. 2000; Schreiber 2002; Nooren and Thornton 2003; Nooren 2003; Ofran and Rost 2003a, b; Bahadur et al. 2004; Keskin et al. 2005; Shenoy and Jayaram 2010).

7.3 Mechanism of PPI

The first and foremost important criterion for protein-protein interactions is the proximity of the protein partners so that they can interact with each other. However, most of the interactions between protein partners are non-covalent interactions with the only exception of covalent disulfide linkages between the cysteine amino acid residues of the interacting partner proteins. Among the non-covalent interactions, the most important one found in PPIs is the hydrogen bonding between the polar atoms in the interacting protein partners. The hydrogen bonding between interacting proteins would involve both the main and side chain atoms of the different amino acid residues in the interacting protein partners. The second most important non-covalent interaction leading to the formation of PPIs is the formation of ion pair. The ion-pair formation occurs mainly between the side chain atoms of an acidic amino acid with that of the basic amino acid in the interacting protein partners. The other important non-covalent interactions involve:

(a) Stacking interactions—between the side chains of the non-polar hydrophobic amino acids in proteins
(b) Cation-pi interactions—between the side chains of the aromatic side amino acid residues like Phe, Tyr, and Trp with the positively charged side chains of basic amino acid residues like Lys and Arg. However, the different protein complexes have different binding interactions (Bogan and Thorn 1998; Faisal et al. 1999; Sheinerman et al. 2000; Schreiber 2002; Nooren and Thornton 2003; Nooren 2003; Ofran and Rost 2003a, b; Bahadur et al. 2004; Keskin et al. 2005; Shenoy and Jayaram 2010).

7.4 Machine Learning and Its Applications in PPI Prediction

Machine Learning (Gallet et al. 2000; Pawson and Nash 2000; Neuvirth et al. 2004; Bradford et al. 2006; Li et al. 2006; Wang et al. 2006; Kushwaha and Shakya 2010; Choong et al. 2013; You et al. 2013; Zahiri et al. 2013; Cukuroglu et al. 2014;

Kobzar et al. 2014; Lage 2014; Lua et al. 2014; Murakami and Mizuguchi 2014): The topic machine learning is a part of computer science. It is derived from pattern recognition. The machine learning algorithms deduce solutions of a problem based on the development of a training model obtained from example inputs. Machine learning can be used to solve various classes of problems like classifications, regressions, clustering, etc. The machine learning techniques can be of the following types:

- Decision tree learning: This is a learning method that uses a treelike architecture which acts as the model for prediction purposes. The new examples are mapped into respective categories with the help of the decision tree. The method is employed in random forest-based classifiers.
- Association rule learning: This is a learning method which extracts some information from input.
- Artificial neural network (ANN): This learning method is inspired by the biological networks of nerve cells, i.e., the neurons. These are the data modeling tools which are based on nonlinear statistics. These methods are used to analyze the so-called complicated relationship problems between inputs and outputs to decipher a suitable rule to find the solution of the given problem.
- Deep learning: This is an extension of ANN. The method tries to mimic the human brain processes by incorporating several hidden layers in the existing ANN.
- Support vector machines (SVMs): This is a supervised machine learning technique. This technique builds a hyperplane to separate the known input data points into different classes. The new data point is then mapped onto the newly generated rule obtained from the hyperplane.
- Bayesian networks: This is a probabilistic method, and it presents a set of variables random in nature and how they are interrelated. A Bayesian network may be used to infer the relationship between diseases and symptoms.
- Genetic algorithm (GA): This method is a method mimicking the process of natural selection. This method generates the methods such as mutation, crossover, etc. to generate new population to find a good solution.

The various machine learning techniques are constantly being used nowadays for the prediction of PPIs. The machine learning tools are mainly used for the classification purposes. In such cases, the input data are used to train a classifier to distinguish between positive (amino acid residues involved in PPIs) and negative (amino acid residues noninvolved in PPIs) examples. The most popular machine learning techniques used for such purposes are the use of classifiers based on random forest and support vector machines. The random forest-based classifiers generally produce results based on domain compositions of interacting and non-interacting protein partners. In other words, such classifiers are trained on information extracted from the domain compositions of interacting and non-interacting protein partners. Given an input of the information of an unknown protein, such classifiers would come up with some likelihood of the amino acids residues in the input protein to be involved in PPIs or not.

The SVM-based predictors would function in a somewhat different way. SVM-based predictors are trained with suitable positive and negative examples. During training, the predictor would create a hyperplane to discriminate between the positive and negative examples. For an input protein, the SVM-based classifier would measure the distances of the amino acid residues from the hyperplane and would thereby come up with some probability values of the amino acid being classified as belonging to PPI or non-PPI class.

However, all the machine learning methods are heavily dependent on the accuracies of training dataset. All the aforementioned tools would generate some features from the training data. The features are needed to discriminate between the positive and negative examples. The resulting classifiers from the machine learning tools would depend on nature of the features. A good feature having a good class discriminating ability would create a better classifier. However, in biological system, it is not always possible to have a good negative example. It cannot be generalized that the amino acid in a protein which is found to be not involved in PPI in one example would do so in all the other protein complexes.

7.5 Conclusion

Identification of amino acid residues involved in PPIs is a very daunting task. As previously mentioned, PPIs exist in all the biochemical reactions. The most important among them is the protease-antiprotease interactions. The abundance of protein sequence information instigated the scientists to come up with protein interaction prediction methodologies that use the protein sequence information only. It is also a well-established fact that sequence is more conserved than structure. So, similar sequences mean similar structures. However, this assumption fails below a sequence similarity level of 30%. So, the sequence-based PPI prediction methods have very low accuracy levels. On the other hand, methods based on protein structures are fairly accurate, but the drawback is there are a very less number of good protein-protein complex structures that are available. Nonetheless, the bioinformatic tools may come up with a firsthand knowledge of PPIs for which experimentation is not yet possible.

Acknowledgment The author would like to acknowledge the help rendered by the DBT-sponsored Bioinformatics Infrastructural Facility of the University of Kalyani. The author would also like to thank the Department of Biotechnology (DBT, India) for the financial support (SAN No. 102/IFD/SAN/1824/2015-2016). The author is grateful to the Virologie et Immunologie Moléculaires, UR892 INRA, Domaine de Vilvert 78352 Jouy-en-Josas, France, for the infrastructural support. The infrastructural help from the Department of Biochemistry and Biophysics, University of Kalyani, is duly acknowledged.

References

Ausubel FM (1987) Current protocols in molecular biology. Wiley, New York/Boston, pp 15.1.1–15.1.14

Bader G et al (2003) BIND the biomolecular interaction network database. Nucleic Acids Res 31:248–250. https://doi.org/10.1093/nar/gkh052

Bahadur RP, Chakrabarti P, Rodier F, Janin J (2004) A dissection of specific and non-specific protein-protein interfaces. J Mol Biol 336:943–955. https://doi.org/10.1016/j.jmb.2003.12.073

Bogan AA, Thorn KS (1998) Anatomy of hot spots in protein interfaces. J Mol Biol 280:1–9. https://doi.org/10.1006/jmbi.1998.1843

Bollag DM, Rozycki MD, Edelstein ST (1996) Protein methods, 2nd edn. Wiley Publishers, New York, pp 1–83

Bourin M, Gautron J, Berges M, Hennequet-Antier C, Cabau C, Nys Y, Réhault-Godbert S (2012) Transcriptomic profiling of proteases and antiproteases in the liver of sexually mature hens in relation to vitellogenesis. BMC Genomics 13:457. https://doi.org/10.1186/1471-2164-13-457

Bradford JR, Needham CJ, Bulpitt AJ, Westhead DR (2006) Insights into protein-protein interfaces using a Bayesian network prediction method. J Mol Biol 362:365–386. https://doi.org/10.1016/j.jmb.2006.07.028

Branden C, Tooze A (2008) Introduction to protein structure, 2nd edn. Garland Publishing Inc, New York, pp 373–392. ISBN 0815304862, 9780815304869

Braun P, Gingras AC (2012) History of protein-protein interactions: from egg-white to complex networks. Proteomics 12:1478–1498. https://doi.org/10.1002/pmic.201100563

Breitkreutz BJ, Stark C, Reguly T, Boucher L, Breitkreutz A, Livstone M, Oughtred R, Lackner DH, Bähler J, Wood V, Dolinski K (2008) The BioGRID interaction database: 2008 update. Nucleic Acids Res 36:D637–D640. https://doi.org/10.1093/nar/gkm1001

Chatr-Aryamontri A, Ceol A, Palazzi LM, Nardelli G, Schneider MV, Castagnoli L, Cesareni G (2007) MINT the molecular interaction database. Nucleic Acids Res 35:D572–D574. https://doi.org/10.1093/nar/gkl950

Choong YS, Tye GJ, Lim TS (2013) Minireview: applied structural bioinformatics in proteomics. Protein J 32:505–511. https://doi.org/10.1007/s10930-013-9514-1

Cox SW, Rodriguez-Gonzalez EM, Booth V, Eley BM (2006) Secretory leukocyte protease inhibitor and its potential interactions with elastase and cathepsin B in gingival crevicular fluid and saliva from patients with chronic periodontitis. J Periodontal Res 41:477–485. https://doi.org/10.1111/j.1600-0765.2006.00891.x

Creighton TE (1992) Proteins: structures and molecular properties, 2nd edn. W.H. Freemann & Company, New York

Cukuroglu E, Engin HB, Gursoy A, Keskin O (2014) Hot spots in protein-protein interfaces: towards drug discovery. Prog Biophys Mol Biol 31:165–173. https://doi.org/10.1016/j.pbiomolbio.2014.06.003

Erickson S (1978) Proteases and protease inhibitors in chronic obstructive lung disease. Acta Med Scand 203:449–455

Faisal M, Oliver JL, Kaattari SL (1999) Potential role of protease-anti-protease interactions in *Perkinsus Marinus* infection in *Crassostrea sp.* Bull Eur Assoc Fish Pathol 19:269–276. https://doi.org/10.1051/alr:2004050

Gallet X, Charloteaux B, Thomas A, Brasseur R (2000) A fast method to predict protein interaction sites from sequences. J Mol Biol 302:917–926. https://doi.org/10.1006/jmbi.2000.4092

Golemis E (2002) Protein-protein interactions: a molecular cloning manual, 2nd edn. Cold Spring Harbor Laboratory Press, Cold Spring Harbor, pp 1–50

Greene CM, McEvanely NG (2009) Proteases and antiproteases in chronic neutrophilic lung disease – relevance to drug discovery. Br J Pharmacol 158:1048–1058. https://doi.org/10.1111/j.1476-5381.2009.00448.x

Hermjakob H, Montecchi-Palazzi L, Lewington C, Mudali S, Kerrien S, Orchard S, Vingron M, Roechert B, Roepstorff P, Valencia A, Margalit H (2004) IntAct an open source molecular interaction database. Nucleic Acids Res 32:D452–D455. https://doi.org/10.1093/nar/gkh052

Kerppola TK (2008) Bimolecular fluorescence complementation: visualization of molecular interactions in living cells. Methods Cell Biol 9:789–798. https://doi.org/10.1016/S0091-679X(08)85019-4

Keskin O, Ma B, Nussinov R (2005) Hot regions in protein-protein interactions: the organization and contribution of structurally conserved hot spot residues. J Mol Biol 345:1281–1294. https://doi.org/10.1016/j.jmb.2004.10.077

Kessel A, Ben-Tal N (2010) Introduction to proteins: structure, function, and motion, 1st edn. Chapman & Hall/CRC, Boca Raton, pp 36–65

Kobzar OL, Trush VV, Tanchuk VY, Zhilenkov AV, Troshin PA, Vovk AI (2014) Fullerene derivatives as a new class of inhibitors of protein tyrosine phosphatases. Bioorg Med Chem Lett 24:3175–3179. https://doi.org/10.1016/j.mencom.2015.05.013

Kohei O (2012) New families of carboxyl peptidases: serine-carboxyl peptidases and glutamic peptidases. J Biochem 151:13–25. https://doi.org/10.1093/jb/mvr129

Kurian J, Conforti B, Wemmer D (2012) The molecules of life: physical and chemical principles, 1st edn. Garland Science, New York, pp 530–787

Kushwaha SK, Shakya M (2010) Protein interaction network analysis-approach for potential drug target identification in mycobacterium tuberculosis. J Theor Biol 262:284–294. https://doi.org/10.1016/j.jtbi.2009.09.029

Lage K (2014) Protein-protein interactions and genetic diseases: the interactome. Biochim Biophys Acta Mol basis Dis 1842:1971–1980. https://doi.org/10.1016/j.bbadis.2014.05.028

Lesk AM (2010) Introduction to protein science: architecture, function, and genomics, 2nd edn. Oxford University Press, New York, pp 17–38. https://doi.org/10.1107/S2059798316018283

Li JJ, Huang DS, Wang B, Chen P (2006) Identifying protein-protein interfacial residues in heterocomplexes using residue conservation scores. Int J Biol Macromol 38:241–247. https://doi.org/10.1016/j.ijbiomac.2006.02.024

Li X, Wu M, Kwoh CK, Ng SK (2010) Computational approaches for detecting protein complexes from protein interaction networks: a survey. BMC Genomics 11:S3. https://doi.org/10.1186/1471-2164-11-S1-S3

Lua RC, Marciano DC, Katsonis P, Adikesavan AK, Wilkins AD, Lichtarge O (2014) Prediction and redesign of protein–protein interactions. Prog Biophys Mol Biol 116:194–202. https://doi.org/10.1186/s12862-016-0608-1

Meyer M, Jaspers I (2015) Respiratory protease/antiprotease balance determines susceptibility to viral infection and can be modified by nutritional antioxidants. Am J Phys Lung Cell Mol Phys 308:L1189–L2010. https://doi.org/10.1152/ajplung.00028.2015

Murakami Y, Mizuguchi K (2014) Homology-based prediction of interactions between proteins using averaged one-dependence estimators. BMC Bioinformatics 15:213. https://doi.org/10.1186/1471-2105-15-213

Nelson DL, Cox MM (2012) Principles of biochemistry, 5th edn. W.H. Freemann & Company, New York, pp 157–237. www.whfreeman.com/lehninger4e

Neuvirth H, Raz R, Schreiber G (2004) ProMate: a structure based prediction program to identify the location of protein-protein binding sites. J Mol Biol 338:181–199. https://doi.org/10.1016/j.jmb.2004.02.040

Nooren IMA (2003) NEW EMBO MEMBER'S REVIEW: diversity of protein-protein interactions. EMBO J 22:3486–3492. https://doi.org/10.1093/emboj/cdg359

Nooren IMA, Thornton JM (2003) Structural characterisation and functional significance of transient protein-protein interactions. J Mol Biol 325:991–1018. https://doi.org/10.1016/S0022-2836(02)01281-0

Ofran Y, Rost B (2003a) Analysing six types of protein-protein interfaces. J Mol Biol 325:377–387. https://doi.org/10.1016/S0022-2836(02)01223-8

Ofran Y, Rost B (2003b) Predicted protein-protein interaction sites from local sequence information. FEBS Lett 544:236–239. https://doi.org/10.1016/S0014-5793(03)00456-3

Park JS, Cochran JR (2009) Protein engineering and design, 1st edn. CRC Press, Boca Raton, pp 131–150. https://doi.org/10.1016/j.chembiol.2010.10.012

Pawson T, Nash P (2000) Protein-protein interactions define specificity in signal transduction. Genes Dev 14:1027–1047. https://doi.org/10.1101/gad.14.9.1027Genes&Dev.2000.14:1027-1047

Peri S, Navarro JD, Kristiansen TZ, Amanchy R, Surendranath V, Muthusamy B, Gandhi TK, Chandrika KN, Deshpande N, Suresh S, Rashmi BP (2004) Human protein reference database as a discovery resource for proteomics. Nucleic Acids Res 32:D497–D501. https://doi.org/10.1093/nar/gkh070

Phizicky EM, Fields S (1995) Protein-protein interactions: methods for detection and analysis. Microbiol Rev 59(1):94–123. http://mmbr.asm.org/content/59/1/94.long

Piehler J (2005) New methodologies for measuring protein interactions in vivo and in vitro. Curr Opin Struct Biol 15:4–14. https://doi.org/10.1016/j.sbi.2005.01.008

Puente XS, López-Otín C (2004) A genomic analysis of rat proteases and protease inhibitors. Genome Res 14:609–622. https://doi.org/10.1101/gr.1946304

Puig O, Caspary F, Rigaut G, Rutz B, Bouveret E, Bragado-Nilsson E, Wilm M, Séraphin B (2001) The tandem affinity purification (TAP) method: a general procedure of protein complex purification. Methods 24:218–229. https://doi.org/10.1006/meth.2001.1183

Rao VS, Srinivas K, Sujini GN, Kumar GN (2014) Protein-protein interaction detection: methods and analysis. Int J Proteomics. https://doi.org/10.1155/2014/147648

Rigaut G, Shevchenko A, Rutz B, Wilm M, Mann M, Séraphin B (1999) A generic protein purification method for protein complex characterization and proteome exploration. Nat Biotechnol 17:1030–1032. https://doi.org/10.1038/13732

Schreiber G (2002) Kinetic studies of protein – protein interactions. Curr Opin Struct Biol:41–47. http://www.biocristalografia.df.ibilce.unesp.br/publications/pdf/bf2003_83.pdf

Sheinerman FB, Norel R, Honig B (2000) Electrostatic aspects of protein – protein interactions. Curr Opin Struct Biol 10:153–159. https://doi.org/10.1016/S0959-440X(02)00287-7

Shenoy RS, Jayaram B (2010) Proteins: sequence to structure and function-current status. Curr Protein Pept Sci 11:498–514. https://doi.org/10.2174/138920310794109094

Shoemaker BA, Panchenko AR (2007) Deciphering protein-protein interactions.Part II. Computational methods to predict protein and domain interaction partners. PLoS Comput Biol 3:e43. https://doi.org/10.1371/journal.pcbi.0030043

Sims GK, Wander MM (2002) Proteolytic activity under nitrogen or sulfur limitation. Appl Soil Ecol 568:1–5. http://citeseerx.ist.psu.edu/viewdoc/download?doi=10.1.1.512.5240&rep=rep1&type=pdf

Skrabanek L, Saini HK, Bader GD, Enright AJ (2008) Computational prediction of protein-protein interactions. Mol Biotechnol 38:1–17. https://doi.org/10.1007/s12033-007-0069-2

Theofilatos K, M Dimitrakopoulos C, K Tsakalidis A, D Likothanassis S, T Papadimitriou S, P Mavroudi S (2011) Computational approaches for the prediction of protein-protein interactions-a survey. Curr Bioinforma 6:398–414. https://doi.org/10.3389/fgene.2015.00289

Tropp BE (2011) Molecular biology: genes to proteins, 4th edn. Jones & Bartlett Learning, Sudbury, pp 27–75

Tuncbag N, Kar G, Keskin O, Gursoy A, Nussinov R (2009) A survey of available tools and web servers for analysis of protein-protein interactions and interfaces. Brief Bioinform 10:217–232. https://doi.org/10.1093/bib/bbp001

Twigg MS, Brockbank S, Lowry P, FitzGerald SP, Taggart C, Weldon S (2015) The role of serine proteases and antiproteases in the cystic fibrosis lung. Mediators Inflamm 293053. https://doi.org/10.1155/2015/293053

van der Hoorn RA (2008) Plant proteases: from phenotypes to molecular mechanisms. Annu Rev Plant Biol 59:191–223. https://doi.org/10.1146/annurev.arplant.59.032607.092835

Walsh G (2002) Proteins: biotechnology and biochemistry, 1st edn. John Wiley & Sons, Chichester, pp 251–278. https://pharmareview.files.wordpress.com/2015/03/biopharmaceuticals-biochemistry-and-biotechnology-walsh-wiley-2e-2003.pdf

Wang B, Chen P, Huang DS, Li JJ, Lok TM, Lyu MR (2006) Predicting protein interaction sites from residue spatial sequence profile and evolution rate. FEBS Lett 580:380–384. https://doi.org/10.1016/j.febslet.2005.11.081

Whiteford D (2005) Proteins: structure and function, 1st edn. Wiley, Chichester, pp 189–244

Woessner FJ (2004) In: Barrett AJ, Rawlings ND (eds) Handbook of proteolytic enzymes, 3rd edn. Elsevier Academic Press, London, pp 1–16

You Z-H, Lei Y-K, Zhu L, Xia J, Wang B (2013) Prediction of protein-protein interactions from amino acid sequences with ensemble extreme learning machines and principal component analysis. BMC Bioinf 14. https://doi.org/10.1186/1471-2105-14-S8-S10

Zahiri J, Yaghoubi O, Mohammad-Noori M, Ebrahimpour R, Masoudi-Nejad A (2013) PPIevo: protein-protein interaction prediction from PSSM based evolutionary information. Genomics 102:237–242. https://doi.org/10.1016/j.ygeno.2013.05.006

Machine Learning Framework: Predicting Protein Structural Features

8

Pramod Kumar, Vandana Mishra, and Subarna Roy

Abstract

Structural biology is a challenging scientific discipline that aims to uncover the topologies and shapes of biomolecules and macromolecules—that is, DNA, RNA, and proteins. Proteins are large macromolecules consisting of more than one chain of amino acids joined together in a linear chain by peptide bonds. Proteins are required in organisms; they help in all biological processes of cells. They catalyze biochemical reactions (enzymes), carry out key roles in cellular processes, and act as structural constituents, catalysis agents, signaling molecules, and molecular machines of every biological system. They are responsible for immune responses, can store molecules (e.g., casein and ovalbumin store amino acids), and are even responsible for cell mechanics (e.g., actin and myosin). The structure prediction of proteins is a difficult task with basic problems in computational biology, structural science, and structural biology. The complex structure of protein prediction has four different levels: (1) - one-dimensional (1D) prediction of different structural features and linear chain of amino acids; (2) two-dimensional (2D) prediction of spatial arrangements between amino acids; (3) three-dimensional (3D) (tertiary) structural features prediction of a protein; and (4) four-dimensional (4D) (quaternary) structure prediction of multicomplex proteins. Researchers have recently used most of the various data mining methods, different scripting-based tools, and machine learning tools for structure prediction of a protein. In this chapter, we provide a comprehensive overview of proteins structure and use different data mining machine learning algorithms for protein structure prediction.

P. Kumar (✉) · S. Roy
Biomedical Informatics Centre, ICMR-National Institute of Traditional Medicine (Formerly Regional Medical Research Centre), Department of Health Research, Belagavi, Karnataka, India
e-mail: pramodbiotech@gmail.com

V. Mishra
CSIR- Unit for Research and Development of Information Products, Pune, Maharashtra, India

H. J. Purohit et al. (eds.), *Soft Computing for Biological Systems*,
https://doi.org/10.1007/978-981-10-7455-4_8

Keywords
Algorithms · Machine learning · Proteins · Structure prediction

8.1 Introduction

8.1.1 Proteins

Proteins are the basic and essential components of living organisms and represent the major class of biomolecules in living things. Proteins play a significant role in all cellular processes and functions of the cell. They provide structural and infra-structure support to hold a creature together. They act as enzymes to make chemical reactions that are essential for life. Proteins also act as switches to control gene expression and as sensors that are involved in taste and smell. They are effectors to make muscles move, act as detectors that make people differentiate self from nonself, and cause immune responses. Some proteins have a globular form of structure because of the assumption of a globe-like shape in a natural water environment. The non-globular proteins are a special class of proteins, represented as globular proteins that depend on shape and interaction with cell membranes.

8.1.2 Primary Structure

Proteins are linear chains of polymers consisting of amino acids. There are 20 amino acids, called small molecules, which are synthesized naturally in organisms. An amino acid consists of an amino group (NH_2), a carboxyl group (COOH), and a hydrogen atom attached to a central (α) carbon. It has a variable side chain (R) group, which is attached to the central (α) carbon. The R group distinguishes one amino acid from another. Amino acids may create bonds between each other through the reaction of carboxyl and amino groups. Such a bond formed between amino acids is called a peptide bond. Peptide bonds hold the amino acid together and form the protein structure. The different parts of the original amino acids in the protein are known as residue. This kind of linear polypeptide chain of amino acids forms the primary structure of proteins.

8.1.3 Secondary Structure

Local conformations of a linear polypeptide chain of amino acids refer to secondary structures which appear repeatedly in the protein's secondary structure. There are different characteristics that are involved in secondary structure responsible for the 3D form of local regions. The α-helix and β-sheets are two types of major dominated conformations found in linear chain polypeptides. There are certain regularities occurring in local structures and these identify hydrogen bonds among the various residues of amino acids.

An α-helix is a rigid, rod-like structure and represents screw-shaped conformations, whereas in a β-sheet there are many parallel strands of residues known as β-pleated. In regular secondary structures there are loose, flexible loops and tight turns showing more consistent elements of secondary structures. There are also random loops and coils included in conformations that are not associated The conformations which do not belong with a methodical structural feature of a secondary structure are referred to as random loops or coils. These can be divided into three different classes –α-helixes, coils, and β-sheets. The sequence annotation for each individual residue represents the secondary structure of the protein. The method used for annotations of secondary structure is covered by the Dictionary of Secondary Structure of Proteins (DSSP) (Kabsch and Sander 1983). The resultant classes of DSSP are H, G, I, E, B, T, S, and C (the letter is actually a "none" assignment represented by white space). Moreover, SSP, a basic and simple version of DSSP and associated with DSSPHEC, is widely used for initial categorization to alpha-helix (H), beta-strand (E), or coil (C). In SSP, a simplified version of DSSP, say DSSPHEC, is typically adopted, and covers and maps each of the eight initial categories to DSSPHEC, related to DSSP in a look-up table. The most significant correspondence is mentioned below:

DSSP	H G I E B T S C.
$DSSP_{HEC}$	H H H E E C C C.

8.1.4 Tertiary Structure

Proteins fold up and form the 3D structure known as the tertiary (3D) structure of a protein. This refers to unique 3D conformations that globular proteins assume as a consequence of the interactions between the side chains in their primary structure. Peptide bonds hold the amino acids in the protein structure, shows two degrees of rotational freedom, φ and ψ angles. The folding (shape) of protein can be summarized as an order of φ/ψ angles, using Cartesian coordinates, of the middle backbone atom (the alpha carbon, written Cα), or using other representational schemes. The atoms are found in specific positions in folded protein known as the tertiary structure.

8.1.5 Quaternary Structure

It is a combination of multiple chains that appears in larger complexes. Interactions between or among the atoms of protein chains occur jointly by non-covalent interaction. Examples include van der Waals interactions, hydrogen bonding, ionic bonding interactions, and disulfide bonding.

Pioneering and experimental research in structural and computational biology, the main objective of structure prediction, refers to the prediction of secondary and tertiary structures using experimental primary sequences or structures of proteins. This is concerned with identifying primary and complex quaternary structures.

A cell consists of different type of proteins and protein complexes, and these interact with each other and neighboring molecules such as DNA, RNA, metabolites which carry different kinds of cellular and biological processes and functions such as enzymatic reactions and catalysis, coordinated motion, immune protection, transport and storage, mechanical support, transmission of nerve impulses, and inhibition of growth and differentiation (Laskowski et al. 2003). Many biochemical experiments are being carried out (Kendrew et al. 1960; Perutz et al. 1960; Travers 1989; Bjorkman and Parham 1990) to determine the native structures of proteins responsible for the key functions of proteins. Therefore, elucidating a native structure of proteins to understand its function plays major roles in many scientific disciplines such as pharmaceutical, biological, biotechnological and medical sciences. The experimental methods being used to derive protein structures include X-ray crystallography (Bragg 1975) and nuclear magnetic resonance (NMR) spectroscopy (Wuthrich 1986; Baldwin et al. 1991). Following the two major protein structures of myoglobin and hemoglobin being determined using X-ray crystallography (Kendrew et al. 1960; Perutz et al. 1960) experimental techniques, the number of proteins with known structures has been speedily enhanced.

At present the Protein Data Bank (PDB) contains details of approximately 125,000 proteins with determined structures (Berman et al. 2000). The available diverse and abundant structures of proteins in the PDB provide invaluable information on exactly how proteins fold into their typical tertiary structure and the prediction of protein structure from its sequence (Chandonia and Brenner 2006). After many precursor techniques and experiments (Sanger and Thompson 1953; Kendrew et al. 1960; Perutz et al. 1960; Anfinsen 1973), it has been shown that the native structure of a protein can be derived or predicted by its amino acid sequence; protein structure prediction from linear sequences has become a difficult challenge and a major task in structural biology.

In genomics, certain methods such as high throughput DNA and protein sequencing have been used to predict or determine protein structure. As the knowledge of protein sequences is increasing exponentially through experiments, experimental determinations of native structures of proteins are still time consuming, labor intensive, expensive, and most of the time impossible to predict. Thus, protein structure prediction from linear sequencing of amino acids is becoming increasingly indispensable and useful. Current structure prediction software is becoming a useful tool to show occurrences in existent molecular and cell biology (Petrey and Honig 2005) and also has significant applications in medical biotechnology, pharmaceutical research, biotechnology, and general medical research tasks such as drug discovery and molecule lead design (Jacobson and Sali 2004) (Fig. 8.1).

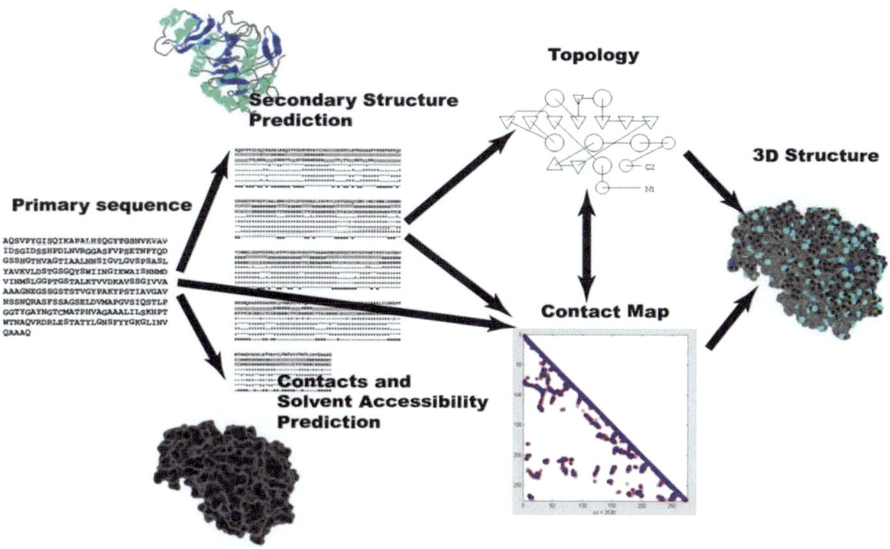

Fig. 8.1 Overall strategy for machine learning protein structures. Example of 1SCJ (subtilisin-propeptide complex) protein. The first stage predicts structural features including secondary structure, contacts, and relative solvent accessibility. The second stage predicts the topology of the protein, using the primary sequence and the structural features. The coarse topology is represented as a cartoon providing the relative proximity of secondary structure elements, such as alpha helices and beta-strands. The high-resolution topology is represented by the contact map between the residues of the protein. The final stage is the prediction of the actual 3D coordinates of all residues and atoms in the structure

Here we have tried to focus on the contributions of machine learning approaches in multilevel protein structure prediction from 1D to 4D (Rost and Chasman 2003; Baldi and Brunak 2001). If we try to predict 1D structure using machine learning techniques we must predict the secondary structural features (Rost and Sander 1993a,b; Jones 1999b; Pollastri et al. 2002b) and relative solvent accessibility (Rost and Sander 1994; Pollastri et al. 2002a) of every residue along the 1D protein sequence (Fig. 8.2). When the 2D structure prediction takes place, this focuses on predicting the spatial arrangements among or between residues, such as distance, contact map prediction (Fariselli et al. 2001; Pollastri and Baldi 2002), and disulfide bond prediction (Fariselli and Casadio 2004; Vullo and Frasconi 2003; Baldi et al. 2005) (Fig. 8.3). A significant and unique feature of 2D representations is that they are independent of translations of protein and can rotate in any direction, therefore being independent of any frame of coordinates, which could be seen only in the 3D level. Tertiary (3D) structure prediction focuses on predicting the 3D coordinates for all residues present or all atoms in a 3D space of a protein. However, the main aim is to predict tertiary (3D) structure, for which1D and 2D structure predictions are frequently used as input, and therefore 1D and 2D predictions are of great interest to biologists as important steps toward tertiary structure prediction

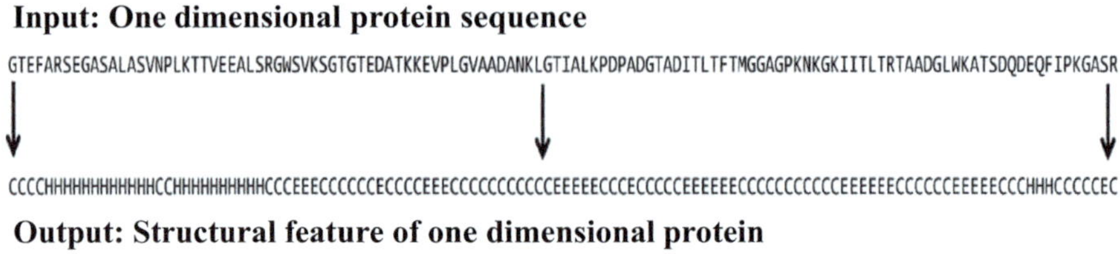

Fig. 8.2 One-dimensional protein structure prediction. Example elected of 1D structure prediction where the input primary sequence of amino acid is "translated" into an output sequence of secondary structure assignments for each amino acid (C= coil; H= helix; E= beeta sheet [extended sheet])

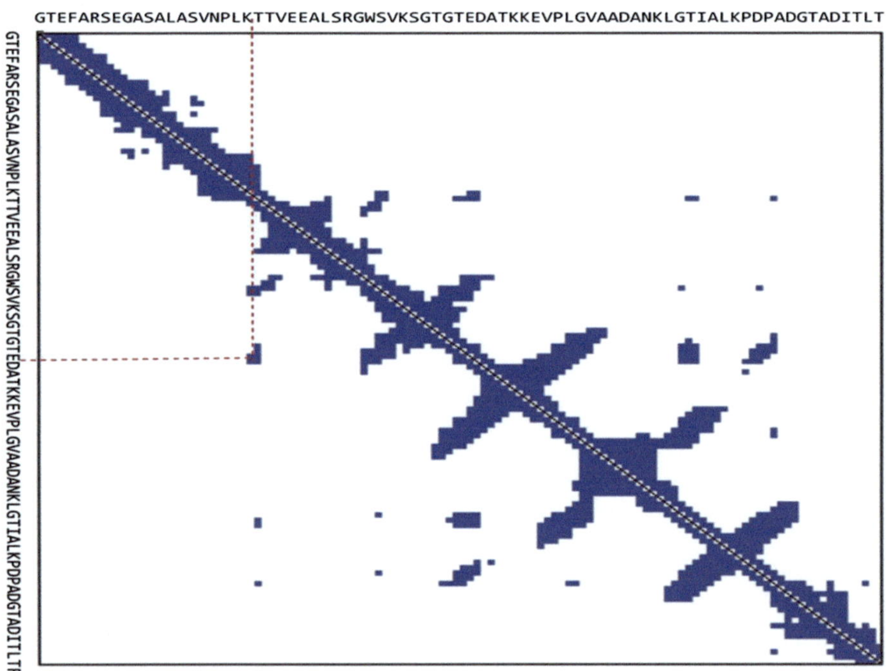

Fig. 8.3 Two-dimensional protein structure prediction. Example depicts a predicted 2D contact map with an 8 Å cutoff. The protein sequence is aligned along the sides of the contact map both horizontally and vertically. Each dot represents a predicted contact, that is, a residue pair whose spatial distance is below 8 Å. For instance, the *red dotted lines* mark a predicted contact associated with the pair (D, T)

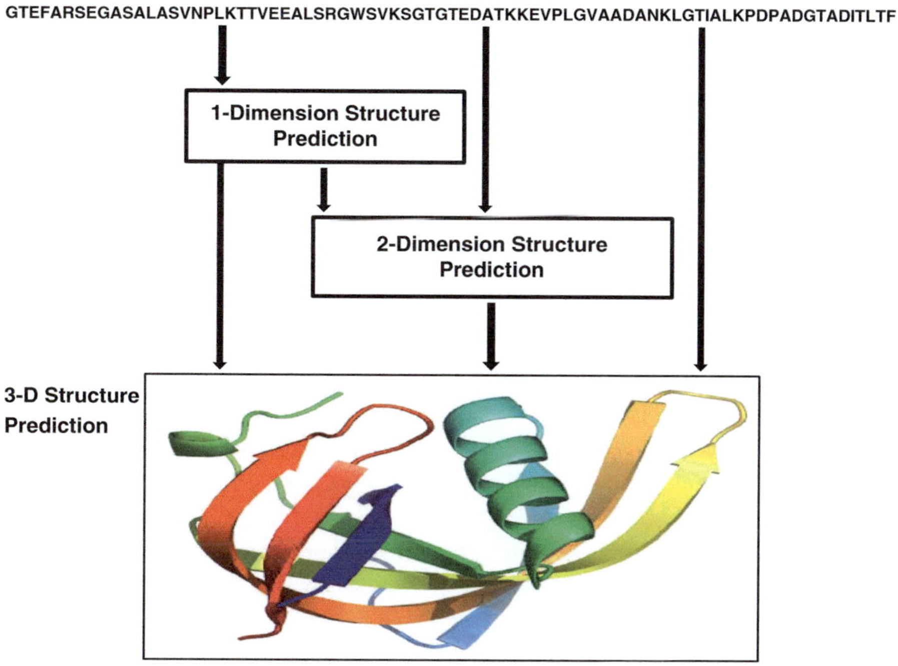

Fig. 8.4 Three-dimensional protein structure prediction. Three-dimensional structure predictors often combine information from the primary sequence and the predicted 1D and 2D structures to produce 3D structure predictions

(Fig. 8.4). Quaternary (4D) structure prediction is aimed at structures and complexes consisting of multiform folded protein chains (Fig. 8.5).

8.2 Computational Approach for Protein Structure Prediction

The deciphering of protein structures is very important in many disciplines of biological sciences and is extremely difficult and challenging, having occupied many researchers for many years. Researchers have tried to achieve significant advancement in computationally solving protein structures for many years. The prediction of 1D structure can be achieved by various paths – methods which rely on amino acid preferences, methods that exploit similar cases, and methods that rely purely on generalizations derived via machine learning.

Two-dimensional structure prediction can be subdivided into methods that exploit co-evolution observed for neighboring residues and machine learning methods. Three-dimensional – also called tertiary –, structure conformation is uniquely determined from amino acid sequence. There is no existing computer algorithm that can accurately map a sequence to tertiary structure; we must rely on experimental techniques, primarily X-ray crystallography, to determine tertiary structure. Advancement in 3D structure investigation is being achieved with

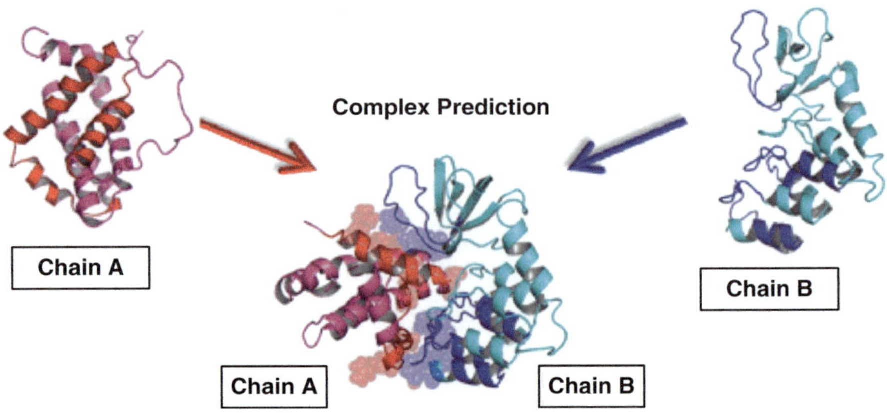

Fig. 8.5 Four-dimensional protein structure prediction. Four-dimensional prediction derived by docking individual protein chains to create a protein complex

many tools, some machine learning programs, neural networks, support vector machines, etc. In addition to classification methods, HMMs are important methods among the machine learning techniques for fold recognition. Earlier HMM approaches, include SAM and HMMer, built an Hiden Markov model (HMM) for a query with its homologous sequences and later used this HMM to score sequences with known structures in the PDB using the Viterbi algorithm, an example of dynamic programming methods. Quaternary structure prediction is very close to achieving 1D, 2D, and 3D structure predictions. If we see an example of protein interaction sites that can be predicted by 1D, the outcome in docking phases, the conformation search space would be reduced very drastically. As this is a problem in 4D prediction, the size of conformation space should be sampled, which is greater in 3D prediction cases, and therefore improving binding sites prediction is a significant and essential step to resolve bottleneck problems. There are certain methods used, such as HMMs, support vector machines, and neural networks, to predict binding site prediction.

In this chapter we have tried to address several methods, primarily unsupervised, and three supervised machine learning methods, including neural network, support vector machines, and HMM methods for 1D to 3D and quaternary structure 4D prediction problems. We have tried to stress the application of the mentioned prediction methods for globular proteins prediction, which is approximately 75% of the distinctive proteome, for which many methods have been found. We have also tried to stress some of the applications for membrane structure prediction, which has less training data in this class.

The first method (Baldi et al. 2002; Pollastr et al. 2002) introduces a novel class of graphical model architectures along with their allied implementations in terms of recurrent neural network architectures. The 1D structures have been introduced to address sequence analysis issues, specifically prediction of structural features of proteins, such as protein secondary structure (Baldi et al. 1999). The main contribution generalizes 1D to 2D, leading to further generalization both to higher dimensions, not necessarily spatial, data structures. Therefore, at this point we tried in the first step (Fig. 8.1) to analyze the 1D style of the structural design and to introduce key generalizations in the second step. Subsequently, the second class of approach (Frasconi and Vullo 2002) involved the learning task that involves predicting a scoring function allied with a hypothetical contact map for supervising a graph search algorithm. For this, recursive neural networks have been stretched to manage undirected (and possibly cyclic) graphs. This was accomplished by taking a lead of the appropriate property of protein contact maps, where vertices are distinctively organized (e.g., from the protein's N- to C-terminus). Instead of getting into the complexity of protein structure prediction, it is enough that the prediction of contact maps is possibly the utmost challenging and important step in the overall approach. Discussion can be in terms of processing architectures with inputs and outputs. As with this strategy, the reader can use or practice similar concepts to create parallel architectures based on inputs only, outputs only (e.g., HMMs), or even no inputs and no outputs (e.g., Markov chains).

8.3 Machine Learning Methods for Protein Structure Prediction

The most important aspect and major task in structural biology is 3D structure prediction from 1D linear protein sequences. Our main goal is to determine the spatial arrangements (shape or fold) that a given protein sequence adopts. The major problems are further divided by specific conditions as to whether the amino acid sequence adopts a new fold or carries a similar existing fold in some other databases. Fold recognition prediction is possible if any sequence query is analogous with known structures (Bourne and Weissig 2003). If the two sequences are similar and share evolutionary ancestry, they are called homologous. This kind of information or sequence similarity of protein sequences can help to build or predict the protein query sequence by choosing the known homologue sequence as a template. This method is known as comparative modeling.

In some cases, if we do not have a homologous template structure for the given query sequence, we can try to build the protein tertiary structure from scratch. Such methods are called *ab initio* methods. Ab initio approaches are based on physiochemical principles or statistical machinery which could help to simulate protein folding. The fundamental feature of ab initio approaches is to try to predict protein structure without referring to any specific template protein sequence with known structure. In the case of fold prediction, it is not necessary that we always have good sequence similarity to known structures, but the template with structural features may exist for a given sequence. Here, if the target structure is available, the template can be identified against the entire structure database. Therefore, it is not necessary that template and target query should be homologous. These two cases represent fold prediction (homologous) and fold prediction (analogous) problems during the critical assessment of protein structure prediction (CASP) competition.

The evaluation of 1D to 3D structure prediction is being examined in critical assessment techniques for the protein structure prediction (CASP) (Moult et al. 2007) and 4D structure prediction methods are evaluated in critical assessment of techniques for protein interaction (CAPRI) (Wodak 2007). To date, knowledge-based methods have been most successful in the structure prediction areas. A knowledge-based method refers to retrieving knowledge from known and available protein structures and inferring about new proteins for which structures are unavailable. Machine learning methods (Baldi and Brunak 2001) represent a specific class of tools which are used in all kinds of protein structure prediction. Here, we explain the 1D, 2D, 3D, and 4D structure prediction using a machine learning approach, and also try to address the advancement and application of machine learning methods. We focus on some of the methods such as unsupervised clustering and supervised machine learning, including support vector machines, neural networks, and HMMs for 1D, 2D, 3D, and 4D structure prediction problems. We also briefly try to deliver the knowledge and explain some applications processes and techniques for structure prediction of membrane proteins.

8.3.1 Machine Learning Methods for 1D Structure Prediction

In many proteins, the general observation made regarding 1D prediction, which corresponds to structural feature problems, includes solvent accessibility prediction, binding site prediction, secondary structure prediction, disordered region prediction, protein domain boundary prediction, functional site prediction, and transmembrane helix prediction (Cheng et al. 2005; Bryson et al. 2007). Problems found during 1D predictions are protein primary sequences which are used as input and outcome leads to predicted features of sequence for each amino acid in the sequence (Fig. 8.2). The main aim of the map is to predict structural features using primary protein sequences of amino acids, mostly the 1D prediction problem seen to classify each of the amino acid involves in protein sequences. In the past, the protein secondary structure prediction was seen to be most studied in the development of protein structure prediction processes and techniques (Rost and Sander 1993a, b; Chou and Fasman 1978; Baldi et al. 1999). Here, we have mainly tried to focus on prediction of globular proteins of secondary structure using machine learning methods. Similarly, the techniques mentioned have been used for other 1D prediction issues. In the early phase, the structure prediction methods (Chou and Fasman 1978) were available for extracting statistically significant correlation between the consecutive amino acids present in protein sequences and secondary structure classification.

This statistical correlation method for amino acids present in protein sequences and structural features of proteins used a certain amount information to give an accuracy of about 50%. With recent advancement in structure prediction, with most progress in strong pattern recognition and nonlinear function fitting techniques, new techniques and methods have being used for structure prediction study. In the late 1980s, for the first time, the feed forward neural networks method was applied to secondary structure prediction and the accuracy of structure prediction increased by 60–70% (Qian and Sejnowski 1988). It was observed that machine learning methods were successfully introduced on a large scale to resolve difficult issues in bioinformatics. In bioinformatics history a third significant breakthrough was observed whereby higher accuracy in structure prediction could be achieved using a richer input, which can be determined from a multiple sequence alignment to its homologues. The fact is that protein secondary structure is more conserved compared to primary sequence from the same protein family, evolved from the same ancestor with different amino acid sequences but usually sustaining the same secondary structure (Crawford et al. 1987; Barton et al. 1991), and this was the first combination using a neural network with multiple sequence alignment that improved secondary structure features accuracy by as much as 70–74%. In this method, instead of encoding each amino acid with a sparse binary vector of length 20 containing a single 1-bit present at a different position for each different amino acid, the empirical probabilities (i.e., normalized frequencies) of the 20 amino acids appearing in the corresponding column of the multiple sequence alignment are used. The positional frequency vector, called the profile of the family at the corresponding position, captures evolutionary information related to the structural properties of the protein family. Under this method, profiles can be created easily and allow one to leverage information contained in the sequence database, for example, SWISSPROT (Bairoch et al. 2005), which is a huge database, and then the PDB.

The profile has been used in all kinds of perception-based protein structure prediction techniques, and it has been refined. For example, PSI-PRED (Jones 1999b) uses PSI-BLAST (Altschul et al. 1997) for deriving new profiles using position specific scoring matrices for subsequent improvement of secondary structure prediction. It is necessary to improve the accuracy of secondary structure prediction for this new algorithm (Baldi et al. 1999). Pollastri et al. was inspired by probabilistic graphical models which helps sophisticated neural network models to improve the accuracy of structure prediction by adding information which elongates beyond the limited size window the input of traditional feed forward neural networks. Hundreds of huge neural network ensembles also have been used (Pollastri and McLysaght 2005). To date, the available technologies and databases have improved structure prediction, helping to reach an accuracy of about 78–80%. In addition, hybrid methods (Cheng et al. 2005; Bondugula and Xu 2007) combine neural network methods, and approaches with homology searches also play a crucial role in secondary structure prediction. As we know, homologous proteins are those usually derived from the same ancestor and they are used to refer to participating structural and functional characteristics. Any protein showing homology with the query protein sequence is likely to share similar features and structure (Berman et al. 2000). Moreover, the existing statistical machine learning techniques such as neural networks and support vector machines (SVMs) help to optimize the accuracy of secondary structure prediction and 1D features of globular protein prediction (Ward et al. 2003).

Machine learning methods (e.g., neural networks and HMMs) have been used to predict 1D features of membrane proteins, which include location specific regions of alpha-helical or beta-strand and localization of intracellular or extracellular fragments of the loop regions (Randall et al. 2008). In structure prediction, the 1D prediction methods have progressed well in the last three decades, but there is still much scope to develop prediction methods and increase the accuracy of these methods. There are several methods with significant roles in structure prediction that need to improve. For instance, the accuracy of secondary structure prediction is 8% below the limit of 88% (Rost and Chasman 2003). However, the prediction of protein domain boundaries (Cheng et al. 2006a; Bryson et al. 2007) and disordered regions (Obradovic et al. 2005; Cheng et al. 2005) are at a preliminary level of expansion, although already showing favorable results. There are a few developments from algorithmic improvements; for example, ensemble and meta learning methods (bagging and boosting) (Freund 1990) have been used to combine classifiers in the improvement of structure accuracy. On the other hand, we need new and up-to-date sources of biological information for structure prediction improvements, such as gene structure information, including alternative splicing sites, to help domain boundary prediction.

8.3.2 Machine Learning Methods for 2D Structure Prediction

Two-dimension structure prediction mainly refers to the prediction of protein contact maps (Olmea and Valencia 1997; Baldi and Pollastri 2002). A protein contact map (Fig. 8.3) is represented by a matrix M which shows a matrix such as M [i,j] which is either 1 (one) or 0 (zero). This depends on the Euclidean distance between two amino acids at linear positions i and j, which is more than a threshold value (e.g., 8 Å) or not. The range between two amino acids, for example, the corresponding backbone carbon atoms, can be measured. Similarly, using secondary structure elements, a coarser contact map could be determined in an identical manner. The appropriate contact maps could be determined by considering all the atoms present in every amino acid. As we mentioned, the contact map uniformly corresponds to rotations and translations. For any standard contact map, the corresponding 3D structure can be modified or reconstructed using several algorithms and methods (Aszodi et al. 1995; Skolnick et al. 1997). As we know, any contact map is essential for representation of a 3D structure, and contact map prediction is a more challenging and difficult task compared to predicting the corresponding 3D structure of a protein. Contact maps could be useful to infer folding rates in a particular protein (Plaxco et al. 1998; Punta and Rost 2005). There are several methods of machine learning with applications, including neural networks (Fariselli et al. 2001; Baldi and Pollastri 2003; Shackelford and Karplus 2007), self-organizing maps (MacCallum 2004), and support vector machines for contact map prediction. The appropriate feed forward and support vector machine approaches try to predict around two amino acids, whether they are in contact or not, and this can be seen in binary classification problems. Every individual place in a window around an amino acid usually relates to a vector containing 20 numbers corresponding to the 20 profile probabilities, as in the 1D prediction problem.

A suitable 1D knowledge can be leveraged, including predicted secondary structures and their relative accessibility of each of the amino acids. As the 1D prediction method refers to local window approaches it does not take any exterior effect from the window. To resolve this issue, a 2D recursive neural network architecture principle may be useful for the complete sequence to determine each prediction created to rectify contact map prediction. In the updated CASP (Moult et al. 2007), there are three methods – neural networks (Shackelford and Karplus 2007), 2D recursive neural networks (Cheng et al. 2005), and support vector machines – to obtain the best authentic results (Izarzugaza et al. 2007). Although progress has been seen in the last few years, the contact map still remains a very challenging and unsolved problem. The observed accuracy and precision of contact prediction is around 28%. However, this percentage sounds quite low, although the accuracy is good enough in comparison to that predicted by different ab initio 3D structure prediction methods. In 3D structure prediction, the predicted contact maps could be helpful for structure prediction problems, and even a small fraction of a correctly predicted contact map may be used to build an accurate protein topology (Wu and Zhang 2008).

In residue contact maps it is important that we pay more attention specifically to contact predictions: beta-strand pairing prediction (Cheng et al. 2005) and disulfide bond prediction (Fariselli et al. 1999; Vullo and Frasconi 2004). Disulfide bonds are covalent bonds that can form between cysteine amino acid residues. These disulfide bonds are significant and play a key role in stabilizing proteins, especially small proteins. The prediction of disulfide bonds forming between any two cysteine residues in a protein can be made if such disulfide bonds exist. Two effective methods – neural networks and support vector machines – are used to predict disulfide bonds. The accuracy and precision are much better (50%) after applying these two methods. Likewise, the prediction can be made with any two amino acids in two different beta-strands or not in the same beta sheet. Generally, two beta residues form a hydrogen bond between them in the protein, helping to stabilize the corresponding beta sheet. The major requirement is imposed by constraints with hydrogen bonding, the stringency in beta sheet containing 41%, which is quite a lot higher than the generic contacts in contact maps. In other 2D prediction tasks, such as beta sheet parings prediction, two methods –feed forward methods and recursive methods – have been widely used for the appropriate beta sheet parings prediction. To date, the most successful method is a 2D recursive neural network method, which usually takes inputs as a grid of beta residues (Cheng et al. 2005), concurrently using graph matching algorithms which help to predict pairings at strand, sheet, and residue, levels.

The above-mentioned applied methods have also been used to predict globular proteins for 2D prediction, these methods helping to making predictions of contacts in transmembrane beta-barrel proteins. The transmembrane helix prediction is used to reconstruct 3D structures with appropriate accuracy (Randall et al. 2008). Two-dimensional prediction is used as input to improve 3D structure prediction;

this requires improvement in 2D prediction accuracy. For 1D prediction, successful improvements may come by improvements in machine learning methods or by adding illuminating features in the inputs; for example, reciprocal selective information has been identified as an impactful feature for 2D structure prediction (Shackelford and Karplus 2007). On the other hand, if we focus on reconstruction of 3D structures, several methods and optimization algorithms exist, and these can play a crucial role in reconstructing 3D structures from contact maps using Monte Carlo methods (Vassura et al. 2008) and experimentally adding contacts into protein structure predictions (Rohl and Baker 2004) or determining protein structures using NMR methods. However, these methods could not reproduce reliable 3D structures from irregular contact maps predicted from the information existing in primary structures alone (Vendruscolo et al. 1997; Vassura et al. 2008). Thus, there are demands and a need to develop 3D construction algorithms that can ignore the noise existing in predicted contact maps.

8.3.3 Machine Learning Methods for 3D Structure Prediction

Machine learning approaches are being used in different aspects of 3D structure predictions, such as fold recognition, model generation, and model evaluation. Fold recognition has as its main objective the determination of a protein, with known features of available structure, presumably homologous to the unknown structure of a query protein. There is always a need to have essential steps to find homologous structural features for most successive template-based 3D protein structure prediction methods. Neural network methods were first used to complete this challenge in combination with threading (Jones 1999a). Recently, researchers proposed a generalized machine learning approach to enrich two important factors, sensitivity and specificity, of fold recognition using homology between query and protein sequences (Cheng et al. 2006a). However, the earlier implemented support vector machines determined the folds, and it is feasible that this could help to elongate other methods of supervised learning. In the classification, HMM is a significant technique which plays a crucial role in fold recognition. Recently, HMM techniques (SAM and HMMer) (Eddy 1998) have been used to build a Markov model for a query sequence with its homologue sequence and this model has been used to score sequences with available structures in the PDB by Viterbi algorithm, for example, dynamic programming methods. This could be an example of profile-sequence alignment. Very recently the profile methods have been considered to be a more significant improvement in the sensitivity of fold recognition in comparison to profile-sequence and sequence-sequence methods (Soeding 2005). The profile-profile method is used in the HMM version from earlier times. This model is used to align a query with the available and known HMMs from the template library. Such a type of profile-profile alignment is computed using standard dynamic programming methods.

Optimization techniques include conjugate gradient descent and are being globally used in statistical machine learning techniques, which are also important techniques for 3D protein structure generation and sampling. Conjugate gradient descent methods are used in neural network and tool (modeler) development for 3D structures prediction, and are widely used in comparative modeling (Sali and Blundell 1993). The lattice Monte Carlo sampling method is used in model generation techniques such as ab initio structure modeling (Zhang and Skolnick 2004), and the globally known ab initio fragment assembly tool Rosetta, which is based on simulated annealing sampling techniques. The machine learning methods used very frequently for model generation to select and evaluate protein models and ab initio structure prediction methods use mostly clustering techniques to choose the appropriate models (Zhang et al. 2004a).

These types of techniques generate a huge population of candidate models and then try to arrange themselves among them based on structure homology into different clusters using means clustering and other algorithms of clustering. The elements described from different clusters show as centroid and then are proposed as potential 3D structures. Usually a centroid is found as a most confident prediction among the largest clusters and this centroid could possibly be closer to the native structure of protein. Further additions to clustering supervised learning methods can be applied directly to the RMSD between the model and native structure of protein (Wallner and Elofsson 2007), support vector machines being used to rank protein models (Qiu et al. 2007).

One major challenge is that current methods used in the chapter cannot select the best model with the lowest RMSD. The model quality could be evaluated between predicted scores and real quality scores for poor models and is still a little low, which means some poor models may have obtained good predicted scores (Cozzetto et al. 2007). In addition, the significant statistical confidence score should be designated to determine quality scores for significant model usage and analysis. There is a demand for additional machine learning approaches and techniques to enhance the quality score and resolve the major challenging problems.

8.3.4 Machine Learning Methods for 4D Structure Prediction

Protein docking is a method for predicting complex protein structure, that is, 4D structure (Fig. 8.5) containing multiple protein chains. The main objective of 4D structure (Fig. 8.5) prediction is to predict the complex structure of proteins consisting of multiple protein chains (Aloy et al. 1998; Gray et al. 2003). As with 3D structure prediction, 4D structure prediction uses energy functions to reduce the problem of confirmation sampling. Three-dimensional grid Fourier transformation methods (Katchalski-Katzir et al. 1992) use small subunits of protein to dock together. RosettaDock uses a simulated annealing method with some adjustment for 4D problem in the same way as Rosetta for 3D structure prediction (Gray et al. 2003). More broadly, some methods are being used which lead to adapting 3D techniques to 4D issues, such as clustering methods used for both cluster docking confirmation and to select the centroid of clusters to generate an outcome of the predictions (Lorenzen and Zhang 2007).

Four-dimensional prediction is very similar and close to 1D, 2D, and 3D predictions; for instance, 1D predictors can easily predict protein interaction sites (Zhou and Qin 2007) and the search space for the docking phase could be drastically changed. Therefore, one of the major problems of 4D prediction is that the size of the confirmation space is simplified, which improves interface (site) prediction. Neural networks, HMMs, and SVM techniques for the prediction of protein interface use few features of 3D structures of protein subunits (Zhou and Shan 2001). Experimentally, in some cases 3D, structures themselves are currently not available, and hence further methods may be developed which can predict interactions from protein sequences alone.

Conformational changes are the biggest bottlenecks in protein docking, making processes even more complex and not easily handled by current techniques (Wodak and Mendez 2004). In protein binding, every protein may undergo substantial or high scale conformational changes rather than little changes, leading to complexity and limitations of current methods. Machine learning methods have been developed to identify several regions, such as flexible hinges, which facilitate major modifications to determine overall complex structures of proteins. However, for these problems the training data are somewhat less, and may not be sufficient to rectify the problems. Finally, machine learning methods show advancement and reliability for accessing the quality of 4D models and the confidence score.

8.4 Conclusion

As discussed above, machine learning methods and approaches have been used globally in protein structure predictions and have significantly contributed to the transformation of amino acid sequence information into structural features. An attempt has been made to provide a glimpse of the approaches and methods of machine learning in structural biology.

Machine learning methods have played a key role in structure prediction over the past few decades and still play an important and significant role in 1D to 4D structure predictions, as well as many structural feature predictions. Machine learning approaches have been applied to many structural feature prediction problems, for instance predicting protein solubility (Smialowski et al. 2007), protein stability (Cheng et al. 2006c), protein signal peptides, protein cellular localization (Emanuelsson et al. 2007), protein post-translation modification sites, such as phosphorylation sites (Blom et al. 1999), and protein epitopes (Andersen et al. 2006; Sweredoski and Baldi 2009). We have tried to cover some of the important approaches and application of machine learning methods to structure predictions and structural features of proteins. A general query asked by students is which method or approach of machine learning is good enough for a given problem? When we look at opinion for this question, it turns out that this is not a fundamental question as it may first seem It is really a challenging question for the researcher and computer scientist to answer.

In upcoming future trends, the machine learning approach has a significant role in structure prediction. The emergence of available training sets coupled with the existing gap between the number of sequences and available structures remains a powerful challenge for further development. In addition, machine learning methods are relatively fast in comparison to other methods. The observation has been made that machine learning methods usually take most time in the learning phase, which is possible to conduct offline. In "production" mode, feed forward neural networks may predict faster. We are aware that accuracy and speed are important features of any structural feature predictions, these features and considerations are likely remain important, and there are challenges and the scope to generate growth or emergence in this area.

Acknowledgements We are pleased to acknowledge the contributions of the authors from Biomedical Informatics Centre of ICMR-NITM, Belagavi and CSIR-URDIP, Pune. We also gratefully acknowledge the Indian Council of Medical Research, Dept. of Health Research, New Delhi, Government of India, for providing infrastructure and funding for the facility.

References

Aloy P, Moont G, Gabb HA, Querol E, Aviles FX, Sternberg MJE (1998) Modelling protein docking using shape complementarity, electrostatics and biochemical information. Proteins 33:535–549. https://doi.org/10.1006/jmbi.1997.1203

Altschul SF, Madden TL, Schaer AA, Zhang J, Zhang Z, Miller W, Lipman DJ (1997) Gapped BLAST and PSI-BLAST: a new generation of protein database search programs. Nucleic Acids Res 25:3389–3402. https://doi.org/10.1093/nar/25.17.3389

Andersen PH, Nielsen M, Lund O (2006) Prediction of residues in discontinuous B-cell epitopes using protein 3D structures. Protein Sci 15:2558–2567. https://doi.org/10.1110/ps.062405906

Anfinsen CB (1973) Principles that govern the folding of protein chains. Science 181:223–230. https://doi.org/10.1126/science.181.4096.223

Aszodi A, Gradwell M, Taylor W (1995) Global fold determination from a small number of distance restraints. J Mol Biol 251:308–326. https://doi.org/10.1006/jmbi.1995.0436

Bairoch A, Apweiler R, Barker CH, Wu WC, Boeckmann B, Ferro S, Gasteiger E, Huang H, Lopez R, Magrane M, Martin MJ, Natale DA, O'Donovan C, Redaschi N, Yeh LS (2005) The universal protein resource (UniProt). Nucleic Acids Res 33:D154–D159. https://doi.org/10.1093/nar/gki070

Baldi P, Brunak S (2001) Bioinformatics: the machine learning approach, 2nd edn. MIT Press, Cambridge, MA. https://mitpress.mit.edu/books/bioinformatics

Baldi P, Pollastri G (2002) Generalized IOHMMs and recurrent neural network architectures. http://www.jsnc.caltech.edu/2002/abstracts02/Baldi-P.pdf

Baldi P, Pollastri G (2003) The principle design of large-scale recursive neural network architectures-DAG-RNNs and the protein structure prediction problem. J Mach Learn Res 4:575–602. https://doi.org/10.1162/153244304773936054

Baldi P, Brunak S, Frasconi P, Soda G, Pollastri G (1999) Exploiting the past and the future in protein secondary structure prediction. Bioinformatics 15:937–946. https://doi.org/10.1093/bioinformatics/15.11.937

Baldi P, Cheng J, Vullo A (2005) Large-scale prediction of disulphide bond connectivity. In: Advances in neural information processing systems, vol 17. MIT Press, Cambridge, MA, pp 97–104. http://papers.nips.cc/paper/2607-large-scale-prediction-of-disulphide-bond-connectivity.pdf

Baldwin EN, Weber IT, Charles RS, Xuan J, Appella E, Yamada M, Matsushima K, Edwards BFP, Clore GM, Gronenborn AM, Wlodawar A (1991) Crystal structure of interleukin 8: symbiosis of NMR and crystallography. Proc Natl Acad Sci 88:502–506. http://www.jstor.org/stable/2355898

Barton GJ, Newman RH, Freemont PS, Crumpton MJ (1991) Amino acid sequence analysis of the annexin supergene family of proteins. Eur J Biochem 198:749–760. https://doi.org/10.1111/j.1432-1033.1991.tb16076.x

Berman HM, Westbrook J, Feng Z, Gilliland G, Bhat TN, Weissig H, Shindyalov IN, Bourne PE (2000) The protein data bank. Nucleic Acids Res 28:235–242. https://www.ncbi.nlm.nih.gov/pmc/articles/PMC102472/

Blom N, Gammeltoft S, Brunak S (1999) Sequence-and structure based prediction of eukaryotic protein phosphorylation sites. J Mol Biol 294:1351–1362. https://doi.org/10.1006/jmbi.1999.3310

Bjorkman PJ, Parham P (1990) Structure, function and diversity of class I major histocompatibility complex molecules. Annu Rev Biochem 59:253–288. https://doi.org/10.1146/annurev.bi.59.070190.001345

Bondugula R, Xu D (2007) MUPRED: a tool for bridging the gap between template based methods and sequence profile based methods for protein secondary structure prediction. Proteins 66:664–670. https://doi.org/10.1002/prot.21177

Bourne P, Weissig H (2003) Structural bioinformatics. Wiley, Hoboken. ftp://ftp.ufv.br/dbg/material%20curso%20bioinfo/Leitura%20Complementar/livros/Wiley-Liss.Structural.Bioinformatics.pdf

Bragg SL (1975) The development of X-ray analysis. G Bell and Sons, London. http://trove.nla.gov.au/work/10865227?selectedversion=NBD729968

Bryson K, Cozzetto D, Jones DT (2007) Computer-assisted protein domain boundary prediction using the DomPred server. Curr Protein Pept Sci 8:181–188. https://doi.org/10.2174/138920307780363415

Chandonia JM, Brenner SE (2006) The impact of structural genomics: expectations and outcomes. Science 311:347–351. https://doi.org/10.1126/science.1121018

Cheng J, Sweredoski MJ, Baldi P (2005) Accurate prediction of protein disordered regions by mining protein structure data. Data Min Knowl Disc 11:213–222. http://download.igb.uci.edu/disorder.pdf

Cheng J, Sweredoski M, Baldi P (2006a) DOMpro: protein domain prediction using profiles, secondary structure, relative solvent accessibility, and recursive neural networks. Data Min Knowl Disc 13:1–10. 10.1007%2Fs10618-005-0023-5

Cheng J, Saigo H, Baldi P (2006b) Large-scale prediction of disulphide bridges using kernel methods, two-dimensional recursive neural networks, and weighted graph matching. Proteins: Struct Funct Bioinf 62:617–629. https://doi.org/10.1002/prot.20787

Cheng J, Randall A, Baldi P (2006c) Prediction of protein stability changes for single site mutations using support vector machines. Proteins 62(4):1125–1132. https://doi.org/10.1002/prot.20810

Chou PY, Fasman GD (1978) Prediction of the secondary structure of proteins from their amino acid sequence. Adv Enzymol 47:45–148. https://doi.org/10.1002/9780470122921

Cozzetto D, Kryshtafovych A, Ceriani M, Tramontano A (2007) Assessment of predictions in the model quality assessment category. Proteins 69:175–183. https://doi.org/10.1002/prot.21669

Crawford IP, Niermann T, Kirchner K (1987) Prediction of secondary structure by evolutionary comparison: application to a subunit of tryptophan synthase. Proteins 2:118–129. https://doi.org/10.1002/prot.340020206

Eddy SR (1998) Profile hidden Markov models. Bioinformatics 14:755–763. https://doi.org/10.1093/bioinformatics/14.9.755

Emanuelsson O, Brunak S, Heijne GV, Nielsen H (2007) Locating proteins in the cell using TargetP, SignalP, and related tools. Nat Protoc 2:953–971. https://doi.org/10.1038/nprot.2007.131

Fariselli P, Riccobelli P, Casadio R (1999) Role of evolutionary information in predicting the disulfide-bonding state of cysteine in proteins. Proteins 36:340–346. https://doi.org/10.1002/(SICI)1097-0134(19990815)36:3<340::AID-PROT8>3.0.CO;2-D

Fariselli P, Casadio R (2004) Prediction of disulfide connectivity in proteins. Bioinformatics 17:957–964. https://doi.org/10.1093/bioinformatics/17.10.957

Fariselli P, Olmea O, Valencia A, Casadio R (2001) Prediction of contact maps with neural networks and correlated mutations. Protein Eng 13:835–843. https://doi.org/10.1093/protein/14.11.835

Frasconi P, Vullo A (2002) Prediction of protein coarse contact maps using recursive neural networks. Proc IEEE-EMBS Conf Mol Cell Tissue Eng. https://doi.org/10.1109/MCTE.2002.1175038

Freund Y (1990) Boosting a weak learning algorithm by majority. Inf Comput 121:256–285. https://doi.org/10.1006/inco.1995.1136

Gray JJ, Moughan SE, Wang C, Schueler-Furman O, Kuhlman B, Rohl CA, Baker D (2003) Protein-protein docking with simultaneous optimization of rigid body displacement and side chain conformations. J Mol Biol 331:281–299. https://doi.org/10.1016/S0022-2836(03)00670-3

Izarzugaza JMG, Graña O, Tress ML, Valencia A, Clarke ND (2007) Assessment of intramolecular contact predictions for CASP7. Proteins 69:152–158. https://doi.org/10.1002/prot.21637

Jacobson M, Sali A (2004) Comparative protein structure modeling and its applications to drug discovery. In: Overington J (ed) Annual reports in medical chemistry. Academic, London, pp 259–276. https://doi.org/10.1016/s0065-7743(04)39020-2

Jones DT (1999a) GenTHREADER: an efficient and reliable protein fold recognition method for genomic sequences. J Mol Biol 287:797–815. https://doi.org/10.1006/jmbi.1999.2583

Jones DT (1999b) Protein secondary structure prediction based on position specific scoring matrices. J Mol Biol 292:195–202. https://doi.org/10.1006/jmbi.1999.3091

Kabsch W, Sander C (1983) Dictionary of protein secondary structure: pattern recognition of hydrogen-bonded and geometrical features. Biopolymers 22:2577–2637. https://doi.org/10.1002/bip.360221211

Katchalski-Katzir E, Shariv I, Eisenstein M, Friesem AA, Aflalo C, VakseI AR (1992) Molecular surface recognition: determination of geometric fit between proteins and their ligands by correlation techniques. Proc Natl Acad Sci 89:2195–2199. https://www.ncbi.nlm.nih.gov/pmc/articles/PMC48623/

Kendrew JC, Dickerson RE, Strandberg BE, Hart RJ, Davies DR, Phillips DC, Shore VC (1960) Structure of myoglobin: a three-dimensional Fourier synthesis at 2°Å resolution. Nature 185:422–427. https://www.ncbi.nlm.nih.gov/pubmed/18990802

Laskowski RA, Watson JD, Thornton JM (2003) From protein structure to biochemical function? J Struct Funct Genom 4:167–177. https://doi.org/10.1023/a:1026127927612

Lorenzen S, Zhang Y (2007) Identification of near-native structures by clustering protein docking conformations. Proteins 68:187–194. https://doi.org/10.1002/prot.21442

MacCallum R (2004) Striped sheets and protein contact prediction. Bioinformatics 20:i224–i231. https://doi.org/10.1093/bioinformatics/bth913

Moult J, Fidelis K, Kryshtafovych A, Rost B, Hubbard T, Tramontano A (2007) Critical assessment methods of protein structure prediction-Round VII. Proteins 29:179–187. https://doi.org/10.1002/prot.24452

Obradovic Z, Peng K, Vucetic S, Radivojac P, Dunker AK (2005) Exploiting heterogeneous sequence properties improves prediction of protein disorder. Proteins 61:176–182. https://doi.org/10.1002/prot.20735

Olmea O, Valencia A (1997) Improving contact predictions by the combination of correlated mutations and other sources of sequence information. Fold Des 2:s25–s32. https://doi.org/10.1016/S1359-0278(97)00060-6

Perutz MF, Rossmann MG, Cullis AF, Muirhead G, Will G, North AT (1960) Structure of haemoglobin: a three-dimensional fourier synthesis at 5.5°Å resolution, obtained by X-ray analysis. Nature 185:416–422. https://doi.org/10.1038/185416a0

Petrey D, Honig B (2005) Protein structure prediction: inroads to biology. Mol Cell 20:811–819. https://doi.org/10.1016/j.molcel.2005.12.005

Plaxco K, Simons K, Baker D (1998) Contact order, transition state placement and the refolding rates of single domain proteins. J Mol Biol 277:985–994. https://doi.org/10.1006/jmbi.1998.1645

Pollastri G, Baldi P (2002) Prediction of contact maps by GIOHMMs and recurrent neural networks using lateral propagation from all four cardinal corners. Bioinformatics 18:S62–S70. https://doi.org/10.1093/bioinformatics/18.suppl_1.S62

Pollastri G, Przybylski D, Rost B, Baldi P (2002a) Improving the prediction of protein secondary structure in three and eight classes using recurrent neural networks and profiles. Proteins 47:228–235. https://doi.org/10.1002/prot.10082

Pollastri G, Baldi P, Fariselli P, Casadio R (2002b) Prediction of coordination number and relative solvent accessibility in proteins. Proteins 47:142–153. https://doi.org/10.1002/prot.10069

Pollastri G, McLysaght A (2005) Porter: a new, accurate server for protein secondary structure prediction. Bioinformatics 21:1719–1720. https://doi.org/10.1093/bioinformatics/bti203

Punta M, Rost B (2005) Protein folding rates estimated from contact predictions. J Mol Biol 348:507–512. https://doi.org/10.1016/j.jmb.2005.02.068

Qian N, Sejnowski TJ (1988) Predicting the secondary structure of globular proteins using neural network models. J Mol Biol 202:265–884. https://doi.org/10.1016/0022-2836(88)90564-5

Qiu J, Sheffler W, Baker D, Noble WS (2007) Ranking predicted protein structures with support vector regression. Proteins 71:1175–1182. https://doi.org/10.1002/prot.21809

Randall A, Cheng J, Sweredoski M, Baldi P (2008) TMBpro: secondary structure, beta- contact, and tertiary structure prediction of transmembrane beta-barrel proteins. Bioinformatics 24:513–520. https://doi.org/10.1093/bioinformatics/btm548

Rohl CA, Baker D (2004) De novo determination of protein backbone structure from residual dipolar couplings using Rosetta. J Am Chem Soc 124:2723–2729. https://doi.org/10.1021/ja016880e

Rost B, Chasman D (2003) Rising accuracy of protein secondary structure prediction. In: Chasman D (ed) Protein structure determination, analysis, and modeling for drug discovery. Marcel Dekker, New York, pp 207–249. https://www.rostlab.org/papers/2003_rev_dekker/paper.html

Rost B, Sander C (1993a) Improved prediction of protein secondary structure by use of sequence profiles and neural networks. Proc Natl Acad Sci 90(16):7558–7562. http://www.pnas.org/content/90/16/7558

Rost B, Sander C (1993b) Prediction of protein secondary structure at better than 70% accuracy. J Mol Bio 232(2):584–599. https://doi.org/10.1006/jmbi.1993.1413

Rost B, Sander C (1994) Conservation and prediction of solvent accessibility in protein families. Proteins 20(3):216–226. https://doi.org/10.1002/prot.340200303

Sali A, Blundell TL (1993) Comparative protein modelling by satisfaction of spatial restraints. J Mol Biol 234:779–815. https://doi.org/10.1006/jmbi.1993.1626

Sanger F, Thompson EO (1953) The amino-acid sequence in the glycyl chain of insulin. 1. The identification of lower peptides from partial hydrolysates. J Biochem 53:353–366. https://www.ncbi.nlm.nih.gov/pmc/articles/pmc1198157

Shackelford G, Karplus K (2007) Contact prediction using mutual information and neural nets. Proteins 69:159–164. https://doi.org/10.1002/prot.21791

Skolnick J, Kolinski A, Ortiz A (1997) MONSSTER: a method for folding globular proteins with a small number of distance restraints. J Mol Biol 265:217–241. https://ub.cbm.uam.es/publications/downloads/pdfs/9020984.pdf

Smialowski P, Martin-Galiano AJ, Mikolajka A, Girschick T, Holak TA, Frishman D (2007) Protein solubility: sequence based prediction and experimental verification. Bioinformatics 23:2536–2542. https://doi.org/10.1093/bioinformatics/btl623

Soeding J (2005) Protein homology detection by HMM-HMM comparison. Bioinformatics 21:951–960. https://doi.org/10.1093/bioinformatics/bti125

Sweredoski MJ, Baldi P (2009) COBEpro: a novel system for predicting continuous B-cell epitopes. Protein Eng Des Sel 22:113–120. https://doi.org/10.1093/protein/gzn075

Travers A (1989) DNA conformation and protein binding. Annu Rev Biochem 58:427–452. https://doi.org/10.1146/annurev.bi.58.070189.002235

Vassura M, Margara L, Di Lena P, Medri F, Fariselli P, Casadio R (2008) FT-COMAR: fault tolerant three-dimensional structure reconstruction from protein contact maps. Bioinformatics 24:1313–1315. https://doi.org/10.1093/bioinformatics/btn115

Vendruscolo M, Kussell E, Domany E (1997) Recovery of protein structure from contact maps. Fold Des 2:295–306. https://doi.org/10.1016/S1359-0278(97)00041-2

Vullo A, Frasconi P (2003) A recursive connectionist approach for predicting disulfide connectivity in proteins. In: Eighteenth annual ACM symposium on applied computing (SAC '03), pp 67–71. https://doi.org/10.1145/952532.952550

Vullo A, Frasconi P (2004) Disulfide connectivity prediction using recursive neural networks and evolutionary information. Bioinformatics 20:653–659. https://doi.org/10.1093/bioinformatics/btg463

Wallner B, Elofsson A (2007) Prediction of global and local model quality in CASP7 using Pcons and ProQ. Proteins 69:184–193. https://doi.org/10.1002/prot.21774

Ward JJ, McGuffin LJ, Buxton BF, Jones DT (2003) Secondary structure prediction using support vector machines. Bioinformatics 19:1650–1655. https://doi.org/10.1093/bioinformatics/btg223

Wodak SJ (2007) From the Mediterranean coast to the shores of Lake Ontario: CAPRI's premiere on the American continent. Proteins 69:687–698. https://doi.org/10.1002/prot.21805

Wodak SJ, Mendez R (2004) Prediction of protein-protein interactions: the CAPRI experiment, its evaluation and implications. Curr Opin Struct Biol 14:242–249. https://doi.org/10.1016/j.sbi.2004.02.003

Wu S, Zhang Y (2008) A comprehensive assessment of sequence-based and template-based methods for protein contact prediction. Bioinformatics 24:924–931. https://doi.org/10.1093/bioinformatics/btn069

Wuthrich K (1986) NMR of proteins and nucleic acids. Wiley, New York. http://as.wiley.com/WileyCDA/WileyTitle/productCd-0471828939.html

Zhang Y, Skolnick J (2004a) Automated structure prediction of weakly homologous proteins on a genomic scale. Proc Natl Acad Sci 101:7594–7599. https://doi.org/10.1073/pnas.0305695101

Zhou HX, Qin S (2007) Interaction-site prediction for protein complexes: a critical assessment. Bioinformatics 23:2203–2209. https://doi.org/10.1093/bioinformatics/btm323

Zhou HX, Shan Y (2001) Prediction of protein interaction sites from sequence profile and residue neighbor list. Proteins 44:336–343. https://doi.org/10.1002/prot.1099

Drug Transporters as Therapeutic Targets: Computational Models, Challenges, and Future Perspective

9

Deepak Singla, Ritika Bishnoi, Sandeep Kumar Dhanda, and Shailendra Asthana

Abstract

Tissue level expression, mutation, and substrate specificity of the transporter proteins have been widely accepted for their usefulness in drug disposition and efficacy. Many transporters play a significant role in normal human physiology as well as in disease conditions. Association of these properties, with systemic plasma concentration of the drug, is the leading reason for adverse drug reactions and drug resistance. The identification and validation of transporter proteins in experiments and their atomic resolution for characterization of structural-functional relationship is a costly, time-consuming, and more tedious process. However, predictive in silico tools claimed well for accurately accessing the pharmacokinetics, pharmacodynamics properties in early drug discovery stage. But the huge amount of data requires the development of reliable computational techniques and databases for the identification and/or prediction of membrane transport proteins as well as their ligands has become essential. Here, we review

D. Singla (✉)
ICMR-National Institute of Pathology, New Delhi, India

Host-Parasite Interaction Biology group, ICMR-National Institute of Malaria Research (NIMR), Sec-8 Dwarka, New Delhi, India
e-mail: deepkumar1983@gmail.com

R. Bishnoi
CSIR-Institute of Microbial Technology, Chandigarh, India

S. K. Dhanda
CSIR-Institute of Microbial Technology, Chandigarh, India

La Jolla Institute for Allergy and Immunology, La Jolla, CA, USA

S. Asthana (✉)
Drug Discovery Research Center, Translational Health Science and Technology Institute (THSTI), NCR Biotech Science Cluster, 3rd Milestone, Faridabad, India
e-mail: sasthana@thsti.res.in

H. J. Purohit et al. (eds.), *Soft Computing for Biological Systems*,
https://doi.org/10.1007/978-981-10-7455-4_9

the available datasets and the computational methods, which put forth more insights for better understanding of human drug transporter proteins.

Keywords
Transporter · Inhibitor · Drug · Database · ADMET, ligand prediction

9.1 Introduction

The transporters (proteins) are the foremost determinants of regulating the in-/outflow of molecules through the plasma membrane and thus controlling the pharmacokinetics behavior of various drugs. The concentration of drug at various sites of action is another crucial element for its desirable as well as undesirable effect. Their plasma level is dependent on the different transporter proteins expressed in various tissues or organs. The importance of drug transporter proteins in pharmacokinetics and the associated absorption, distribution, metabolism, excretion, and toxicity (ADMET) properties are well described in literature (Saier 1998, 2000; Saier et al. 2014). For example, metformin a substrate of organic cation transporters OCTs (Fig. 9.1a), when co-administrated with another drug cimetidine (an inhibitor of OCTs), increased the concentration of metformin (Fig. 9.1d) (Viereck et al. 2014).

According to Food and Drug Administration (FDA) guideline, each investigational drug should be evaluated in vitro for their function as P-glycoprotein (P-gp/BCRP) substrate/inhibitor. Furthermore, expression of drug transporter proteins from bacteria to mammals indicates its immense value in each domain of life (Saier 1998). Previous studies suggested that eukaryotes have more number of transporters as compared to prokaryotes (Quentin and Fichant 2000; Ren and Paulsen 2005). In view of its significance, various classification schemes have been adopted to understand their function and distribution (Saier 2000; Saier et al. 2014). In 2012, Viereck et al. conducted a comparative study on different classification schemes adopted so far and recognized different pitfall in a classification system (Viereck et al. 2014). Although the importance of drug transporter proteins in other species/organism can not be ignored, but it's difficult to cover all these in comprehensive ways. Overall, several key questions associated with drug development need to be addressed such as clinically relevant drug transporters and a protocol for validation of drug transport interactions. Therefore in this chapter, we will summarize the database of human drug transporter proteins and their ligands as well as computational models developed so far (Fig. 9.2).

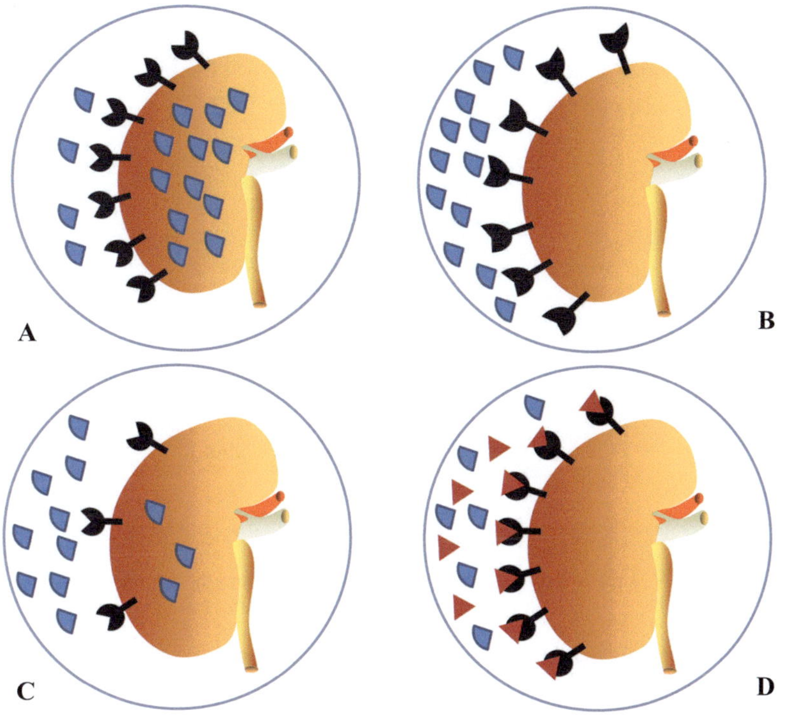

Fig. 9.1 Transport of molecule in the kidney through drug transporter protein. The black color depicts as black, transporter protein; blue, a substrate; and red, inhibitor. (**a**) High-level expression of transporter protein, (**b**) mutated transporter protein, (**c**) poorly expressed transporter protein, (**d**) transporter protein bound with inhibitors

9.2 Database of Human Drug Transporter Proteins

In previous studies, a number of databases have been developed for a different purpose that provides information on transporter proteins (Fig. 9.2). The complete drug transporter proteins could be represented by two major families, namely, ATP-binding cassette (ABC) and solute carrier (SLC) family (Zhao et al. 2011). Additionally, ABC transporters play a crucial role in bioavailability and toxicity of drugs (DeGorter et al. 2012; Hee Choi and Yu 2014; Szakács et al. 2008). Recently, Lin et al. have reported that SLC transporters are the important therapeutic targets by playing a major role in regulating the physiological process by controlling the cellular uptake of molecules (César-Razquin et al. 2015; Lin et al. 2015). Therefore, we mainly focused on these two transporter families. The first database specifically designed for human ABC transporters proteins (http://nutrigene.4t.com/humanabc.html) provides information of 49 protein sequences (Quentin and Fichant 2000). Along with basic information like gene/protein sequence, this database also has information on the disease associated with a mutation in a particular transporter

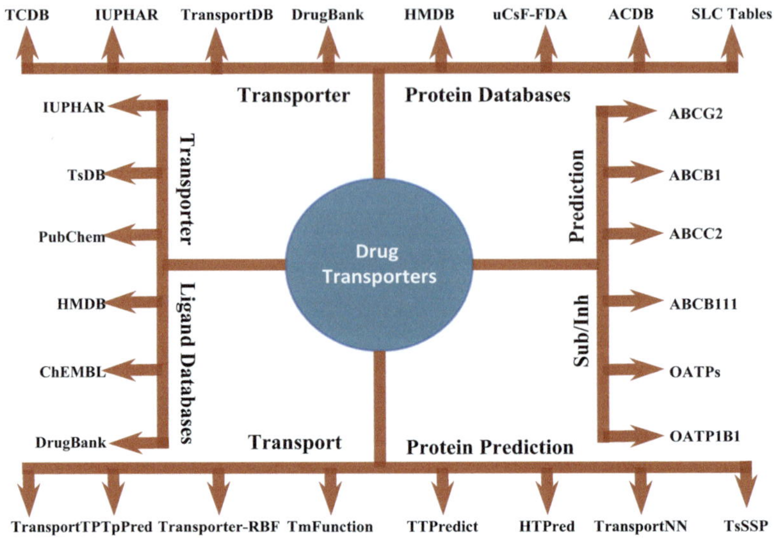

Fig. 9.2 Overview of the transporter protein, ligand database, and their prediction methods

protein. Similarly, SLC table (http://www.bioparadigms.org/slc/) has about 400 solute carrier genes which are further categorized into 52 subfamilies along with their tissue level expression and associated diseases gathered from the literature.

Later on, a common database of transporter proteins (TransportDB http://www.membranetransport.org/) was developed which enclosed the transporter protein information for 365 organisms, of which human transporter proteins were classified into four classes: (a) ATP-dependent, (b) ion channels, (c) secondary transporters, and (d) unclassified (Ren et al. 2007). The 53 ABC genes belonged to ATP-dependent category and ~250 genes belonged to SLC family, the majority of which were secondary transporters. This database provides the information of tissue in which these genes are expressed. This database also integrates a tool, TransAAP, for whole genome-based annotation of transporter proteins. However, the major limitation of this tool is its applicability on prokaryotic genomes only. Based on the functional and phylogenetic relationship, another database TCDB (transporter classification database http://www.tcdb.org/) has been developed for classification of transporter proteins. This database was extensively used for various analysis and predictions (Saier et al. 2014). This database has a record of ~10,000 proteins of which 1555 protein belongs to human transporter proteins. At present, it encapsulates the information of 48/454 genes that belongs to ABC/SLC family, respectively. The advantage of this database involves the integration of protein analysis tools such as transmembrane segment prediction and hydropathy/amphipathicity prediction.

Human metabolome database (http://www.hmdb.ca/), a database of human metabolites, was developed for accessing the information like spectral data, bio-fluid concentration, and location of metabolites (Wishart et al. 2013). Although, this database is primarily designed for human metabolites, yet we also extracted the information of about 12 ABC and 58 SLC genes (Wishart et al. 2013). In 2014, a guide to pharmacology holding the information about drugs and various targets has been developed. This web service (http://www.guidetopharmacology.org/) supports approximately 500 human transporter proteins which are comprised of 42 ABC, 206 SLC, and 204 others transporter proteins (Alexander et al. 2013). Similarly, DrugBank v4.0 (http://www.drugbank.ca/) has information on about 18 ABC and 84 SLC genes that are involved in the transport of various drugs (Law et al. 2014). As shown in Fig. 9.2, UCSF-FDA (http://bts.ucsf.edu/fdatransportal) transport database has been developed for storing the information of almost 30 drug transporter proteins along with their clinically relevant drug interactions (Ye et al. 2014). Recently, a human transporter protein-specific database HumanTDB (http://htd.cbi.pku.edu.cn) has been developed that contains information of a total of 1555 proteins of which 48 belong to ABC and 454 belong to SLC transporter families, respectively (Ye et al. 2014).

9.3 Ligand Database of Transporter Proteins

With the advent of high-throughput chemical screening technology, a huge amount of chemical activity data is produced and stored in different databases. The PubChem BioAssay is a large repository of high-throughput chemical screening data coming from various sources and research labs (Table 9.1). Our search found that at present, it holds the BioAssay information of 50 human transporter proteins of which only four belong to ABC family (Wang et al. 2012). Similarly, ChEMBL is another database having chemical activity data of small molecules. Although, in ChEMBL and PubChem BioAssay, some overlapping chemical data is also present, yet a major difference in both is that the ChEMBL database also has information from previous literature studies (Zhao et al. 2011). This database covers a total of 87 human transporter proteins with more than 12,000 ligands for ABC family and ~1400 for SLC family (Table 9.1). Similarly, the IUPHAR has only ~200 ligands targeted by 77 transporter proteins (Alexander et al. 2013). The updated version of DrugBank4.0 holds about ~650 approved drugs that act as substrate, inhibitor, or inducer for 117 transporter proteins (Law et al. 2014). As shown in Table 9.1, human metabolome database covers more than 1200 ABC and 15,000 SLC metabolites that are interacting with 70 transporter proteins (Wishart et al. 2013). TSDB is a database of transporter protein substrates that captures information for 105 ABC and 96 SLC substrates, respectively (Zhao et al. 2011). Recently, Mak and colleagues created a database of human drug transporter ligands collected from previous studies. Presently, this database has information for 6 ABC and 14 SLC transporter protein small molecules. It also represents a unique resource for ~3500 unique compounds with a different mechanism of action such as substrate/inhibitor/modulator, etc. Besides that, it also includes information for non-substrate and non-inhibitor molecules that would be useful for developing robust prediction

Table 9.1 Databases for ligands of transporter proteins in literature

Name	Target covered (ABC + SLC)	ABC	SLC	Total
PubChem BioAssay	50 (4 + 46)	–	–	–
ChEMBL	87 (10 + 75)	12,466	1402	13,868
DrugBank	117 (18 + 84)	246	406	652
HMDB	84 (12 + 58)	1229	15,169	16,604
IUPHAR	(2 + 75)	2	197	199
TSDB	(19 + 128)	105	96	201
Metrabase	(6 + 14)	2572	2254	4826

models. In addition to this, it also compiled the ligand molecules for a CYP450 enzyme that plays important role in drug metabolism.

9.4 Computational Model for Prediction of Transporter Proteins

In the current scenario, the application of next-generation sequencing (NGS) is being widely used for sequencing the new or previously unknown organisms. A huge amount of data generated from sequencing projects is required to be annotated and functionally characterized. With a view of its importance, a number of computational tools have been developed for predicting the transporter proteins and their families (Table 9.2). In 2008, a nearest-neighbor (NN) based method has been developed to predict the family of transporter proteins (Table 9.2). This method used the dataset of ~3800 proteins and showed 72.3% accuracy on independent datasets (Li et al. 2008). In 2010–2011, two models were developed for prediction of a subfamily of transporter proteins based on the position-specific scoring matrix (PSSM) and biochemical properties (Chen et al. 2011b; Li et al. 2009; Ou et al. 2010). The radial basis function-based network model identified hydrophobic residues (Leu, Ile, and Lys) in electrochemical transporter proteins, Glutamic acid (Glu) in active transporters, and Aspartic acid (Asp) in channel transporter proteins. The only limitation of the above-mentioned model is its ability to assign the new transporter protein to either of these three classes: (a) active transporters, (b) electrochemical, or (c) channels/pores. In 2009, a support vector machine (SVM)- and hidden Markov model (HMM)-based two-phase classification model with ~81% accuracy was developed to predict the transporter proteins (Li et al. 2009). Gromiha et al. developed another model to predict the functional residue in membrane proteins (Gromiha et al. 2009). This method is useful for identification of critical residue whose change results in loss of function of that particular protein. A classification model based on physicochemical properties was also developed to predict transporter proteins with 65% accuracy (Table 9.2). In 2014, Mishra et al. developed a tool TrSSP (Transporter Substrate Specificity Prediction) for prediction of transporter proteins and their substrate specificities (Mishra et al. 2014). The model developed in this study is able to discriminate the transporter proteins into

Table 9.2 Computational methods developed for the prediction of transporter proteins and their families

Name	Methodology	Dataset	Performance (%)	Reference
Transporter family prediction	Nearest neighbor	TS-3899	72.3%	2008
TransportTP (http://bioinfo3.noble.org/transporter)	HMM and SVM	–	81.8%	2009
Functional residue prediction	Evolutionary features	–	–	2009
Transporters, their class, family prediction (http://rbf.bioinfo.tw $ sachen/tcrbf.html)	PSSM and biochemical properties	TS-693	Transporter, 76% Class, 73% Family, 69%	2010
Transporter family prediction (http://rbf.bioinfo.tw/~sachen/ttrbf.html)	PSSM and biochemical properties	TS-651	ET, 90.1% Protein/mRNA, 80.1% Ion transporters, 70.3% Others, 82.3%	2011
Transporters, their class, subclass prediction (http://www.juit.ac.in/attachments/tppred/Home.html)	Physiochemical properties	TS-5359 NTS-2907	65%	2012
Human transporter protein prediction	Physiochemical properties	HTS-728 NHTS-4258	~86%	2014
TrSSP (http://bioinfo.noble.org/TrSSP/)	PSSM, physiochemical and biochemical properties	TS-900 NTS-660	76.69%	2014

seven classes (amino acid transporters, anion transporters, cation transporters, electron transporters, protein/mRNA transporters, sugar transporters, and other transporters) based on their substrate specificity. The models developed so far were not species specific. To answer this problem, Huang et al. developed a method for prediction of human transporter proteins with an accuracy of ~84% on the independent dataset (Huang et al. 2014). This model used the AAindex database for computation of physicochemical properties of proteins and used it as input for support vector machine (SVM) based classification. The only limitation of the model is its inability to categorize the predicted protein to their respective family or class (Table 9.2).

9.5 Computational Models for Transporter Proteins Ligand Prediction

In the past, numerous reviews have been published summarizing the machine learning and structure-based models for drug transporter proteins (Chen et al. 2012; Montanari and Ecker 2015; Tao et al. 2015). In this study, we are only describing the methods that were not reported before (Table 9.3). Besides that, we also compiled a comprehensive list of in silico models developed so far (Table 9.3). In 2013, Tan et al. developed an inhibitor prediction model for P-glycoprotein. The SVM-based linear model by using only three descriptors predicted the overall accuracy of 86.8% on an independent dataset. Furthermore, the docking-based model also showed 82.3% accuracy on the test set. Analysis of molecular properties and structural properties of the binding site identified aromaticity, molecular volume, and lipophilicity as important determinants of P-gp inhibitors.

For solute carrier (SLC) protein family, Karlgren et al. compiled a list of 65 OATP1B1 inhibitors from in vitro screening. This dataset comprised of 98 training and 48 test set molecules used for computational modeling. Examination of molecular properties showed that a high value of molecular size, polarity, and logP was favored in inhibitors, while a high value of shape descriptor (MSD) was responsible for the non-inhibitory action. Based on these descriptors, OPLS-DA-based single component-based model leads to an overall accuracy of 87% on the test set (Table 9.4). In the same year, authors also reported another in silico model for OATP1B1, OATB1B3, and OATP2B1 hepatic transporters. A multivariate PLS models predicted overall accuracies of 79%, 92%, and 75% for OATP1B1, OATP1B3, and OATP2B1 on test sets, respectively (Table 9.4). In addition to common features such as lipophilicity and polar surface area, a high count of hydrogen bond donor (HBD) was found to be important for OATP1B3 inhibitors than non-inhibitors. Similarly, high polarity was found to be the main characteristic of OATP1B1/OATP1B3 inhibitor, with a small influence on OATP2B1 inhibitor.

Recently, You et al. compiled a dataset of 284 compounds to predict the binder among the four representative hepatic importers (OATP1B1, OATP1B3, OAT2, and OCT1). Support vector machine (SVM)-based algorithm is able to differentiate binder from non-binder with 76.38%, 77.72%, 84.31%, 84.21%, and 76.38% accuracies for OCT1, OATP1B1, OATP1B3, and OAT2, respectively (Table 9.4). Analysis of descriptors revealed the importance of hydrophobicity, H-bonding, and charged molecules common among all the four hepatic importers. However, logD, acidity, and basicity were found to be specific for OATP1B1, OATP1B3, and OCT1, respectively. The major pitfall of the model is its incapability of predicting the action of molecules like as a substrate, inhibitor, or inducer.

In 2014, Garg et al. compiled a dataset of BCRP substrate from previous studies, comprising a total of 234 compounds. Their prediction approach was based on greedy stepwise algorithm and SVM for descriptors selection and model building. The final SVM-based model was based on eight molecular descriptors that resulted in 65% accuracy on the test set. The specificity of molecules was further studied by docking the selected substrate into the binding site of BCRP protein, and it was

Table 9.3 Computational models developed for ABC transporter genes

Target	Method	Action	Descriptors	Dataset			Performance (accuracy)
				Train	Test	Total	
P-gp (Penzotti et al. 2002)	CONAN	Substrate	Pharmacophore	144	45	189	Test 63%
P-gp (Gombar et al. 2004)	LDA	Substrate	Electrotopological, shape indices, and molecular properties	95	58	153	Test 86.2%
P-gp (Xue et al. 2004)	SVM	Substrate	159	74	25	99	Acc 80%
P-gp (Crivori et al. 2006)	PLSD	Substrate	VolSurf	53	272	325	Test 72.4%
P-gp (Sun 2005)	Bayes	Substrate	Atom type and fingerprints	424	185	609	Test 82.2%
P-gp (Cabrera et al. 2006)	TOPS-MODE	Substrate	TOPS-MODE	163	40	203	Test 77.5%
P-gp (De Cerqueira Lima et al. 2006)	SVM, KNN, DT, binary QSAR	Substrate	MolconnZ, VolSurf, MOE	144	51	195	Test 81%
P-gp (Huang et al. 2007)	SVM, PS	Substrate	Molecular descriptors 79	163	40	203	Test 90%
P-gp (Chen et al. 2011a, b)	RP, NBC	Inhibitor	Fingerprints and molecular properties 13	973	300	1273	Test 81.2%
P-gp (Li et al. 2007)	DT	Substrate	Pharmacophore	163	97	260	Test 87.6%
P-gp (Wang et al. 2011)	SVM	Substrate	ADRIANA.Code, MOE, ECFP4	212	120	332	Test 88%
P-gp (Broccatelli et al. 2011)	Combined	Inhibitor	Molecular field, pharmacophore, molecular properties	857	418	1275	Test 86%
P-gp (Li et al. 2014)	NB	Substrate	Molecular properties, topological and fingerprint	723	200	923	Test 83.5%
P-gp (Poongavanam et al. 2012)	RF	Substrate	Checkmol fingerprints	282	202	484	Test 70%
P-gp (Poongavanam et al. 2012)	RF	Inhibitor	Checkmol fingerprints	1268	667	1935	Test 75%

(continued)

Table 9.3 (continued)

Target	Method	Action	Descriptors	Dataset			Performance (accuracy)
				Train	Test	Total	
P-gp (Demel et al. 2010)	RuleFit	Substrate	11 physiochemical and pharmacophore	1877	1436	3313	Acc 90%
P-gp (Klepsch et al. 2014)	SVM, KNN, RF	Inhibitor	62 (2D), 166 MACCS keys, 307 SubFP	1608	346	1954	Test 75%
P-gp (Sedykh et al. 2013)	Consensus (KNN, RF, SVM)	Substrate	286–650 Dragon or 136–148 MOE	544	–	544	$Train_{CV}$ 76%
P-gp (Sedykh et al. 2013)	Consensus (KNN, RF, SVM)	Inhibitor	286–650 Dragon or 136–148 MOE	1571	–	1571	$Train_{CV}$ 94%
P-gp (Hammann et al. 2009)	DT	Substrate	–	206	23	229	Test 78%
P-gp (Hammann et al. 2009)	DT	inhibitor	–	268	30	298	Test 87%
P-gp (Cabrera et al. 2006)	LDA	Substrate	TOPS-MODE	163	40	203	Test 75%
P-gp (Tan et al. 2013)	SVM	Inhibitor	Molecular descriptors 87	857	418	1275	Test 86.8%
P-gp (Thai et al. 2015)	CPG-NN	Inhibitor	MOE, PaDEL	223	26	249	Test 0.47%–0.82%
MRP1 (Sedykh et al. 2013)	Consensus (KNN, RF, SVM)	Inhibitor	286–650 Dragon or 136–148 MOE	418	–	418	$Train_{CV}$ 90%
MRP1 (Sedykh et al. 2013)	Consensus (KNN, RF, SVM)	Substrate	286–650 Dragon or 136–148 MOE	168	–	168	$Train_{CV}$ 91%
MRP1 (Lather and Madan 2005)	Linear model	Inhibitor	Wiener index	82	–	82	$Train_{CV}$ 88%
MRP2 (Pedersen et al. 2008)	OPLS-DA	Inhibitor	Molecular properties 5	76	39	115	Test 72%
MRP2 (Pinto et al. 2012)	RF	Substrate	Molecular properties 16	1204	44	1248	$Train_{CV}$ 75%

MRP2 (Zhang et al. 2009)	Pharmacophore	Inhibitor	HBA and hydrophobicity	9	318	318	Acc 74%
MRP2 (Sedykh et al. 2013)	Consensus (KNN, RF, SVM)	Substrate	Dragon descriptors (286–650), MOE (136–148)	188	–	188	$Train_{CV}$ 87%
MRP2 (Sedykh et al. 2013)	Consensus (KNN, RF, SVM)	Inhibitor	Dragon descriptors (286–650), MOE (136–148)	96	–	96	$Train_{CV}$ 89%
MRP2 (Zhang et al. 2009)	SVM	Inhibitor	Molecular descriptors 16 and pharmacophore	257	61	318	77.1%
MRP2 (Pedersen et al. 2008)	OPLS-DA	Inhibitor	Molecular descriptor 240	79	39	118	72%
MRP3 (Sedykh et al. 2013)	Consensus	Substrate	Dragon descriptors (286–650), MOE (136–148)	62	–	62	$Train_{CV}$ 98%
MRP4 (Sedykh et al. 2013)	Consensus	Substrate	Dragon descriptors (286–650), MOE (136–148)	92	–	92	$Train_{CV}$ 92%
MRP4 (Sedykh et al. 2013)	Consensus	Inhibitor	Dragon descriptors (286–650), MOE (136–148)	64	–	64	$Train_{CV}$ 70%
MRP4 (Welch et al. 2015)	Bayesian	Inhibitor	Molecular descriptor 8, ECFP_6, FCFP_6	57	29	86	Test 83.8%
MRP4 (Welch et al. 2015)	Pharmacophore	Inhibitor	Pharmacophore	9	77	86	Test Sen 71.4%, Spec 62.8%
PEPT1 (Sedykh et al. 2013)	Consensus	Substrate	Dragon descriptors (286–650), MOE (136–148)	158	–	158	$Train_{CV}$ 84%
PEPT1 (Sedykh et al. 2013)	Consensus	Inhibitor	Dragon descriptors (286–650), MOE (136–148)	80	–	80	$Train_{CV}$ 72%
PEPT1 (Kamphorst et al. 2007)	Bayesian	Inhibitor	19	138	46	184	Test 87%
BSEP (Warner et al. 2012)	SVM	Inhibitor	Molecular properties (196 2D/3D)	437	187	624	Test 87%

(continued)

Table 9.3 (continued)

Target	Method	Action	Descriptors	Dataset			Performance (accuracy)
				Train	Test	Total	
BSEP (Welch et al. 2015)	Bayesian	Inhibitor	Molecular descriptors 8, ECFP_6, FCFP_6	171	86	257	Test 87%
BSEP (Welch et al. 2015)	Pharmacophore	Inhibitor	Pharmacophore	9	247	256	Test Sen 82.7% and Spec 37.2%
BCRP (Montanari and Ecker 2014)	NB	Inhibitor	ECFP_6 fingerprints	780	198	978	$Train_{CV}$ 92%
BCRP (Pan et al. 2013)	Pharmacophore	Inhibitor	Pharmacophore	30	79	109	Test 66%
BCRP (Pan et al. 2013)	Bayesian	Inhibitor	Molecular Properties 62, ECFP_6, FCFP_6	124	79	203	Test 90%
BCRP (Hazai et al. 2013)	SVM	Substrate	AAC, SPH, Mor17m, Mor25m, R2m	263	40	303	Test 73%
BCRP (Zhong et al. 2011)	GA-CG-SVM	Substrate	Molecular descriptors (17)	137	40	177	Test 85%
BCRP (Gantner et al. 2013)	LDA	Substrate	Molecular descriptor (867)	164	98	262	Test 74.5%
BCRP (Sedykh et al. 2013)	Consensus (KNN, RF, SVM)	Substrate	Dragon descriptors (286–650), MOE (136–148)	146	–	146	$Train_{CV}$ 80%
BCRP (Sedykh et al. 2013)	Consensus (KNN, RF, SVM)	Inhibitor	Dragon descriptors (286–650), MOE (136–148)	382	–	382	$Train_{CV}$ 83%
BCRP (Matsson et al. 2007)	OPLS-DA	Inhibitor	Molecular descriptors 2	80	43	123	Test 79%
BCRP (Garg et al. 2014)	SVM	Substrate	Molecular descriptors 8	160	74	234	Test 65%
P-gp (Palmeira et al. 2011)	PharmaGist	Inhibitor	Pharmacophore descriptor	26	–	26	–

P-gp (Langer et al. 2004)	Catalyst	Inhibitor	Pharmacophore	15	105	155	Test 80% (top 35 and bottom 35 compound)
P-gp (Pajeva and Wiese 2002)	GASP	Substrate	Pharmacophore	18	–	18	–
P-gp (Cianchetta et al. 2005)	PLS	Inhibitor	Almond and VolSurf	109	20	129	Test R^2 0.72
P-gp (Müller et al. 2008)	PLS	Inhibitor	CoMFA and CoMSIA	28	30	58	Test R^2 0.6
P-gp (Wu et al. 2009)	MLR, SVM	Modulator	423 CODESSA	56	14	70	Test R^2 0.81
P-gp (Wang et al. 2005)	BRNN	Inhibitor	249	43	14	57	Test R^2 0.728
P-gp (Ekins 2002)	Catalyst	Inhibitor	Pharmacophore	27	19	46	Train R^2 0.77
				21	19	40	Train R^2 0.88
				17	19	36	Train R^2 0.86
P-gp (Leong et al. 2012)	PhE/SVM	Inhibitor	Pharmacophore	31	88 + 11	130	R^2 0.87
P-gp (Shen et al. 2014)	BP-ANN	Inhibitor	PaDEL descriptor 118 and pharmacophore	71	17	88	R^2 0.81–0.87
MRP2 (Ng et al. 2005)	SA-PLA	Binding affinity	Molecular descriptors 71 and pharmacophore	20	5	25	R^2 0.82
PEPT1 (Larsen et al. 2008)	PLS	Binding affinity	VolSurf 110, MOE 204, GRIND 500	76	38	114	R^2 0.72
PEPT1 (Biegel et al. 2005)	COMSIA	Inhibitor	COMSIA	98	–	98	Q^2 0.83
BSEP (Ritschel et al. 2014)	Pharmacophore	Inhibitor	MOE	5	59	64	MCC 0.52
BCRP (Ding et al. 2014)	PhE/SVM	Inhibitor	Pharmacophore	22	97 + 16	135	Q^2 0.75–0.89
BCRP (Pick et al. 2011)	MLR/PLS	Inhibitor	CoMFA and CoMSIA	28	13	41	Q^2 0.63

SE sensitivity, *SP* specificity, Q^2 coefficient of determination

Table 9.4 Computational models developed for SLC transporter genes

Target	Method	Action	Descriptors	Dataset			Performance (accuracy)
				Train	Test	Total	
Classification							
OATPs (Karlgren et al. 2012b)	PLS	Inhibitor	93	83	42	125	OATP1B1–79%
							OATP1B3–92%
							OATP2B1–75%
OATP1B1 (Karlgren et al. 2012a)	OPLS-DA	Inhibitor	91 descriptors	98	48	146	Test 87%
OATP2B1 (Sedykh et al. 2013)	Consensus	Substrate	2030 Dragon descriptors and 185 MOE descriptors	53	–	53	$Train_{CV}$ 75%
OATP2B1 (Sedykh et al. 2013)	Consensus	Inhibitor	2030 Dragon descriptors and 185 MOE descriptors	136	–	136	$Train_{CV}$ 80%
OCT1 (Sedykh et al. 2013)	Consensus	Substrate	Dragon descriptors (286–650) and MOE descriptors (136–148)	78	–	78	$Train_{CV}$ 89%
OCT1 (Sedykh et al. 2013)	Consensus	Inhibitor	Dragon descriptors (286–650) and MOE descriptors (136–148)	199	–	199	$Train_{CV}$ 92%
ASBT (Sedykh et al. 2013)	Consensus	Substrate	Dragon descriptors (286–650) and MOE descriptors (136–148)	100	–	100	$Train_{CV}$ 93%
ASBT (Sedykh et al. 2013)	Consensus	Inhibitor	Dragon descriptors (286–650) and MOE descriptors (136–148)	150	–	150	$Train_{CV}$ 92%
ASBT (Zheng et al. 2009)	Bayesian and 3D QSAR	Inhibitor	FCFP_6, pharmacophore, molecular descriptors	38	19–30	57–68	Test 54–88%
MCT1 (Sedykh et al. 2013)	Consensus	Inhibitor	Dragon descriptors (286–650) and MOE descriptors (136–148)	67	–	67	$Train_{CV}$ 100%
OATP1B1 (Van de Steeg et al. 2015)	Bayesian	Inhibitor	Molecular descriptors, FCFP4, ECFP4	437	155	592	Test Sen 100%, Spec 85%

OATP1B1*15 (Van de Steeg et al. 2015)	Bayesian	Inhibitor	Molecular descriptors, FCFP4, ECFP4	452	146	598	Test Sen 100%, Spec 87%
OATP1B1 (You et al. 2015)	SVM	Binder	Molecular descriptors 102, SFED descriptors 16	284	1360	1644	Test 77.72%
OATP1B3 (You et al. 2015)	SVM	Binder	Molecular descriptors 102, SFED descriptors 16	284	1539	1823	Test 84.31%
OAT2 (You et al. 2015)	SVM	Binder	Molecular descriptors 102, SFED descriptors 16	284	19	303	Test 84.21%
OCT (You et al. 2015)	SVM	Binder	Molecular descriptors 102, SFED descriptors 16	284	199	483	Test 76.38%
Regression							
ASBT (Rais et al. 2010)	Multivariate regression	Inhibitor	Molecular descriptors	32	–	32	Q^2 0.89

observed that His457 and Arg465 are involved in specificity (Table 9.3). Ding and colleagues also developed a regression-based model using 135 BCRP inhibitors collected from previous studies. A pharmacophore model was developed based on 22 selected molecules whose IC_{50} values varied from 4 logarithmic unit and the prediction value was used for SVM model development. The final PhE-/SVM-based model showed Q^2 of 0.75–0.89 on the test set and 0.72–0.91 on outlier dataset. A high correlation value R^2 of 0.83 on HIV protease inhibitors also shows the applicability of the model.

Recently, Welch et al. built a computational algorithm for the prediction of inhibitors against multiple drug-resistant protein 4 (MRP4) and bile salt export pump (BSEP). The Bayesian-based classification model showed accuracies of 83.8% and 87% on test dataset for MRP4 and BSEP, respectively (Table 9.3). The pharmacophore model based on nine inhibitors from each class showed the importance of hydrophobicity and H-bond acceptor in inhibitor prediction. The importance of same descriptors for two transporters partially explains the affinity of various drugs for both proteins. Analysis of dataset also revealed that higher logP value positively contributes in inhibitory activity of compounds.

However, the major challenge in the field of drug transporter modeling is the molecular basis for the poly-specificity. To overcome this issue, Thai et al. developed a counter-propagation neural network (CPG-NN) based multi-label classification system. In this study, 223 compounds were used to develop three models (SION, SIO, SIN) in order to classify specific "true" P-gp inhibitors, and three other models (CPBN, CPB1, CPN) were developed in order to distinguish between CYP3A, P-gp inhibitors, and co-inhibitors of these proteins. The overall accuracies of SIN, SIO, and SION model on diverse independent test set were 82%, 65%, and 55%, respectively.

In 2015, Steeg and colleagues screened 640 FDA drugs for their inhibitory action on OATP1B1, OATP1B1*15, and OATP1B3 (Table 9.4). Based on more than 60% inhibition cutoff value, 8%, 7%, and 1% drugs were found to act as inhibitors for OATP1B1, OATP1B1*15, and OATP1B3, respectively. In vitro screening also showed that all OATP1B3 inhibitors also inhibited OATP1B1 but not vice versa. Due to a low number of OATP1B3 inhibitors, only OATB1B1 and OATB1B1*15 inhibitors were selected for computational modeling. The Bayesian-based model showed the overall performance of 80% on the test set. Analyses of molecular properties of compounds described the function of mol. wt., logP, molecular surface area, rotatable bond, and ring count for inhibiting the OATP1B1 transporter. However, positively charged alkylamine group showed negative contribution on the binding. This might be due to the absence of counteracting negatively charged amino acids in OATP1B1 transporter.

9.6 Genomic Alteration in Drug Resistance and Diseases

Drug transporter proteins act as gatekeeper for influx/efflux of molecules; therefore, any kind of genomic alterations like single nucleotide polymorphisms (SNPs), indels, expression, etc. has been associated with various kinds of diseases (César-Razquin et al. 2015; Honjo et al. 2002; Kim et al. 2002; Mizuarai et al. 2004). As evident from Table 9.5, transporter proteins have been implicated in a number of neurological disorders such as Alzheimer disease, Huntington disease, schizophrenia, Parkinson disease, etc. (Ashraf et al. 2012; Dean et al. 2001; Hediger et al. 2013). These kinds of alterations have a major impact on drug efficacy and its potency. For example, the overexpression of ABC family of protein and the underexpression of SLC family of protein were found to be associated with drug resistance particularly in cancer (Table 9.6). Likewise, single nucleotide polymorphism (SNP) either may lead to change in the promoter region, non-synonymous mutation, or may lead to stop codon. These type of changes cause variation in the structure of proteins or might result in loss of function of the protein

Table 9.5 Association of transporter proteins in various diseases

Gene	Disease	Gene	Disease
SLC1	ALS, AD, schizophrenia	SLC2	Early-onset nephropathy
SLC6	Depression, addiction, aggression, PTSD, anxiety, OCD, ADHD, Autism	SLC13	Nephrolithiasis, GA1, CD
SLC17	Gout, schizophrenia, ALS, AD, HD	SLC26	Chondrodysplasias, chloride diarrhea, oxalate urolithiasis, gastric hypochlorhydria, distal renal tubular acidosis, male infertility
SLC30	Transient neonatal zinc deficiency	SLC52	MADD, Brown-Vialetto-Van Laere syndrome
ABCA	Tangier disease T1, HDL deficiency, Stargardt disease-1, age-related macular degeneration, and retinitis pigmentosa	ABCB	AS, type 2 diabetes, celiac disease, lethal neonatal syndrome, X-linked sideroblastic anemia with ataxia
ABCC	MDR, Dubin-Johnson syndrome, congenital bilateral aplasia of the vas deferens, type 2 diabetes, paroxysmal kinesigenic choreoathetosis, cystic fibrosis, pseudoxanthoma elasticum	ABCD	ALD, Zellweger syndrome
ABCG	Sitosterolemia		

ALS amyotrophic lateral sclerosis, *AD* Alzheimer disease, *PTSD* post-traumatic stress disorder, *OCD* obsessive compulsive disorder, *ADHD* attention deficit hyperactivity disorder, *GA1* glutaric aciduria type 1, *CD* Canavan disease, *HD* Huntington disease, *MADD* multiple acyl-CoA dehydrogenase deficiency, *HDL* Familial high-density lipoprotein deficiency, *MDR* multidrug resistance, *AS* ankylosing spondylitis

Table 9.6 Effect of alteration in gene expression on drug transport

Drug	Transporter	Type	Expression	Class	References
Paclitaxel	ABCC3	Efflux	High	Antimitotic	Dong et al. (2014)
Gemcitabine	ABCG2	Efflux	High	Anticancer	Huang and Sadée (2006)
Nucleoside drugs	SLC29A1	Uptake	Low	Anti-HIV	O'Brien et al. (2008)
5-Fluorouracil	ABCC11	Efflux	High	Anticancer	Hauswald et al. (2009)
2′, 3′-Dideoxycytidine	ABCC11	Efflux	High	Anti-HIV	Hauswald et al. (2009)
Monomethyl auristatin E (MMAE)	ABCC3	Efflux	High	Antimitotic	Hauswald et al. (2009)
9′-(2′-Phosphonylmethoxynyl) adenine (PMEA)	ABCC11	Efflux	High	Anti-hepatitis B	Guo et al. (2003)
Fluoropyrimidines	ABCC11, ABCG2	Efflux	High	Antiviral	Guo et al. (2003)
Doxorubicin	ABCB1	Efflux	High	Anticancer	Okabe et al. (2008) and Schlessinger et al. (2013)
	SLC19A3	Uptake	Low		
	SLC22A4	Uptake	Low		
Mitoxantrone	ABCB1	Efflux	High	Anticancer	Huang and Sadée (2006) and Okabe et al. (2008)
	SLC22A4	Uptake	Low		
Cisplatin	ABCB9	Uptake	Low	Anticancer	Huang and Sadée (2006) and Nieth and Lage (2005)
	SLC31A1	Influx	Downregulate		

Table 9.7 Role of SNP in transporters proteins with possible outcomes

Gene	SNP	Action	Reference
ABCC1	2965G>A	Substrate affinity decreases	Cascorbi (2006)
ABCC1	2012G>T	Doxorubicin-induced cardiomyopathy	
ABCC2	24C>T	Decrease expression	
ABCC2	2366C>T	Decrease expression	
ABCC2	4348G>A	Decrease expression	
ABCG2	376C>T	Loss of function	
ABCG2	421C>A	Decrease expression	

(Table 9.7). Thus, this might ultimately lead to various kinds of diseases and/or disorders, including drug resistance.

9.7 Conclusion and Future Perspectives

In the past, significant efforts have been done to compile the computational framework of transporter proteins (Guo et al. 2003; Kruh et al. 2007; Okabe et al. 2008). However, the major attention was paid to a single family of proteins such as ABC and SLC. These two transporters ideally represent the human transportome. Albeit, there is limited information reported till date. It is clear from Table 9.1 that different databases have different numbers of human transporter proteins. Till date, HumanTDB is the updated database with a maximum number of transporter proteins available (Ye et al. 2014). With the advancement of curated dataset, qualitative and quantitative accessibility of biological data, along with the significant progress of structural biology, we can explore our understanding of the molecular basis of ligand-transporter interactions. Furthermore, to determine how transporters may affect variation in drug response, drug levels, and their endogenous role as drug transporters, it is also interesting to have a comparative analysis of these transporters with and without drugs. Since, the genetic variants in transporters at genome-wide level significance have been shown to impact the risk for various human diseases, such as cancers, diabetes, and cardiovascular disease (César-Razquin et al. 2015). Therefore, there is an urgent need to analyze these datasets and search for novel transporter proteins by developing a unified computational model.

We also highlighted that now a lot of datasets are available for active ligand against human transporter proteins. But one must have to check the redundancy and authenticity of the dataset, and in the future, availability of such dataset can help in the advancement of better models for predictions. Indeed, another serious issue pointed out here is the availability of a reliable and comprehensive database for substrate and inhibitors of transporters. As previously reported that most of the models developed were based on similar compounds, thus this kind of new additional information will be helpful in developing the global prediction models and in the identification of key structural features, thereby playing role in structural-functional relationship. We entirely supported the Matsson et al. (2013) report that rather than developing the individual model, more emphasis has to be given towards integrating the protein and ligand chemistry for developing more robust models. This review summarizes the knowledge on different databases, small molecules (ligands and/or substrates), and computational tools for qualitative and quantitative analysis, along with the chemoinformatics methods, for understanding the structural-functional relationship and molecular recognition of transporter proteins at the microscopic level. The more attention will be focused on the endogenous role of drug transporters at atomic level resolution. In comparison to available online databases, a little progress has been observed on the availability of prediction methods. Therefore, it is an important and urgent task to develop prediction methods that would be helpful in advancing drug discovery projects and will also contribute to the comprehensive understanding of their mechanism and function at an atomic level resolution for therapeutic intervention.

Authors' Contributions DS, RB, and SKD compiled the information from the literature. DS and SA analyze the results and wrote the article. RB and SKD helped in editing the article. DS conceived and coordinated the project. The manuscript has been read and approved by all authors.

Acknowledgments The authors are thankful to the Indian Council of Medical Research (ICMR) and the Department of Biotechnology (DBT), Government of India, for financial assistance.

References

Alexander SPH, Benson HE, Faccenda E, Pawson AJ, Sharman JL, McGrath JC, Catterall WA, Spedding M, Peters JA, Harmar AJ, Abul-Hasn N, Anderson CM, Anderson CMH, Araiksinen MS, Arita M, Arthofer E, Barker EL, Barratt C, Barnes NM, Bathgate R, Beart PM, Belelli D, Bennett AJ, Birdsall NJM, Boison D, Bonner TI, Brailsford L, Bröer S, Brown P, Calo G, Carter WG, Catterall WA, Chan SLF, Chao MV, Chiang N, Christopoulos A, Chun JJ, Cidlowski J, Clapham DE, Cockcroft S, Connor MA, Cox HM, Cuthbert A, Dautzenberg FM, Davenport AP, Dawson PA, Dent G, Dijksterhuis JP, Dollery CT, Dolphin AC, Donowitz M, Dubocovich ML, Eiden L, Eidne K, Evans BA, Fabbro D, Fahlke C, Farndale R, Fitzgerald GA, Fong TM, Fowler CJ, Fry JR, Funk CD, Futerman AH, Ganapathy V, Gaisnier B, Gershengorn MA, Goldin A, Goldman ID, Gundlach AL, Hagenbuch B, Hales TG, Hammond JR, Hamon M, Hancox JC, Hauger RL, Hay DL, Hobbs AJ, Hollenberg MD, Holliday ND, Hoyer D, Hynes NA, Inui K-I, Ishii S, Jacobson KA, Jarvis GE, Jarvis MF, Jensen R, Jones CE, Jones RL, Kaibuchi K, Kanai Y, Kennedy C, Kerr ID, Khan AA, Klienz MJ, Kukkonen JP, Lapoint JY, Leurs R, Lingueglia E, Lippiat J, Lolait SJ, Lummis SCR, Lynch JW, MacEwan D, Maguire JJ, Marshall IL, May JM, McArdle CA, McGrath JC, Michel MC, Millar NS, Miller LJ, Mitolo V, Monk PN, Moore PK, Moorhouse AJ, Mouillac B, Murphy PM, Neubig RR, Neumaier J, Niesler B, Obaidat A, Offermanns S, Ohlstein E, Panaro MA, Parsons S, Pwrtwee RG, Petersen J, Pin J-P, Poyner DR, Prigent S, Prossnitz ER, Pyne NJ, Pyne S, Quigley JG, Ramachandran R, Richelson EL, Roberts RE, Roskoski R, Ross RA, Roth M, Rudnick G, Ryan RM, Said SI, Schild L, Sanger GJ, Scholich K, Schousboe A, Schulte G, Schulz S, Serhan CN, Sexton PM, Sibley DR, Siegel

JM, Singh G, Sitsapesan R, Smart TG, Smith DM, Soga T, Stahl A, Stewart G, Stoddart LA, Summers RJ, Thorens B, Thwaites DT, Toll L, Traynor JR, Usdin TB, Vandenberg RJ, Villalon C, Vore M, Waldman SA, Ward DT, Willars GB, Wonnacott SJ, Wright E, Ye RD, Yonezawa A, Zimmermann M (2013) The concise guide to PHARMACOLOGY 2013/14: overview. Br J Pharmacol 170:1449–1458. https://doi.org/10.1111/bph.12444

Ashraf T, Kis O, Banerjee N, Bendayan R (2012) Drug transporters at brain barriers: expression and regulation by neurological disorders. Adv Exp Med Biol 763:20–69

Biegel A, Gebauer S, Hartrodt B, Brandsch M, Neubert K, Thondorf I (2005) Three-dimensional quantitative structure-activity relationship analyses of beta-lactam antibiotics and tripeptides as substrates of the mammalian H+/peptide cotransporter PEPT1. J Med Chem 48:4410–4419. https://doi.org/10.1021/jm048982w

Broccatelli F, Carosati E, Neri A, Frosini M, Goracci L, Oprea TI, Cruciani G (2011) A novel approach for predicting P-glycoprotein (ABCB1) inhibition using molecular interaction fields. J Med Chem 54:1740–1751. https://doi.org/10.1021/jm101421d

Cabrera MA, González I, Fernández C, Navarro C, Bermejo M (2006) A topological substructural approach for the prediction of P-glycoprotein substrates. J Pharm Sci 95:589–606. https://doi.org/10.1002/jps.20449

Cascorbi I (2006) Role of pharmacogenetics of ATP-binding cassette transporters in the pharmacokinetics of drugs. Pharmacol Ther 112:457–473. https://doi.org/10.1016/j.pharmthera.2006.04.009

César-Razquin A, Snijder B, Frappier-Brinton T, Isserlin R, Gyimesi G, Bai X, Reithmeier RA, Hepworth D, Hediger MA, Edwards AM, Superti-Furga G (2015) A call for systematic research on solute carriers. Cell 162:478–487. https://doi.org/10.1016/j.cell.2015.07.022

Chen L, Li Y, Zhao Q, Peng H, Hou T (2011a) ADME evaluation in drug discovery. 10. Predictions of P-glycoprotein inhibitors using recursive partitioning and naive Bayesian classification techniques. Mol Pharm 8:889–900. https://doi.org/10.1021/mp100465q

Chen S-A, Y-Y O, Lee T-Y, Gromiha MM (2011b) Prediction of transporter targets using efficient RBF networks with PSSM profiles and biochemical properties. Bioinformatics 27:2062–2067. https://doi.org/10.1093/bioinformatics/btr340

Chen L, Li Y, Yu H, Zhang L, Hou T (2012) Computational models for predicting substrates or inhibitors of P-glycoprotein. Drug Discov Today 17:343–351. https://doi.org/10.1016/j.drudis.2011.11.003

Cianchetta G, Singleton RW, Zhang M, Wildgoose M, Giesing D, Fravolini A, Cruciani G, Vaz RJ (2005) A pharmacophore hypothesis for P-glycoprotein substrate recognition using GRIND-based 3D-QSAR. J Med Chem 48:2927–2935. https://doi.org/10.1021/jm0491851

Crivori P, Reinach B, Pezzetta D, Poggesi I (2006) Computational models for identifying potential P-glycoprotein substrates and inhibitors. Mol Pharm 3:33–44. https://doi.org/10.1021/mp050071a

De Cerqueira Lima P, Golbraikh A, Oloff S, Xiao Y, Tropsha A (2006) Combinatorial QSAR modeling of P-glycoprotein substrates. J Chem Inf Model 46:1245–1254. https://doi.org/10.1021/ci0504317

Dean M, Rzhetsky A, Allikmets R (2001) The human ATP-binding cassette (ABC) transporter superfamily. Genome Res 11:1156–1166. https://doi.org/10.1101/gr.184901

DeGorter MK, Xia CQ, Yang JJ, Kim RB (2012) Drug transporters in drug efficacy and toxicity. Annu Rev Pharmacol Toxicol 52:249–273. https://doi.org/10.1146/annurev-pharmtox-010611-134529

Demel MA, Kraemer O, Ettmayer P, Haaksma E, Ecker GF (2010) Ensemble rule-based classification of substrates of the human ABC-transporter ABCB1 using simple physicochemical descriptors. Mol Inform 29:233–242. https://doi.org/10.1002/minf.200900079

Ding Y-L, Shih Y-H, Tsai F-Y, Leong MK (2014) In silico prediction of inhibition of promiscuous breast cancer resistance protein (BCRP/ABCG2). PLoS One 9:e90689. https://doi.org/10.1371/journal.pone.0090689

Dong Z, Zhong Z, Yang L, Wang S, Gong Z (2014) MicroRNA-31 inhibits cisplatin-induced apoptosis in non-small cell lung cancer cells by regulating the drug transporter ABCB9. Cancer Lett 343:249–257. https://doi.org/10.1016/j.canlet.2013.09.034

Ekins S (2002) Application of three-dimensional quantitative structure-activity relationships of P-glycoprotein inhibitors and substrates. Mol Pharmacol 61:974–981. https://doi.org/10.1124/mol.61.5.974

Gantner ME, Emiliano M, Ianni D, Ruiz ME, Talevi A, Bruno-blanch LE (2013) Development of conformation independent computational models for the early recognition of breast cancer resistance protein substrates. Biomed Res Int. https://doi.org/10.1155/2013/863592

Garg P, Dhakne R, Belekar V (2014) Role of breast cancer resistance protein (BCRP) as active efflux transporter on blood-brain barrier (BBB) permeability. Mol Divers 19:163–172. https://doi.org/10.1007/s11030-014-9562-2

Gombar VK, Polli JW, Humphreys JE, Wring SA, Serabjit-Singh CS (2004) Predicting P-glycoprotein substrates by a quantitative structure-activity relationship model. J Pharm Sci 93:957–968. https://doi.org/10.1002/jps.20035

Gromiha MM, Yabuki Y, Suresh MX, Thangakani AM, Suwa M, Fukui K (2009) TMFunction: database for functional residues in membrane proteins. Nucleic Acids Res 37:D201–D204. https://doi.org/10.1093/nar/gkn672

Guo Y, Kotova E, Chen Z-S, Lee K, Hopper-Borge E, Belinsky MG, Kruh GD (2003) MRP8, ATP-binding cassette C11 (ABCC11), is a cyclic nucleotide efflux pump and a resistance factor for fluoropyrimidines 2′,3′-dideoxycytidine and 9′-(2′-phosphonylmethoxyethyl)adenine. J Biol Chem 278:29509–29514. https://doi.org/10.1074/jbc.M304059200

Hammann F, Gutmann H, Jecklin U, Maunz A, Helma C, Drewe J (2009) Development of decision tree models for substrates, inhibitors, and inducers of P-glycoprotein. Curr Drug Metab 10:339–346. https://doi.org/10.2174/138920009788499021

Hauswald S, Duque-Afonso J, Wagner MM, Schertl FM, Lübbert M, Peschel C, Keller U, Licht T (2009) Histone deacetylase inhibitors induce a very broad, pleiotropic anticancer drug resistance phenotype in acute myeloid leukemia cells by modulation of multiple ABC transporter genes. Clin Cancer Res 15:3705–3715. https://doi.org/10.1158/1078-0432.CCR-08-2048

Hazai E, Hazai I, Ragueneau-Majlessi I, Chung SP, Bikadi Z, Mao Q (2013) Predicting substrates of the human breast cancer resistance protein using a support vector machine method. BMC Bioinformatics 14:130. https://doi.org/10.1186/1471-2105-14-130

Hediger MA, Clémençon B, Burrier RE, Bruford EA (2013) The ABCs of membrane transporters in health and disease (SLC series): introduction. Mol Asp Med 34:95–107. https://doi.org/10.1016/j.mam.2012.12.009

Hee Choi Y, Yu A-M (2014) ABC transporters in multidrug resistance and pharmacokinetics, and strategies for drug development. Curr Pharm Des 20:793–807. https://doi.org/10.2174/138161282005140214165212

Honjo Y, Morisaki K, Mickley Huff L, Robey RW, Hung J, Dean M, Bates SE (2002) Single-Nucleotide Polymorphism (SNP) analysis in the ABC half-transporter ABCG2 (MXR/BCRP/ABCP1). Cancer Biol Ther 1:696–702. https://doi.org/10.4161/cbt.322

Huang Y, Sadée W (2006) Membrane transporters and channels in chemoresistance and -sensitivity of tumor cells. Cancer Lett 239:168–182. https://doi.org/10.1016/j.canlet.2005.07.032

Huang J, Ma G, Muhammad I, Cheng Y (2007) Identifying P-glycoprotein substrates using a support vector machine optimized by a particle swarm. J Chem Inf Model 47:1638–1647. https://doi.org/10.1021/ci700083n

Huang H-L, Li M-C, Vasylenko T, Ho S-Y (2014) Computational prediction and analysis of human transporters using physicochemical properties of amino acids. Int J Eng Tech Res 2:180–187. https://doi.org/10.1186/1741-7007-7-50

Kamphorst J, Cucurull-Sanchez L, Jones B (2007) A performance evaluation of multiple classification models of human PEPT1 inhibitors and non-inhibitors. QSAR Comb Sci 26:220–226. https://doi.org/10.1002/qsar.200630025

Karlgren M, Ahlin G, Bergström CAS, Svensson R, Palm J, Artursson P (2012a) In vitro and in silico strategies to identify OATP1B1 inhibitors and predict clinical drug-drug interactions. Pharm Res 29:411–426. https://doi.org/10.1007/s11095-011-0564-9

Karlgren M, Vildhede A, Norinder U, Wisniewski JR, Kimoto E, Lai Y, Haglund U, Artursson P (2012b) Classification of inhibitors of hepatic organic anion transporting polypeptides (OATPs): influence of protein expression on drug-drug interactions. J Med Chem 55:4740–4763. https://doi.org/10.1021/jm300212s

Kim M, Turnquist H, Jackson J, Sgagias M, Yan Y, Gong M, Dean M, Sharp JG, Cowan K (2002) The multidrug resistance transporter ABCG2 (Breast Cancer Resistance Protein 1) Effluxes Hoechst 33342 and is overexpressed in hematopoietic stem cells. Clin Cancer Res 8:22–28

Klepsch F, Vasanthanathan P, Ecker GF (2014) Ligand and structure-based classification models for prediction of P-glycoprotein inhibitors. J Chem Inf Model 54:218–229. https://doi.org/10.1021/ci400289j

Kruh GD, Guo Y, Hopper-Borge E, Belinsky MG, Chen Z-S (2007) ABCC10, ABCC11, and ABCC12. Pflugers Arch 453:675–684. https://doi.org/10.1007/s00424-006-0114-1

Langer T, Eder M, Hoffmann RD, Chiba P, Ecker GF (2004) Lead identification for modulators of multidrug resistance based on in silico screening with a pharmacophoric feature model. Arch Pharm (Weinheim) 337:317–327. https://doi.org/10.1002/ardp.200300817

Larsen SB, Jørgensen FS, Olsen L (2008) QSAR models for the human H(+)/peptide symporter, hPEPT1: affinity prediction using alignment-independent descriptors. J Chem Inf Model 48:233–241. https://doi.org/10.1021/ci700346y

Lather V, Madan AK (2005) Topological model for the prediction of MRP1 inhibitory activity of pyrrolopyrimidines and templates derived from pyrrolopyrimidine. Bioorg Med Chem Lett 15:4967–4972. https://doi.org/10.1016/j.bmcl.2005.08.011

Law V, Knox C, Djoumbou Y, Jewison T, Guo AC, Liu Y, Maciejewski A, Arndt D, Wilson M, Neveu V, Tang A, Gabriel G, Ly C, Adamjee S, Dame ZT, Han B, Zhou Y, Wishart DS (2014) DrugBank 4.0: shedding new light on drug metabolism. Nucleic Acids Res 42:D1091–D1097. https://doi.org/10.1093/nar/gkt1068

Leong MK, Chen H-B, Shih Y-H (2012) Prediction of promiscuous p-glycoprotein inhibition using a novel machine learning scheme. PLoS One 7:e33829. https://doi.org/10.1371/journal.pone.0033829

Li W-X, Li L, Eksterowicz J, Ling XB, Cardozo M (2007) Significance analysis and multiple pharmacophore models for differentiating P-glycoprotein substrates. J Chem Inf Model 47:2429–2438. https://doi.org/10.1021/ci700284p

Li H, Dai X, Zhao X (2008) A nearest neighbor approach for automated transporter prediction and categorization from protein sequences. Bioinformatics 24:1129–1136. https://doi.org/10.1093/bioinformatics/btn099

Li H, Benedito VA, Udvardi MK, Zhao PX (2009) TransportTP: a two-phase classification approach for membrane transporter prediction and characterization. BMC Bioinformatics 10:418. https://doi.org/10.1186/1471-2105-10-418

Li D, Chen L, Li Y, Tian S, Sun H, Hou T (2014) ADMET evaluation in drug discovery. 13. Development of in silico prediction models for P-glycoprotein substrates. Mol Pharm 11:716–726. https://doi.org/10.1021/mp400450m

Lin L, Yee SW, Kim RB, Giacomini KM (2015) SLC transporters as therapeutic targets: emerging opportunities. Nat Rev Drug Discov 14:543–560. https://doi.org/10.1038/nrd4626

Matsson P, Englund G, Ahlin G, Bergström CAS, Norinder U, Artursson P (2007) A global drug inhibition pattern for the human ATP-binding cassette transporter breast cancer resistance protein (ABCG2). J Pharmacol Exp Ther 323:19–30. http://doi:10.1124/jpet.107.124768

Matsson P, Artursson P (2013) Computational prospecting for drug–transporter interactions. Clin Pharmacol Ther 94:30–32. https://doi.org/10.1038/clpt.2013.67

Mishra NK, Chang J, Zhao PX (2014) Prediction of membrane transport proteins and their substrate specificities using primary sequence information. PLoS One 9:e100278. https://doi.org/10.1371/journal.pone.0100278

Mizuarai S, Aozasa N, Kotani H (2004) Single nucleotide polymorphisms result in impaired membrane localization and reduced atpase activity in multidrug transporter ABCG2. Int J Cancer 109:238–246. https://doi.org/10.1002/ijc.11669

Montanari F, Ecker GF (2014) BCRP Inhibition: from data collection to ligand-based modeling. Mol Inform 33:322–331. https://doi.org/10.1002/minf.201400012

Montanari F, Ecker GF (2015) Prediction of drug-ABC-transporter interaction – recent advances and future challenges. Adv Drug Deliv Rev 86:17–26. https://doi.org/10.1016/j.addr.2015.03.001

Müller H, Pajeva IK, Globisch C, Wiese M (2008) Functional assay and structure-activity relationships of new third-generation P-glycoprotein inhibitors. Bioorg Med Chem 16:2448–2462. https://doi.org/10.1016/j.bmc.2007.11.057

Ng C, Xiao Y-D, Lum BL, Han Y-H (2005) Quantitative structure-activity relationships of methotrexate and methotrexate analogues transported by the rat multispecific resistance-associated protein 2 (rMrp2). Eur J Pharm Sci 26:405–413. https://doi.org/10.1016/j.ejps.2005.07.008

Nieth C, Lage H (2005) Induction of the ABC-transporters Mdr1/P-gp (Abcb1), mrpl (Abcc1), and bcrp (Abcg2) during establishment of multidrug resistance following exposure to mitoxantrone. J Chemother 17:215–223. https://doi.org/10.1179/joc.2005.17.2.215

O'Brien C, Cavet G, Pandita A, Hu X, Haydu L, Mohan S, Toy K, Rivers CS, Modrusan Z, Amler LC, Lackner MR (2008) Functional genomics identifies ABCC3 as a mediator of taxane resistance in HER2-amplified breast cancer. Cancer Res 68:5380–5389. https://doi.org/10.1158/0008-5472.CAN-08-0234

Okabe M, Szakács G, Reimers MA, Suzuki T, Hall MD, Abe T, Weinstein JN, Gottesman MM (2008) Profiling SLCO and SLC22 genes in the NCI-60 cancer cell lines to identify drug uptake transporters. Mol Cancer Ther 7:3081–3091. https://doi.org/10.1158/1535-7163.MCT-08-0539

Ou Y-Y, Chen S-A, Gromiha MM (2010) Classification of transporters using efficient radial basis function networks with position-specific scoring matrices and biochemical properties. Proteins 78:1789–1797. https://doi.org/10.1002/prot.22694

Pajeva IK, Wiese M (2002) Pharmacophore model of drugs involved in P-glycoprotein multidrug resistance: explanation of structural variety (hypothesis). J Med Chem 45:5671–5686. https://doi.org/10.1021/jm020941h

Palmeira A, Rodrigues F, Sousa E, Pinto M, Vasconcelos MH, Fernandes MX (2011) New uses for old drugs: pharmacophore-based screening for the discovery of P-glycoprotein inhibitors. Chem Biol Drug Des 78:57–72. https://doi.org/10.1111/j.1747-0285.2011.01089.x

Pan Y, Chothe PP, Swaan PW (2013) Identification of novel breast cancer resistance protein (BCRP) inhibitors by virtual screening. Mol Pharm 10:1236–1248. https://doi.org/10.1021/mp300547h

Pedersen JM, Matsson P, Bergström CAS, Norinder U, Hoogstraate J, Artursson P (2008) Prediction and identification of drug interactions with the human ATP-binding cassette transporter multidrug-resistance associated protein 2 (MRP2; ABCC2). J Med Chem 51:3275–3287. https://doi.org/10.1021/jm7015683

Penzotti JE, Lamb ML, Evensen E, Grootenhuis PDJ (2002) A computational ensemble pharmacophore model for identifying substrates of P-glycoprotein. J Med Chem 45:1737–1740. https://doi.org/10.1021/jm0255062

Pick A, Müller H, Mayer R, Haenisch B, Pajeva IK, Weigt M, Bönisch H, Müller CE, Wiese M (2011) Structure-activity relationships of flavonoids as inhibitors of breast cancer resistance protein (BCRP). Bioorg Med Chem 19:2090–2102. https://doi.org/10.1016/j.bmc.2010.12.043

Pinto M, Trauner M, Ecker GF (2012) An in silico classification model for putative ABCC2 substrates. Mol Inform 31:547–553. https://doi.org/10.1002/minf.201200049

Poongavanam V, Haider N, Ecker GF (2012) Fingerprint-based in silico models for the prediction of P-glycoprotein substrates and inhibitors. Bioorg Med Chem 20:5388–5395. https://doi.org/10.1016/j.bmc.2012.03.045

Quentin Y, Fichant G (2000) ABCdb: an ABC transporter database. J Mol Microbiol Biotechnol 2:501–504

Rais R, Acharya C, Tririya G, Mackerell AD, Polli JE (2010) Molecular switch controlling the binding of anionic bile acid conjugates to human apical sodium-dependent bile acid transporter. J Med Chem 53:4749–4760. https://doi.org/10.1021/jm1003683

Ren Q, Paulsen IT (2005) Comparative analyses of fundamental differences in membrane transport capabilities in prokaryotes and eukaryotes. PLoS Comput Biol 1:e27. https://doi.org/10.1371/journal.pcbi.0010027

Ren Q, Chen K, Paulsen IT (2007) TransportDB: a comprehensive database resource for cytoplasmic membrane transport systems and outer membrane channels. Nucleic Acids Res 35:D274–D279. https://doi.org/10.1093/nar/gkl925

Ritschel T, Hermans SMA, Schreurs M, van den Heuvel JJMW, Koenderink JB, Greupink R, Russel FGM (2014) In silico identification and in vitro validation of potential cholestatic compounds through 3D ligand-based pharmacophore modeling of BSEP inhibitors. Chem Res Toxicol 27:873–881. https://doi.org/10.1021/tx5000393

Saier MH (1998) Molecular phylogeny as a basis for the classification of transport proteins from bacteria, archaea and eukarya. Adv Microb Physiol 40:81–136. https://doi.org/10.1016/S0065-2911(08)60130-7

Saier MH (2000) A functional-phylogenetic classification system for transmembrane solute transporters. Microbiol Mol Biol Rev 64:354–411. https://doi.org/10.1128/MMBR.64.2.354-411.2000

Saier MH, Reddy VS, Tamang DG, Västermark A (2014) The transporter classification database. Nucleic Acids Res 42:D251–D258. https://doi.org/10.1093/nar/gkt1097

Schlessinger A, Khuri N, Giacomini KM, Sali A (2013) Molecular modeling and ligand docking for solute carrier (SLC) transporters. Curr Top Med Chem 13:843–856. https://doi.org/10.2174/1568026611313070007

Sedykh A, Fourches D, Duan J, Hucke O, Garneau M, Zhu H, Bonneau P, Tropsha A (2013) Human intestinal transporter database: QSAR modeling and virtual profiling of drug uptake, efflux and interactions. Pharm Res 30:996–1007. https://doi.org/10.1007/s11095-012-0935-x

Shen J, Cue Y, Gy J, Li Y, Li L (2014) A genetic algorithm- back propagation artificial neural network model to quantify the affinity of flavonoids toward P-glycoprotein. Comb Chem High Throughput Screen 17:162–172. https://doi.org/10.2174/1386207311301010002

Sun H (2005) A naive bayes classifier for prediction of multidrug resistance reversal activity on the basis of atom typing. J Med Chem 48:4031–4039. https://doi.org/10.1021/jm050180t

Szakács G, Váradi A, Ozvegy-Laczka C, Sarkadi B (2008) The role of ABC transporters in drug absorption, distribution, metabolism, excretion and toxicity (ADME-Tox). Drug Discov Today 13:379–393. https://doi.org/10.1016/j.drudis.2007.12.010

Tan W, Mei H, Chao L, Liu T, Pan X, Shu M, Yang L (2013) Combined QSAR and molecule docking studies on predicting P-glycoprotein inhibitors. J Comput Aided Mol Des 27:1067–1073. https://doi.org/10.1007/s10822-013-9697-8

Tao L, Zhang P, Qin C, Chen SY, Zhang C, Chen Z, Zhu F, Yang SY, Wei YQ, Chen YZ (2015) Recent progresses in the exploration of machine learning methods as in-silico ADME prediction tools. Adv Drug Deliv Rev 86:83–100. https://doi.org/10.1016/j.addr.2015.03.014

Thai K-M, Huynh N-T, Ngo T-D, Mai T-T, Nguyen T-H, Tran T-D (2015) Three- and four-class classification models for P-glycoprotein inhibitors using counter-propagation neural networks. SAR QSAR Environ Res 26:139–163. https://doi.org/10.1080/1062936X.2014.995701

Van de Steeg E, Venhorst J, Jansen HT, Nooijen IHG, DeGroot J, Wortelboer HM, Vlaming MLH (2015) Generation of Bayesian prediction models for OATP-mediated drug–drug interactions based on inhibition screen of OATP1B1, OATP1B1∗15 and OATP1B3. Eur J Pharm Sci 70:29–36. https://doi.org/10.1016/j.ejps.2015.01.004

Viereck M, Gaulton A, Digles D, Ecker GF (2014) Transporter taxonomy: a comparison of different transport protein classification schemes. Drug Discov Today Technol 12:e37–e46. https://doi.org/10.1016/j.ddtec.2014.03.004

Wang Y-H, Li Y, Yang S-L, Yang L (2005) Classification of substrates and inhibitors of P-glycoprotein using unsupervised machine learning approach. J Chem Inf Model 45:750–757. https://doi.org/10.1021/ci050041k

Wang Z, Chen Y, Liang H, Bender A, Glen RC, Yan A (2011) P-glycoprotein substrate models using support vector machines based on a comprehensive data set. J Chem Inf Model 51:1447–1456. https://doi.org/10.1021/ci2001583

Wang Y, Xiao J, Suzek TO, Zhang J, Wang J, Zhou Z, Han L, Karapetyan K, Dracheva S, Shoemaker BA, Bolton E, Gindulyte A, Bryant SH (2012) PubChem's BioAssay database. Nucleic Acids Res 40:D400–D412. https://doi.org/10.1093/nar/gkr1132

Warner DJ, Chen H, Cantin L-D, Kenna JG, Stahl S, Walker CL, Noeske T (2012) Mitigating the inhibition of human bile salt export pump by drugs: opportunities provided by physicochemical property modulation, in silico modeling, and structural modification. Drug Metab Dispos 40:2332–2341. https://doi.org/10.1124/dmd.112.047068

Welch MA, Köck K, Urban TJ, Brouwer KLR, Swaan PW (2015) Toward predicting drug-induced liver injury : parallel computational approaches to identify multidrug resistance protein 4 and bile salt export pump inhibitors. Drug Metab Dispos 43:725–734. https://doi.org/10.1124/dmd.114.062539

Wishart DS, Jewison T, Guo AC, Wilson M, Knox C, Liu Y, Djoumbou Y, Mandal R, Aziat F, Dong E, Bouatra S, Sinelnikov I, Arndt D, Xia J, Liu P, Yallou F, Bjorndahl T, Perez-Pineiro R, Eisner R, Allen F, Neveu V, Greiner R, Scalbert A (2013) HMDB 3.0--the human metabolome database in 2013. Nucleic Acids Res 41:D801–D807. https://doi.org/10.1093/nar/gks1065

Wu J, Li X, Cheng W, Xie Q, Liu Y, Zhao C (2009) Quantitative Structure Activity Relationship (QSAR) approach to Multiple Drug Resistance (MDR) Modulators based on combined hybrid system. QSAR Comb Sci 28:969–978. https://doi.org/10.1002/qsar.200860134

Xue Y, Yap CW, Sun LZ, Cao ZW, Wang JF, Chen YZ (2004) Prediction of P-glycoprotein substrates by a support vector machine approach. J Chem Inf Comput Sci 44:497–1505. https://doi.org/10.1021/ci049971e

Ye AY, Liu Q-R, Li C-Y, Zhao M, Qu H (2014) Human transporter database: comprehensive knowledge and discovery tools in the human transporter genes. PLoS One 9:e88883. https://doi.org/10.1371/journal.pone.0088883

You H, Lee K, Lee S, Hwang SB, Kim K-Y, Cho K-H, No KT (2015) Computational classification models for predicting the interaction of compounds with hepatic organic ion importers. Drug Metab Pharmacokinet 30(5):347–351. https://doi.org/10.1016/j.dmpk.2015.06.004

Zhang H, Xiang M-L, Zhao Y-L, Wei Y-Q, Yang S-Y (2009) Support vector machine and pharmacophore-based prediction models of multidrug-resistance protein 2 (MRP2) inhibitors. Eur J Pharm Sci 36:451–457. https://doi.org/10.1016/j.ejps.2008.11.014

Zhao M, Chen Y, Qu D, Qu H (2011) TSdb: a database of transporter substrates linking metabolic pathways and transporter systems on a genome scale via their shared substrates. Sci China Life Sci 54:60–64. https://doi.org/10.1007/s11427-010-4125-y

Zheng X, Ekins S, Raufman J-P, Polli JE (2009) Computational models for drug inhibition of the human apical sodium-dependent bile acid transporter. Mol Pharm 6:1591–1603. https://doi.org/10.1021/mp900163d

Zhong L, Ma C-Y, Zhang H, Yang L-J, Wan H-L, Xie Q-Q, Li L-L, Yang S-Y (2011) A prediction model of substrates and non-substrates of breast cancer resistance protein (BCRP) developed by GA-CG-SVM method. Comput Biol Med 41:1006–1013. https://doi.org/10.1016/j.compbiomed.2011.08.009

Module-Based Knowledge Discovery for Multiple-Cytosine-Variant Methylation Profile

10

Saurav Mallik and Ujjwal Maulik

Abstract

Methylation-based study is currently a popular ongoing research topic. The researchers generally use 5-methylcytosine (5-mC) samples for their study since this category of samples is the highest stable methylation cytosine variant, and the impact of 5-mC methylation on different diseases is known to the common people. But, through recent studies, it has been observed that other cytosine variants (e.g., 5-hmC) have also high impact on those diseases. Therefore, in this chapter, we firstly demonstrate the abovementioned different cytosine variants. In the second part of the chapter, we describe a framework of identifying co-methylated gene modules on a methylation profile having multiple cytosine variants (viz., 5-hmC and 5-mC samples). For this, at first we determine significant genes using statistical method. Thereafter, weighted topological overlap matrix (weighted TOM) measure and average linkage method are applied, consecutively on the resultant significant genes. Then dynamic tree cut method with color thresholding is utilized, and co-methylated gene modules are identified from it. The resultant gene modules are then validated biologically by KEGG pathway and gene ontology analyses. Moreover, regulatory transcription factors (TFs) and targeter miRNAs connected with the genes belonging to the different modules are found, and further biological validation has been carried out on them. Finally, other related module-based and correlation-based popular computation methodologies and applications are also shortly demonstrated.

S. Mallik (✉)
Machine Intelligence Unit, Indian Statistical Institute, Kolkata, India

Department of Computer Science and Engineering, Jadavpur University, Kolkata, India
e-mail: sauravmtech2@gmail.com; chasaurav_r@isical.ac.in

U. Maulik
Department of Computer Science and Engineering, Jadavpur University, Kolkata, India

H. J. Purohit et al. (eds.), *Soft Computing for Biological Systems*,
https://doi.org/10.1007/978-981-10-7455-4_10

Keywords
Methylation · Multiple cytosine variants · Gene modules · Module discovery · Correlation-based approaches

10.1 Introduction

Methylation-based study is a latest popular research domain. DNA methylation at fifth position of cytosine (i.e., 5-methylcytosine or 5-mC) is one of the epigenetic factors (Wu et al. 2012; Tan and Shi 2012; Mallik et al. 2013, 2014, 2015). DNA methylation generally reduces the expression level of gene, and hence it plays a key role for silencing of the gene (Tan and Shi 2012; Mallik et al. 2013, 2015; Bhadra et al. 2013). It is observed that the researchers generally use 5-methylcytosine (5-mC)-related samples for their study since this category of samples is the highest stable methylation cytosine variant, and the impact of 5-mC methylation on different diseases is known to the common people. But, through recent studies, it has been observed that other cytosine variants (e.g., 5-hmC) have also high impact on those diseases. Therefore, in this chapter, we firstly describe the relation between gene expression and methylation and then demonstrate fundamental information about different cytosine variants.

A gene regulatory network (i.e., GRN) is a group of different regulators (viz., RNAs, DNAs, proteins, and their corresponding complexes) that link with each other and with other biomolecules in the cell in order to control over the gene expression levels of genes and proteins. In system biology, it is known that the proteins and their transcripts (i.e., genes) carry out cellular processes in the background of some modules, whereas a gene module is a cluster/group of some tightly interconnected genes in the network (Hartwell et al. 1999). The gene module detection in the network is a significant task for deep understanding of the architecture of the whole network. Notably, co-expression (Bhattacharyya and Bandyopadhyay 2009) is a useful term in gene module identification. Co-expression between genes signifies that these expression profiles can together rise and drop in between a subrange of time series rather than complete time series. In a similar fashion, co-methylation states that the methylation profiles of these genes might rise and fall simultaneously in between a subrange of time series rather than the complete time series.

The second half of this chapter deals with the framework of determining co-methylated gene modules on a methylation profile having multiple cytosine variants (viz., 5-mC and 5-hmC samples) or equivalent other microarray profile having two groups, experimental/diseased and normal/control. For this regard, we firstly identify differentially 5-hmC methylated (significant) genes using statistical method. Weighted topological overlap matrix (weighted TOM) measure (Ravasz et al. 2002) is then applied on the significant genes. The corresponding dissimilarity values are calculated from the resultant TOM values. Thereafter, average linkage method is then utilized on these significant genes using the dissimilarity values, and we get corresponding dendrogram of the cluster analysis. Dynamic tree cut method (Langfelder et al. 2007) with color thresholding is then applied for identifying co-methylated gene modules. The resultant gene modules are then validated biologically through KEGG pathway and gene ontology analyses. The transcription factors (TFs), which can regulate the genes belonging to the different gene modules, are accumulated. The miRNAs, which can target the genes belonging to the different gene modules, are collected. Thereafter, the top five transcription factors that connect with the highest number of resultant modules and highest number of genes from the different modules are determined. Subsequently, the top five miRNAs, which associate with the highest number of resultant modules and highest number of genes from the different modules, are identified. Finally, these top five TFs/miRNAs are validated through literature search in order to find the diseased-related TFs/miRNAs among them.

Furthermore, other related module-based and correlation-based popular computation methodologies and applications are shortly demonstrated year by year. Finally, conclusion of this chapter is provided at the end.

10.2 Relation Between Gene Expression and DNA Methylation

DNA methylation (Wagner et al. 2014) is an epigenetic factor which occurs through the inclusion of a methyl group (i.e., CH3) to the fifth place of cytosine pyrimidine ring or sixth nitrogen place of adenine purine ring in the genomic DNA. Methylation generally decreases the regulatory function of the genes. It has been observed that when the presence of methylation is taken place near the place of transcription start site (TSS) in a gene (i.e., for the case of promoter methylation) (Kass et al. 1997), then the relationship between the gene expression and methylation becomes inverse. Otherwise (i.e., for the case of elevated gene body methylation) (Jones 1999), the relationship between them is heterogeneous. Notably, methylation of CpG dinucleotides has an essential role in the inactivation of X chromosome (Payer and Lee 2008), imprinting of genes (Li et al. 1993), and transcriptional inactivation of the foreign DNA elements, whereas aberrant DNA methylation causes many categories of cancer (Baylin et al. 1998).

Methylation is extremely variable over the cell types with different sites. It falls into two major types: (1) those which have inverse correlation in between DNA methylation and chromatin accessibility and (2) those that are with the constitutive DNA hypomethylation as well as variable chromatin accessibility (Thurman et al. 2012). Furthermore, according to Cedar and Bergman (Cedar and Bergman 2009), histone modification and DNA methylation have different associations from the starting time of the embryonic development. For example, when DNA methylation is involved in active promoters, then the hypothesized functions of the DNA methylation prevent tri-methylation of the histone 3 lysine 4 (viz., H3K4me3), whereas in other time H3K4me3 blocks DNA methylation (Hashimshony et al. 2003).

Notably, according to the literature, many integrated analyses that include gene expression and methylation data together have been performed. Mallik et al. carry out such integrative analysis for identifying gene markers that have inverse relationship between their expression level and methylation level (Mallik et al. 2013, 2015; Mallik and Maulik 2015). Thereafter, another integrative study has been performed using intrinsically disordered proteins and differentially expressed and methylated genes (Mallik et al. 2016).

10.3 Fundamental Description About Multiple Cytosine Variants

The most popular DNA methylation methodology is the inclusion of the methyl group in the fifth carbon place of the cytosine ring that results in the formation of 5-methylcytosine (viz., 5-mC). This methyl group hampers transcription. 5-mC generally takes place within CpG dinucleotide motifs, although non-CpG methylation is recognized in the embryonic stem cells (Ramsahoye et al. 2000).

Except DNA methylation, another important process is DNA demethylation (i.e., the elimination of a methyl group). The demethylation is required in order to do epigenetic reprogramming of genes. It is directly associated with characterization of several important diseases like tumor progression. Demethylation of DNA might be either active or passive or integration of the both.

The passive demethylation of DNA generally occurs on newly synthesized DNA strands through DNA (cytosine-5)-methyltransferase 1 (i.e., DNMT1) (Latham et al. 2008) at the period of replication rounds. Active DNA demethylation basically exists by the elimination of 5-methylcytosine through the sequential updating of cytosine bases which is converted by ten-eleven translocation (i.e., TET) enzyme-mediated oxidation.

TET family of 5-mC hydroxylases consists of TET methylcytosine dioxygenase 1 (viz., TET1), TET methylcytosine dioxygenase 2 (i.e., TET2), and TET methylcytosine dioxygenase 3 (viz., TET3). These three proteins might raise DNA demethylation through binding into the CpG regions for stopping the undesirable methyltransferase action of DNA and through transforming the 5-mC to the 5-hydroxymethylcytosine (5-hmC), the 5-hmC to the 5-formylcytosine (5-fC), and the 5-fC to the 5-carboxylcytosine (5-caC) via hydroxylase activity (Fig. 10.1). TET1 protein involves in transcriptional activation and repression, and TET2 protein associates with tumor suppression, whereas TET3 protein is connected with DNA methylation reprogramming procedures.

5-hydroxymethylcytosine (i.e., 5-hmC) is basically a modification of DNA methylation which is created through the enzymatic oxidation of the 5-mC made by the TET family of the iron-based dioxygenases (Tahiliani et al. 2009). 5-hmC was discovered in the T-even bacteriophage (Wyatt and Cohen (1953)). It was later found in the vertebrate brain as well as several other tissues (Kriaucionis and Heintz 2009). 5-hmC is mostly identified in some mammalian tissues like mouse Purkinje cells and granule neurons (Kriaucionis and Heintz 2009). Besides that, the 5-hmC is

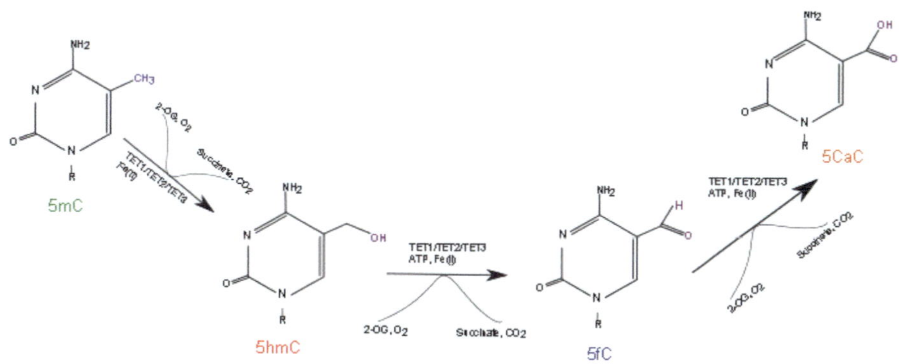

Fig. 10.1 Transformation from 5-methylcytosine (viz., 5-mC) to other cytosine variants

identified in the embryonic stem cells of mouse. By recent studies (about 2009), the presence of the 5-hmC in the mouse embryonic stem (ES) cells and mammalian brain cells is found. 5-hmC is identified in zygotes of bovines, mice, and rabbits. 5-hmC is accumulated for paternal pronucleus concurring with decreasing of 5mC. However, according to several evidences, it has been observed that 5-hmC mainly takes place within promoter regions of gene, and it is connected with the transcriptionally activated genes. 5-hmC also plays a key role in chromatin remodeling, DNA demethylation, and brain-related gene regulation. Additionally, it is especially important to determine the hydroxymethylation status in human tissues/cells with and without disease if 5-hmC could be proven to have a relation between cancer and the DNA demethylation procedure.

Through recent studies, some more cytosine variants have been discovered. 5-formylcytosine (5-fC) is one of them. It is created when TET enzymes work on 5-hmC (Ito et al. 2011; He et al. 2011). Further oxidation of the variant 5-fC by the TET enzyme will create another variant 5-carboxylcytosine (i.e., 5-caC) (Ito et al. 2011; He et al. 2011). Hence, oxidation of the 5-mC via several DNA methylation variants illustrates a procedure of the DNA demethylation of which associated pathway acts throughout the development and the programming related to the germ cell. The 5-fC is found in the major mouse organs and the mouse embryonic stem (ES) cells (Ito et al. 2011), whereas the 5-caC is identified in mouse embryonic stem (ES) cells (Ito et al. 2011). However, the 5-caC is extracted from the genomic DNA through the help of the thymine DNA glycosylase (TDG), which transforms that cytosine variant into its (previous) unchanged state.

Besides that, 3-methylcytosine (3-mC) is the other methylation variant, but it is not involved in the oxidative pathway of TET family proteins. 3-mC is produced through the automatic exposure of nitrogen-three base of the cytosine to the endogenous S-adenosyl methionine. The 3-mC is mutative. It has been recreated either via dealkylation through human homologues of *E. coli* AlkB protein or via base excision repair in humans. If cells lose ALKBH3, level of 3-mC increases as well as cell proliferation decreases (Dango et al. 2011).

The major biological significance of 5-mC is recognized as major epigenetic change in gene expression as well as phenotype. For example, DNA hypomethylation happened by deficiency of methyl due to various environmental impacts. Thus, it can be stated as a molecular marker in various biological processes like cancer. The measurement on 5-mC content (or global methylation) in diseased cells might provide significant clue for detection of the corresponding disease. Additionally, the identification of 5-fC in different cells/tissues might be utilized as a marker for specifying active DNA demethylation.

10.4 Module-Based Knowledge Extraction for Methylation Data

In this section, we demonstrate the method of detecting significant gene modules from the gene regulatory network using a well-known connectivity measure (viz., weighted topological overlap matrix) for a multiple-cytosine-variant methylation dataset. The step-by-step description is provided in the following.

10.4.1 Identification of Statistically Significant Genes

First of all, remember that there might be multiple gene probes for several genes in methylation dataset. In that case, there are several methods to eliminate redundant probes of each gene and keep only a single probe for each gene. For this, at first the median absolute deviation (MAD) of the raw methylation values over all samples for each probe of each gene is computed. Only the probe of the gene of which MAD value is maximum is chosen. The remaining probes of the gene are discarded from the methylation dataset. This process is repeated for the other genes belonging to the methylation dataset. Now, each gene of the dataset consists of only a single probe.

The methylation data is thereafter normalized gene-wise since normalization transforms the data from various scales into a common scale. For this purpose, several normalization techniques are available, viz., min-max, zero-mean, median, sigmoid, statistical column normalization, etc. (Bolstad et al. 2003; Bandyopadhyay et al. 2013). Now, after normalization, normality test is required to apply on each gene-wise data for determining whether the data follows a normal (Gaussian) distribution or not for each population/group. The well-known normality tests are Jarque-Bera test (Thadewald and Buning 2007), Lilliefors test (Razali and Wah 2011), Anderson-Darling test (Razali and Wah 2011), and Shapiro-Wilk test (Razali and Wah 2011).

After normality test, a parametric test can be applied on the data of the gene that follows normal distribution. Related popular parametric tests are two-sample t-test (Sreekumar and Jose 2008; Bandyopadhyay et al. 2013), ANOVA-1 (Sreekumar and Jose 2008; Bandyopadhyay et al. 2013), etc. Similarly, a nonparametric test might be utilized on the data of the gene which does not follow normal distribution.

Popular nonparametric tests for this regard are SAM (significance analysis of microarray) (Tusher et al. 2001; Bandyopadhyay et al. 2013), Limma (linear models for microarray data) (Smyth 2004; Bandyopadhyay et al. 2013), permuted t-test (Anderson 2001; Bandyopadhyay et al. 2013), etc. Notably, for using any statistical test, one cytosine variant (e.g., 5-hmC samples) is utilized as experimental samples, whereas another cytosine variant (e.g., 5-mC samples) is considered as control samples for the corresponding multiple-cytosine-variant methylation datasct.

For each gene of the dataset, individual statistical test is applied. Each statistical test provides a t-value from which corresponding p-value is computed using cumulative distribution function (cdf). The genes of which p-values are less than 0.05 are called as statistically significant (i.e., differentially 5-hmC methylated) genes. The significant (nonredundant) genes are then ranked with respect to their p-values in ascending order.

10.4.2 Computing Co-methylation and Identifying Gene Modules

For measuring the correlation in terms of methylation values, Pearson's correlation score is utilized between pairwise genes belonging to the set of significant (nonredundant) genes. Thereafter, a well-known connectivity measure, namely, weighted topological overlap matrix (viz., weighted TOM) (Ravasz et al. 2002), is applied on the above pairwise genes where above calculated Pearson's correlation scores are considered here as the weighted values for the corresponding adjacency matrix. The corresponding dissimilarity value for each gene is then calculated for each resultant weighted TOM value (similarity value). Average linkage clustering technique is thereafter applied on the genes using the resultant dissimilarity scores. The corresponding dendrogram of the resultant clusters is recognized. Dynamic tree cut (Langfelder et al. 2007) is then performed on the dendrogram through color thresholding methodology, and as a result, co-methylated gene modules are detected.

10.4.3 Trimolecular Network Analysis and Biological Validation

After finding co-methylated gene modules, KEGG pathway and gene ontology analyses should be carried out individually on the genes belonging to each resulting module in order to validate the modules. For KEGG pathway and gene ontology (GO) analyses, several online databases are available such as DAVID (Huang et al. 2008), Enrichr (Chen et al. 2013), GSEA (Gene Set Enrichment Analysis) (Subramanian et al. 2005), etc. The pathways or the GO terms of which corresponding p-values are less than the traditional cutoff 0.05 are considered as statistically significant (enriched) KEGG pathways or GO terms. The significant pathways and GO terms are identified for each module. The modules can be ranked according to maximum number of participating genes belonging to each module.

Besides that, trimolecular network can be formed. For this regard, transcription factors (TFs) which regulate the genes belonging to the each module are determined. For accumulating TFs, there are available several online databases like TRANSFAC (Wingender et al. 1996), ITFP (Zheng et al. 2008), JASPAR (Sandelin et al. 2004), etc. Similarly, miRNAs that target the genes belonging to each module are found. For collecting miRNAs, some online databases are available such as miRWalk (Dweep et al. 2011), miRTarBase (Chou et al. 2016), PITA (Kertesz et al. 2007), DIANA-microT (Maragkakis et al. 2011), miRanda (John et al. 2004), RNAhybrid (Kruger and Rehmsmeier 2006), PicTar (Krek et al. 2005), TargetScan (Lewis et al. 2003), miRNA_Targets (Kumar et al. 2012), etc.

Thereafter, a TF-miRNA-gene network can be built, and further network analysis for each participating biomolecule (e.g., gene, miRNA, and TF) can be performed. In the network analysis, various topological measures can be utilized in different prospects. These measures are degree centrality (Freeman 1977; Ozgur et al. 2008), betweenness centrality (Freeman 1977; Ozgur et al. 2008), k-coreness centrality (Batagelj and Zavernik 2011), closeness centrality (Freeman 1979; Ozgur et al. 2008), subgraph centrality (Estrada and Rodrguez-Velzquez 2005), clustering coefficient-based centrality (Barrat and Weigt 2000; Newman 2003), weighted clustering coefficient-based centrality (Barrat et al. 2004), eigenvector centrality (Bonacich and Lloyd 2001; Mallik and Maulik 2015), etc.

In addition, the top five regulator TFs, which are connected to the highest number of gene modules and highest number of participating genes from the different modules, are determined. Similarly, top five targeter miRNAs, which are linked with the highest number of gene modules and highest number of genes from the different modules, are identified. Finally, the top resulting TFs/miRNAs are validated through literature search for understanding how many of these TFs/miRNAs have a strong connection with the corresponding disease. Notably, the whole framework of gene module-based knowledge discovery is provided in Fig. 10.2.

The above-described methodology is not only valid for multiple-cytosine-variant methylation dataset, but it is also useful for any microarray dataset like gene expression dataset having two populations/groups (viz., one group consisting of only experimental/diseased samples and another group having only control/normal samples).

10.5 Other Module Discovery and Correlation-Based Approaches

According to the literature, there are different other related methodologies in different perspectives except the above-described integrative method. Among these, some are computational methodologies, whereas others are application-based studies.

In the first category (i.e., in the case of computational methodologies), several techniques are already developed. For example, Shen et al. have proposed a new method, namely, "ELDA" (eigengene-based linear discriminant analysis), in which

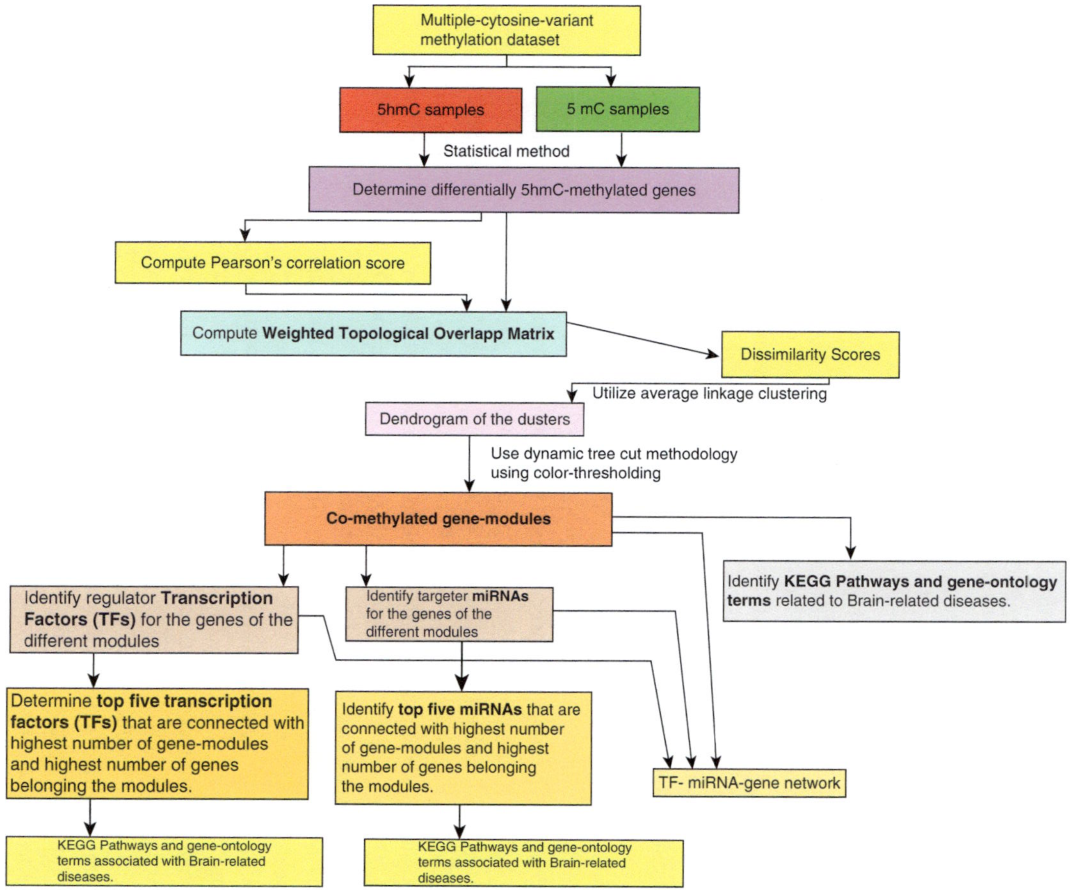

Fig. 10.2 Flowchart of framework of module-based knowledge discovery for multiple-cytosine-variant methylation dataset

a modified rotated spectral decomposition (SpD) approach is applied in order to determine the hub genes that are connected to the most important eigenvectors (Shen et al. 2006). In the year 2007, Langfelder and Horvath have developed an eigengene network-based computational work that signifies the associations between the co-expression modules (Langfelder and Horvath 2007). Thereafter, in 2008, Langfelder and Horvath made a new popular tool "WCGNA" for weighted correlation network analysis that is applicable for finding co-expression for gene expression data as well as retrieving co-methylation from methylation data (Langfelder and Horvath 2008). Bandyopadhyay and Bhattacharyya have proposed a novel similarity measure, namely, "BioSim," in which the resulting correlation values lie in between "−1" and "+1" (Bandyopadhyay and Bhattacharyya 2011) for gene expression data. Here, "−1," "0," and "+1" stand for negative correlation, independency, and positive correlation, respectively. Thereafter, Gevaert et al. have produced a new algorithm, namely, "AMARETTO," that is prepared for determining cancer driver genes as well as gene modules through integrating different omics data across cancer and normal tissues/samples (Gevaert et al. 2013). In the same year (i.e., 2013), two new related techniques are proposed. One of these is developed by Saas et al. Here, a novel methodology is introduced for analyzing combined dataset across the multi-omics levels in order to assess their biological prospects simultaneously. A model-dependent Bayesian approach is here included in order to infer interpretable term probabilities in a module-based framework (Sass et al. 2013). Second technique is developed by Bhattacharyya et al. in which a novel integrative measure is proposed through integrating several co-expression networks that might be important for determining the dependency between co-expression and functional similarity (Bhattacharyya et al. 2013). In 2015, an integrative network-oriented methodology is developed through combining information from miRNA expression, gene expression, DNA methylation, and somatic mutation data (Hamed et al. 2015). Table 10.1 provides a list of the above-elaborated module/correlation-based computational methodologies year by year.

In the second category, some important module and correlation-based applications/studies are included. For example, Liu et al. provide a correlation-based approach between pairwise genes belonging to the same (validated) yeast protein complex (Liu et al. 2009). Van Eijk et al. perform a study (1) to create link between methylation and expression levels, (2) to find relation between co-methylation modules and co-expression modules, and (3) to analyze the association between the genetic markers, expression, and methylation (Van Eijk et al. 2012). Bhattacharyya carries out a new analysis on co-expression toggling of the miRNAs in the brain of Alzheimer patients (Bhattacharyya and Bandyopadhyay 2009). The hypothesis regarding this is that the differential co-expression might recognize the changing patterns in different phenotypes (Bhattacharyya 2012a, b). Again in 2013, Bhattacharyya and Bandyopadhyay have proposed a new computational study indicating the crucial role of the white matter (WM) in the early stage of Alzheimer's disease progression (Bhattacharyya and Bandyopadhyay 2013). Another related work regarding miRNA, methylation and Alzheimer's disease is performed by Roy and Bhattacharyya (Roy and Bhattacharyya 2016). In 2014, Aqil et al. perform first

Table 10.1 Some important module-based and correlation-based different methodologies for biological datasets

Name of algorithms or title of the works	Short description regarding the work	References
ELDA	An eigengene-based linear discriminant analysis (ELDA) for the gene identification in the basis of a multivariate framework. Here, a modified rotated spectral decomposition (SpD) method is utilized for determining the hub genes which is related to the most important eigenvectors	Shen et al. (2006)
"Eigengene networks for studying the relationships between co-expression modules"	A new computational (eigengene network-based) methodology signifying the associations between the resultant co-expression modules	Langfelder and Horvath (2007)
WGCNA	A novel tool for the weighted correlation network-based analysis (i.e., co-expression/co-methylation network-based analysis of the gene expression/methylation dataset	Langfelder and Horvath (2008)
BioSim	A novel similarity measure assuming values in between −1 and +1 as negative and positive dependencies, respectively, whereas 0 for independency	Bandyopadhyay and Bhattacharyya (2011)
AMARETTO	An algorithm for determining cancer driver genes as well as gene modules through integrating different omics data across cancer and normal tissues/samples	Gevaert et al. (2013)
"A modular framework for gene set analysis integrating multilevel omics data"	A novel methodology for analyzing combined dataset across multi-omics levels in order to assess their biological prospects simultaneously. Here, a model-based Bayesian technique is included in order to infer interpretable term probabilities in a module-based framework	Sass et al. (2013)
"A new approach for combining knowledge from multiple coexpression networks of microRNAs"	A novel integrative measure through integrating several co-expression networks which might be important for determining the dependency between co-expression and functional similarity	Bhattacharyya et al. (2013)

(continued)

Table 10.1 (continued)

Name of algorithms or title of the works	Short description regarding the work	References
"Identifying epigenetic biomarkers using maximal relevance and minimal redundancy-based feature selection for multi-omics data"	A novel framework of epigenetic marker discovery through integral study of mutual information-based feature selection, data distribution, and statistical hypothesis test for multi-omics data	Mallik et al. (2017)
"Integrating multiple data sources for combinatorial marker discovery: A study in tumorigenesis"	A novel method of identifying combinatorial markers depending upon the intrinsic relationship between the expression and methylation for the multi-omics data	Bandyopadhyay and Mallik (2016)
"Integrative network-based approach identifies key genetic elements in breast invasive carcinoma"	An integrative network-oriented methodology through combining information from the miRNA expression, DNA methylation, gene expression, and somatic mutation datasets	Hamed et al. (2015)

miRnome study between U937 cells expressing HIV-1 Nef-EYFP samples (i.e., experimental samples) and U937 cells expressing EYFP samples (i.e., control samples) for novel wet laboratory-made cellular and exosomal miRNA datasets (Aqil et al. 2014). From the miRnome analysis, several miRNAs are determined which are selectively secreted into Nef exosomes. Some miRNAs are also recognized that are retained selectively in the Nef-expressing cells. Aqil et al. carry out a new transcriptomic analysis as well as network analysis of the human monocytic cells that express HIV-1 Nef protein and the corresponding exosomes in 2015 (Aqil et al. 2015). Here, some mRNAs that are retained in the Nef-expressing cells, but whose targeting miRNAs are exported out in the exosomes, are identified. Additionally, some mRNAs that are preferentially secreted in exosomes, but whose targeting miRNAs are retained in Nef-expressing monocytes, are found. Table 10.2 provides the list of the above-elaborated module/correlation-based applications year by year.

10.6 Conclusion

Nowadays, the study of methylation is highly appreciable in disease discovery or characterization. Generally, impact of 5-mC methylation on different diseases is known to the common people. But, through recent studies, it has been observed that other less stable cytosine variants (e.g., 5-hmC) have also high impact on those diseases. Therefore, in this chapter, multiple cytosine variants (5-mC, 5-hmC, etc.) are firstly described. Then the framework of module-based knowledge discovery for the multiple-cytosine-variant-based methylation dataset or other equivalent

Table 10.2 Some important module-based and correlation-based several applications/studies for biological datasets

Name of algorithms or title of the articles	Short description regarding the work	References
"Patterns of co-expression for protein complexes by size in *Saccharomyces cerevisiae*"	Computation correlation-based approach between pairwise genes belonging to the same (validated) yeast protein complex	Liu et al. (2009)
"Genetic analysis of DNA methylation and gene expression levels in whole blood of healthy human subjects"	A study (1) to create link between expression and methylation levels, (2) to find relation between co-expression modules and co-methylation modules, and (3) to analyze the association between genetic markers, expression, and methylation	Van Eijk et al. (2012)
"Co-expression toggling of microRNAs in Alzheimer's brain"	Differential co-expression might recognize the changing patterns in different phenotypes	Bhattacharyya (2012a, b)
"Studying the differential co-expression of microRNAs reveals significant role of white matter in early Alzheimer's progression"	A new computational study indicating the crucial role of white matter in early stage of Alzheimer's disease progression	Bhattacharyya and Bandyopadhyay (2013)
"The HIV Nef protein modulates cellular and exosomal miRNA profiles in human monocytic cells"	First-time miRnome study in both HIV Nef-expressing monocytes and their exosomes. Some miRNAs are determined which are selectively secreted into Nef exosomes. Also some miRNAs are recognized that are selectively retained in Nef-expressing cells	Aqil et al. (2014)
"Transcriptomic analysis of mRNAs in human monocytic cells expressing the HIV-1 Nef protein and their exosomes"	A new transcriptomic as well as network analyses of human monocytic cells expressing the HIV-1 Nef protein and the corresponding exosomes. Some mRNAs that are retained in Nef-expressing cells, but whose targeting miRNAs are exported out in the exosomes, are identified. Additionally, some mRNAs that are preferentially secreted in exosomes, but whose targeting miRNAs are retained in Nef-expressing monocytes, are found	Aqil et al. (2015)

microarray dataset having two populations (one for 5-hmC samples and other for 5-mC samples) is demonstrated. From the above method, some co-methylated gene modules can be obtained which are validated by KEGG pathway and gene ontology analyses. Additionally, transcription factors (TFs), which can regulate the genes belonging to the different gene modules, are accumulated, whereas the miRNAs that can target the genes of the modules are identified. Thereafter, TF-miRNA-gene network has been formed, and further network analyses have been performed. At the end, the other related works are also described as much as possible.

10.7 Opinion

Methylation-based study is a current popular ongoing research topic. The effect of 5-methylcytosine (5-mC) methylation is already known to all researchers. Currently, it is observed that other less stable cytosine variants (e.g., 5-hmC) have also high impact on different diseases especially brain-related diseases. Thus, in this chapter, a framework of the module-based knowledge discovery for the multiple-cytosine-variant-based methylation dataset having two populations (one for 5-hmC samples and other for 5-mC samples) is described. We expect that this framework will enlighten a new direction in disease discovery.

Acknowledgments The authors wish to thank Prof. Sanghamitra Bandyopadhyay, Director, Indian Statistical Institute, Kolkata, India, and Dr. Hemant J. Purohit, Chief Scientist and Head, Environmental Genomics Division, National Environmental Engineering, Research Institute, NEERI, CSIR, Nagpur, India.

References

Anderson M (2001) Permutation tests for univariate or multivariate analysis of variance and regression. Can J Fish Aquat Sci 58:626–639

Aqil M, Naqvi AR, Mallik S, Bandyopadhyay S, Maulik U, Jameel S (2014) The HIV Nef protein modulates cellular and exosomal miRNA profiles in human monocytic cells. J Extracell Vesicles 3:1–11. https://doi.org/10.3402/jev.v3.23129

Aqil M, Mallik S, Bandyopadhyay S, Maulik U, Jameel S (2015) Transcriptomic analysis of mRNAs in human Monocytic cells expressing the HIV-1 Nef protein and their exosomes. Biomed Res Int 2015(492395):1–10. https://doi.org/10.1155/2015/492395

Bandyopadhyay S, Bhattacharyya M (2011) A biologically inspired measure for coexpression analysis. IEEE/ACM Trans Comput Biol Bioinform 8:929–942. https://doi.org/10.1109/TCBB.2010.106

Bandyopadhyay S, Mallik S (2016) Integrating multiple data sources for combinatorial marker discovery: a study in tumorigenesis. IEEE/ACM Trans Comput Biol Bioinform. https://doi.org/10.1109/TCBB.2016.2636207

Bandyopadhyay S, Mallik S, Mukhopadhyay A (2013) A survey and comparative study of statistical tests for identifying differential expression from microarray data. IEEE/ACM Trans Comput Biol Bioinform 11:95–115. https://doi.org/10.1109/TCBB.2013.147

Barrat A, Weigt M (2000) On the properties of small world networks. Eur Phys J B 13:547–560

Barrat A, Barthelemy M, Pastor-Satorras R, Vespignani A (2004) The architecture of complex weighted networks. PNAS 101:3747–3752

Batagelj V, Zavernik M (2011) Fast algorithms for determining (generalized) core groups in social networks. Adv Data Anal Classif 5:129–145

Baylin SB, Herman JG, Graff JR, Vertino PM, Issa JP (1998) Alterations in DNA methylation: a fundamental aspect of neoplasia. Adv Cancer Res 72:141–196

Bhadra T, Bhattacharyya M, Feuerbach L, Lengauer T, Bandyopadhyay S (2013) DNA methylation patterns facilitate the identification of microRNA transcription start sites: a brain-specific study. PLoS One 8:1–7. https://doi.org/10.1371/annotation/dd8f4acc-3859-46e2-9136-20b6b4d08d21

Bhattacharyya M (2012a) Mining co-expression graphs: applications to microRNA regulation and disease analysis. Nat Precedings. https://doi.org/10.1038/npre.2012.7119.1

Bhattacharyya M (2012b) Co-expression toggling of microRNAs in Alzheimer's brain. Nat Precedings. https://doi.org/10.1038/npre.2012.7123.1

Bhattacharyya M, Bandyopadhyay S (2009) Integration of co-expression networks for gene clustering. Seventh international conference on advances in pattern recognition, pp 355–358. doi: https://doi.org/10.1109/ICAPR.2009.55

Bhattacharyya M, Bandyopadhyay S (2013) Studying the differential co-expression of microRNAs reveals significant role of white matter in early Alzheimer's progression. Mol BioSyst 9:457–466. https://doi.org/10.1039/C2MB25434D

Bhattacharyya M, Das M, Bandyopadhyay S (2013) A new approach for combining knowledge from multiple Coexpression networks of microRNAs. IEEE Trans Biomed 60:2167–2173. https://doi.org/10.1109/TBME.2013.2250285

Bolstad BM, Irizarry RA, Astrand M, Speed T (2003) A comparison of normalization methods for high density oligonucleotide array data based on variance and bias. Bioinformatics 19:185–193

Bonacich P, Lloyd P (2001) Eigenvector-like measures of centrality for asymmetric relations. Soc Networks 23:191–201

Cedar H, Bergman Y (2009) Linking DNA methylation and histone modification: patterns and paradigms. Nat Rev Genet 10:295–304. https://doi.org/10.1038/nrg2540

Chen EY, Tan CM, Kou Y, Duan Q, Wang Z, Meirelles GV, Clark NR, Ma'ayan A (2013) Enrichr: interactive and collaborative HTML5 gene list enrichment analysis tool. BMC Bioinf 14:128. https://doi.org/10.1186/1471-2105-14-128

Chou CH, Chang NW, Shrestha S, Hsu SD, Lin YL, Lee WH, Yang CD, Hong HC, Wei TY, SJ T, Tsai TR, Ho SY, Jian TY, HY W, Chen PR, Lin NC, Huang HT, Yang TL, Pai CY, Tai CS, Chen WL, Huang CY, Liu CC, Weng SL, Liao KW, Hsu WL, Huang HD (2016) miRTarBase 2016: updates to the experimentally validated miRNA-target interactions database. Nucleic Acids Res 44:D239–D247. https://doi.org/10.1093/nar/gkv1258

Dango S et al (2011) DNA unwinding by ASCC3 helicase is coupled to ALKBH3 dependent DNA alkylation repair and cancer cell proliferation. Mol Cell 44:373–384. https://doi.org/10.1016/j.molcel.2011.08.039

Dweep H, Sticht C, Pandey P, Gretz N (2011) miRWalk--database: prediction of possible miRNA binding sites by "walking" the genes of three genomes. J Biomed Inform 44:839–847. https://doi.org/10.1016/j.jbi.2011.05.002

Estrada E, Rodrguez-Velzquez JA (2005) Subgraph centrality in complex networks. Phys Rev E 71:1–9

Freeman LC (1977) A set of measures of centrality based on betweenness. Sociometry 577:35–41

Freeman LC (1979) Centrality in social networks: conceptual clarification. Sociometry 1:215–239

Gevaert O, Villalobos V, Sikic BI, Plevritis SK (2013) Identification of ovarian cancer driver genes by using module network integration of multi-omics data. Interface Focus 3 (4):20130013. https://doi.org/10.1098/rsfs.2013.0013

Hamed M, Spaniol C, Zapp A, Helms V (2015) Integrative network-based approach identifies key genetic elements in breast invasive carcinoma. BMC Genomics 16:S2. https://doi.org/10.1186/1471-2164-16-S5-S2

Hartwell LH, Hopfield JJ, Leibler S, Murray AW (1999) From molecular to modular cell biology. Nature 402:C47–C52
Hashimshony T, Zhang JM, Keshet I, Bustin M, Cedar H (2003) The role of DNA methylation in setting up chromatin structure during development. Nat Genet 34:187–192. https://doi.org/10.1038/ng1158
He YF et al (2011) Tet-mediated formation of 5-carboxylcytosine and its excision by TDG in mammalian DNA. Science 333:1303–1307. https://doi.org/10.1126/science.1210944
Huang DW, Sherman BT, Lempicki RA (2008) Systematic and integrative analysis of large gene lists using DAVID bioinformatics resources. Nat Protoc 4:44–57. https://doi.org/10.1038/nprot.2008.211
Ito S et al (2011) Tet proteins can convert 5-methylcytosine to 5-formylcytosine and 5-carboxylcytosine. Science 333(6047):1300–1303. https://doi.org/10.1126/science.1210597
John B, Enright AJ, Aravin A, Tuschl T, Sander C, Marks DS (2004) Human MicroRNA targets. PLoS Biol 2:1862–1879
Jones PA (1999) The DNA methylation paradox. Trends Genet 15:34–37. https://doi.org/10.1016/S0168-9525(98)01636-9
Kass SU, Landsberger N, Wolffe AP (1997) DNA methylation directs a time-dependent repression of transcription initiation. Curr Biol 7:157–165. https://doi.org/10.1016/S0960-9822(97)70086-1
Kertesz M, Iovino N, Unnerstall U, Gaul U, Segal E (2007) The role of site accessibility in microRNA target recognition. Nat Genet 39:1278–1284
Krek A, Grun D, Poy MN, Wolf R, Rosenberg L, Epstein EJ, MacMenamin P, Piedade ID, Gunsalus KC, Stoffel M, Rajewsky N (2005) Combinatorial microRNA target predictions. Nat Genet 37:495–500
Kriaucionis S, Heintz N (2009) The nuclear DNA base 5-hydroxymethylcytosine is present in Purkinje neurons and the brain. Science 324(5929):929–930. https://doi.org/10.1126/science.1169786
Kruger J, Rehmsmeier M (2006) RNAhybrid: microRNA target prediction easy, fast and flexible. Nucleic Acids Res 34:W451–W454
Kumar A, Wong AKL, Tizarda ML, Moorea RJ, Lefèvreb C (2012) miRNA_Targets: a database for miRNA target predictions in coding and non-coding regions of mRNAs. Genomics 100:352–356. https://doi.org/10.1016/j.ygeno.2012.08.006
Langfelder P, Horvath S (2007) Eigengene networks for studying the relationships between co-expression modules. BMC Syst Biol 1(54):1–17. http://www.biomedcentral.com/1752-0509/1/54
Langfelder P, Horvath S (2008) WGCNA: an R package for weighted correlation network analysis. BMC Bioinf 9:559. https://doi.org/10.1186/1471-2105-9-559
Langfelder P, Zhang B, Horvath S (2007) Defining clusters from a hierarchical cluster tree: the dynamic tree cut package for R. Bioinformatics 24(5):719–720
Latham T, Gilbert N, Ramsahoye B (2008) DNA methylation in mouse embryonic stem cells and development. Cell Tissue Res 331:31–55
Lewis BP, Shih IH, Jones-Rhoades MW, Bartel DP, Burge CB (2003) Prediction of mammalian microRNA targets. Cell 115:787–798
Li E, Beard C, Jaenisch R (1993) Role for DNA methylation in genomic imprinting. Nature 366:362–365. https://doi.org/10.1038/366362a0
Liu CT, Yuan S, Li KC (2009) Patterns of co-expression for protein complexes by size in Saccharomyces cerevisiae. Nucleic Acids Res 37:526–532. https://doi.org/10.1093/nar/gkn972
Mallik S, Maulik U (2015) MiRNA-TF-gene network analysis through ranking of biomolecules for multi-informative uterine leiomyoma dataset. J Biomed Inform 57:308–319. https://doi.org/10.1016/j.jbi.2015.08.014
Mallik S, Mukhopadhyay A, Maulik U, Bandyopadhyay S (2013) Integrated analysis of gene expression and genome-wide DNA methylation for tumor prediction: an association rule mining-based approach. Proc IEEE symposium on Computational Intelligence in

Bioinformatics and Computational Biology (CIBCB), IEEE Symposium Series on Computational Intelligence – SSCI, Singapore, pp 120–127. doi:https://doi.org/10.1109/CIBCB.2013.6595397

Mallik S, Mukhopadhyay A, Maulik U (2014) Integrated statistical and rule- mining techniques for DNA methylation and gene expression data analysis. JAISCR 3:101–115. https://doi.org/10.2478/jaiscr-2014-0008

Mallik S, Mukhopadhyay A, Maulik U (2015) RANWAR: rank-based weighted association rule mining from gene expression and methylation data. IEEE Trans Nanobiosci 14:59–66. https://doi.org/10.1109/TNB.2014.2359494

Mallik S, Sen S, Maulik U (2016) IDPT: insights into potential intrinsically disordered proteins through transcriptomic analysis of genes for prostate carcinoma epigenetic data. Gene 586 (2016):87–96. https://doi.org/10.1016/j.gene.2016.03.056

Mallik S, Bhadra T, Maulik U (2017) Identifying epigenetic biomarkers using maximal relevance and minimal redundancy based feature selection for multi-omics data. IEEE Trans Nanobiosci. https://doi.org/10.1109/TNB.2017.2650217

Maragkakis M, Vergoulis T, Alexiou P, Reczko M, Plomaritou K, Gousis M, Kourtis K, Koziris N, Dalamagas T, Hatzigeorgiou AG (2011) DIANA-microT Web server upgrade supports Fly and Worm miRNA target prediction and bibliographic miRNA to disease association. Nucleic Acids Res 39:W145–W148

Maulik U, Mallik S, Mukhopadhyay A, Bandyopadhyay S (2015) Analyzing gene expression and methylation data profiles using StatBicRM: statistical biclustering-based rule mining. PLoS One 10(4):e0119448. https://doi.org/10.1371/journal.pone.0119448

Newman MEJ (2003) The structure and function of complex networks. SIAM Rev 45(2):167–256

Ozgur A, Vu T, Erkan G, Radev DR (2008) Identifying gene-disease associations using centrality on a literature mined gene-interaction network. Bioinformatics 24:i277–i285. https://doi.org/10.1093/bioinformatics/btn182

Payer B, Lee JT (2008) X chromosome dosage compensation: how mammals keep the balance. Annu Rev Genet 42:733–772. https://doi.org/10.1146/annurev.genet.42.110807.091711

Ramsahoye B et al (2000) Non-CpG methylation is prevalent in embryonic stem cells and may be mediated by DNA methyltransferase 3a. PNAS 97:5237–5242

Ravasz E, Somera AL, Mongru DA, Oltvai ZN, Barabasi AL (2002) Hierarchical organization of modularity in metabolic networks. Science 297:1551–1555

Razali N, Wah Y (2011) Power comparisons of Shapiro-Wilk, Kolmogorov-Smirnov, Lilliefors and Anderson-Darling tests. J Stat Model Anal 2:21–33

Roy A, Bhattacharyya M (2016) Identifying microRNAs related to Alzheimer's disease from differential methylation signatures. Gene Rep 4:104–111. https://doi.org/10.1016/j.genrep.2016.04.006

Sandelin A, Alkema W, Engström P, Wasserman WW, Lenhard B (2004) JASPAR: an open-access database for eukaryotic transcription factor binding profiles. Nucleic Acids Res 32: D91–D94. https://doi.org/10.1093/nar/gkh012

Sass S, Buettner F, Mueller NS, Theis FJ (2013) A modular framework for gene set analysis integrating multilevel omics data. Nucleic Acids Res 41:9622–9633. https://doi.org/10.1093/nar/gkt752

Shen R, Ghosh D, Chinnaiyan A, Meng Z (2006) Eigengene-based linear discriminant model for tumor classification using gene expression microarray data. Bioinformatics 22:2635–2642. https://doi.org/10.1093/bioinformatics/btl442

Smyth G (2004) Linear models and empirical Bayes methods for assessing differential expression in microarray experiments. Stat Appl Genet Mol Biol 3:Article3.

Sreekumar J, Jose KK (2008) Statistical tests for identification of differentially expressed genes in cDNA microarray experiments. Indian J Biotechnol 7:423–436

Subramanian A, Tamayo P, Mootha VK, Mukherjee S, Ebert BL, Gillette MA, Paulovich A, Pomeroyh SL, Golub TR, Lander ES, Mesirov JP (2005) Gene set enrichment analysis: a knowledge-based approach for interpreting genome-wide expression profiles. BMC Bioinf 102:15545–15550. https://doi.org/10.1073/pnas.0506580102

Tahiliani M et al (2009) Conversion of 5-methylcytosine to 5-hydroxymethylcytosine in mammalian DNA by MLL partner TET1. Science 324:930–935. https://doi.org/10.1126/science.1170116

Tan L, Shi YG (2012) Tet family proteins and 5-hydroxymethylcytosine in development and disease. Development 139:1895–1902. https://doi.org/10.1242/dev.070771

Thadewald T, Buning H (2007) Jarque-Bera test and its competitors for testing normality. J Appl Stat 34:87–105

Thurman RE, Rynes E, Humbert R, Vierstra J, Maurano MT, Haugen E, Sheffield NC, Stergachis AB, Wang H, Vernot B, Garg K, John S, Sandstrom R, Bates D, Boatman L, Canfield TK, Diegel M, Dunn D, Ebersol AK, Frum T, Giste E, Johnson AK, Johnson EM, Kutyavin T, Lajoie B, Lee BK, Lee K, London D, Lotakis D, Neph S et al (2012) The accessible chromatin landscape of the human genome. Nature 489:75–82. https://doi.org/10.1038/nature11232

Tusher VG, Tibshirani R, Chu G (2001) Significance analysis of microarrays applied to the ionizing radiation response. Proc Natl Acad Sci USA 98:5116–5121

Van Eijk KR, de Jong S, Boks MP et al (2012) Genetic analysis of DNA methylation and gene expression levels in whole blood of healthy human subjects. BMC Genomics 13:636. https://doi.org/10.1186/1471-2164-13-636

Wagner JR, Busche S, Ge B, Kwan T, Pastinen T, Blanchette M (2014) The relationship between DNA methylation, genetic and expression inter-individual variation in untransformed human fibroblasts. Genome Biol 15:R37. https://doi.org/10.1186/gb-2014-15-2-r37

Wingender E, Dietze P, Karas H, Knuppel R (1996) TRANSFAC: a database on transcription factors and their DNA binding sites. Nucleic Acids Res 24:238–241. https://doi.org/10.1093/nar/24.1.238

Wu H, Tao J, Sun YE (2012) Regulation and function of mammalian DNA methylation patterns: a genomic perspective. Brief Funct Genomics 11:240–250

Wyatt GR, Cohen SS (1953) The bases of the nucleic acids of some bacterial and animal viruses: the occurrence of 5-hydroxymethylcytosine. Biochem J 55(5):774–782. PMID: 13115372 PMCID: PMC1269533.

Zheng G, Tu K, Yang Q, Xiong Y, Wei C, Xie L, Zhu Y, Li Y (2008) ITFP: an integrated platform of mammalian transcription factors. Bioinformatics 24:2416–2417. https://doi.org/10.1093/bioinformatics/btn439

Outlook of Various Soft Computing Data Preprocessing Techniques to Study the Pest Population Dynamics in Integrated Pest Management

11

M. Pratheepa and J. Cruz Antony

Abstract

Agriculture is the backbone of Indian economy. The crop loss has been estimated around US$ 36 billion in India in post-green revolution era. The rationale for crop loss is due to damage from pests, diseases and weeds. Big data have been generated every day in agriculture especially in integrated pest management. Pest control strategy is a serious concern in crop production. Robust model in pest prediction is needed to take up the pest control measures beforehand to avoid yield loss. Data preprocessing is an important aspect to derive reliable results for decision-making process in integrated pest management. Pest population dynamics in different crop ecosystem is due to biotic and abiotic factors, and they have innate adaptive capacity with the environment. There is a gap in selection of suitable data preprocessing techniques in agricultural domain. Several soft computing techniques are available, but the usage of these techniques in agricultural field is at minimum level. Hence, the panorama of various soft computing data preprocessing techniques in this field is essential to develop robust models and decision support systems in crop-pest advisory system.

Keywords

Agricultural data · Pest prediction · Data preprocessing · Data mining · Decision support system

---

M. Pratheepa (✉) · J. Cruz Antony
ICAR-National Bureau of Agricultural Insect Resources, Bengaluru, India
e-mail: mpratheepa@gmail.com

H. J. Purohit et al. (eds.), *Soft Computing for Biological Systems*,
https://doi.org/10.1007/978-981-10-7455-4_11

11.1 Introduction

The world population increases day by day and it reaches 10 billion by 2050 (Singh 2005). The additional population of 4.3 billion by 2050 will be living in developing countries, which is roughly three-fourths of the global population. Accordingly the food demand is likely to double by 2025 in comparison with present production. Therefore, it is important to produce enough food to feed everyone adequately, and hence, agriculture plays a role on it. Agriculture is the backbone of Indian economy, and India ranks second in the world in farm production (Limbore and Khillare 2015). The crop loss has been estimated around US$ 36 billion in India in post-green revolution era (Dhaliwal et al. 2015).There are various factors responsible for the low productivity of crops, of which damage from insect pests, diseases and weeds are accountable for the huge losses on major agricultural crops (Singh 2005; Kumar and Parikh 1998). Insect problems have been increased due to the unprecedented increase in area during last four decades. The pest damage varies largely from one crop to the other on different seasons (Trivedi et al. 2005), and the pest control strategy in crop production is a serious concern. Therefore, there is a need to develop robust pest prediction models for emerging pests by considering (i) reliable data on pest population for long periods at definite time intervals, (ii) weather records, (iii) crop phenology and (iv) relative abundance of natural enemies (Trivedi et al. 2005). Pest populations like all animal populations are influenced by various abiotic and biotic factors due to their innate capacity to adapt to the environments (Singh 2005; Southwood 1977).

An expert system is developed for the pest and disease management to assist the coffee industry board of Jamaica (Mansingh et al. 2007). Several expert systems developed in Indian agriculture for crop protection like Pesticide Advisor, Expert System for Pest and Disease on Different Field Crops in India (ESPDDFCI), Indian Cotton Insect Pest Management (ICOTIPM), and Expert System for Management of Malformation Disease of Mango (ESMMDM) are classified as crop specific, crop nonspecific, disease specific and disease nonspecific (Chakraborty and Chakrabarti 2008).The pest population dynamics have been studied by using several methods like artificial neural networks and statistical analysis like correlation analysis and linear regression, but the data mining techniques help to extract more hidden knowledge for the pest prediction (Pratheepa et al. 2016). Robust models are necessary to derive accurate decisions for forewarning the farmers to implement the timely pest management so that crop loss can be reduced. The generic pest dynamics model is built up with five steps: (i) data collection, (ii) data preprocessing, (iii) model development/analytical models, (iv) report generation and (v) decision-making. Among all steps, data preprocessing is very important since pest surveillance data comprises many data types and huge amount as well. The data could be generated from different sources like India Meteorological Department (IMD) and from the farmer's fields. IMD gives weather prediction in the form of *AGROMET* bulletins, and pest surveillance data from farmer's fields contains (i) crop phenology, (ii) natural enemy population, (iii) pest damage level and (iv) biology of the pest. The schematic diagram for pest prediction model

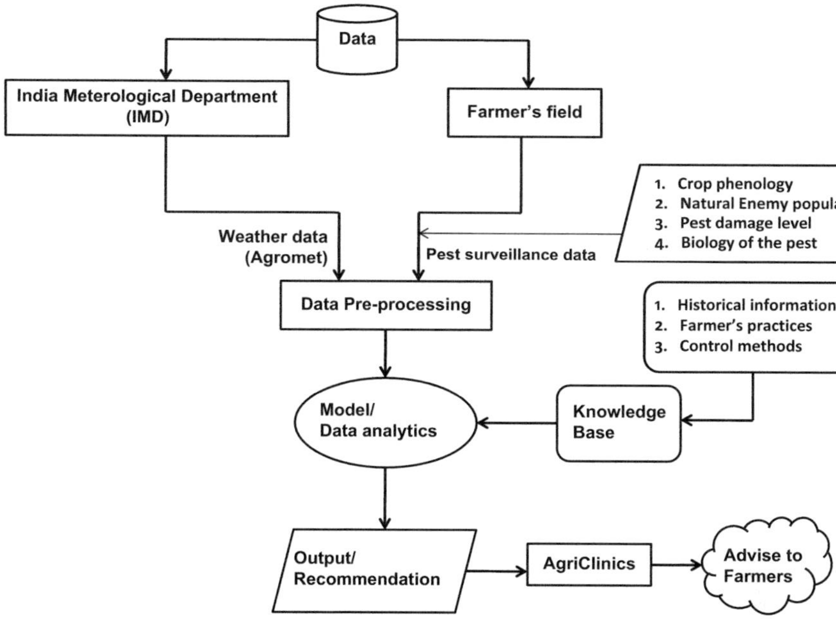

Fig. 11.1 Block diagram of pest prediction model

consists of the modules, data collection from different sources, data preprocessing, model for data analysis and recommendations to agri-clinics and farmers (Fig. 11.1).

11.2 Data Set

The data set for the pest *Helicoverpa armigera* (Hübner) occurrence was acquired from the All India Coordinated Cotton Improvement Project (AICCIP), and the experiment was conducted at the Regional Agricultural Research Station (16°0.21′N/77°0.34′E), Raichur, Karnataka, on non-Bt cotton NCS-145 variety in unsprayed condition (Table 11.1).

The pest incidence was recorded in weekly basis throughout the crop season. The mean value of pest incidence was combined with the mean value of previous week abiotic factors like maximum temperature (MaxT) and minimum temperature (MinT), morning relative humidity (RH1), evening relative humidity (RH2), rainfall (RF) and number of rainy days in a week (RFD) and with biotic factors like spiders (NE1) and *Chrysoperla carnea* sillemi (NE2) (Henry et al. 2010). The season and crop stage of cotton crop are also included in the data set. The selection of variables for generic pest prediction model is important for model accuracy (Table 11.2).

Table 11.1 Partial data set for the pest – *Helicoverpa armigera* – occurrence with biotic and abiotic factors, season and crop stage

PI (pest incidence)	Class	Year	Crop stage	Season	NE1	NE2	MaxT (°C)	MinT (°C)	RF (mm)	RFD	RH1 (%) (morning RH)	RH2 (%) (evening RH)
0	Low	2005	1	Monsoon	0	0	31.81	23.24	5.8	1	84	63
0	Low	2005	1	Monsoon	0	0	31.38	22.58	62.4	3	90	73
0	Low	2005	1	Monsoon	0	0	30.67	22.67	47.6	2	88	56
0	Low	2005	1	Monsoon	0	0	31.5	22.27	6	1	87	60
0	Low	2005	2	Monsoon	0	0	32.45	23.08	59.2	3	91	66
0	Low	2005	2	Monsoon	0	0	31.15	22.9	101	1	84	59
0.25	High	2005	2	Monsoon	0.22	0	31.34	22.25	0	0	87	64
0.4	High	2005	2	Monsoon	0.26	0	29.9	22.04	50.2	2	86	55
0.36	High	2005	3	Postmonsoon	0.32	0.2	31.3	22.5	18.2	1	87	48
0.65	High	2005	3	Postmonsoon	0.36	0.26	33.1	22.72	8.6	1	84	56
0.45	High	2005	3	Postmonsoon	0.39	0.2	32.2	21.24	46.8	1	91	62
0.7	High	2005	3	Postmonsoon	0.41	0.26	30.21	22.58	81.4	3	85	51
0.58	High	2005	3	Postmonsoon	0.45	0.22	31.11	21.45	1.4	1	90	61
0.65	High	2005	3	Postmonsoon	0.46	0.28	28.3	17.8	21.4	1	84	33
0.8	High	2005	4	Postmonsoon	0.47	0.31	30.11	16.25	0	0	84	29
1.06	High	2005	4	Postmonsoon	0.52	0.36	30.22	13.41	0	0	83	35
1.15	High	2005	4	Postmonsoon	0.58	0.39	30.48	16.28	0	0	73	37
0.62	High	2005	4	Postmonsoon	0.59	0.49	29.85	16.51	0	0	81	29
0.56	High	2005	4	Postmonsoon	0.62	0.58	31.04	17.08	0	0	81	41
0.38	High	2005	5	Postmonsoon	0.71	0.61	30.81	14.42	0	0	67	24

Source: All India Coordinated Cotton Improvement Project (AICCIP), Agricultural research station (ARS), Raichur, Karnataka
NE1, spiders; NE2, *Chrysoperla zastrowi* sillemi; MaxT, maximum temperature; MinT, minimum temperature; RF, rainfall, RFD, number of rainy days in a week; RH1, morning relative humidity;
RH2 = Evening relative humidity

Table 11.2 Selection of variables for generic pest prediction model

Variables/ attributes	Explanation of the variable	Type of the variable
Pest incidence (PI) or 'Y'	Response variable/outcome/dependent variable	Continuous
Crop stage	Observation taken when the cotton crop is on different stages	Nominal
Season	Observation taken on that season	Nominal
RFD	Number of rainy days in a week	Continuous or nominal
	Value ranges from 0 to 7	
Biotic		
NE1	No. of spiders per plant (natural enemy – predator)	Continuous (numerical data type)
NE2	No. of *Chrysoperla zastrowi* sillemi (natural enemy – predator)	
Abiotic		Continuous (numerical data type)
MaxT	Maximum temperature	
MinT	Minimum temperature	
RF	Rainfall	
RH1	Morning relative humidity	
RH2	Evening relative humidity	

11.3 Data Preprocessing Technique for the Dependent Variable

The value of 'Y' will be determined when it is of any type either 'nominal' or 'continuous'. In linear models like regression, the values of target variable can be in the form of 'continuous' numerical values. But for non-linear models like logistic regression, inductive models similar to rule learners, Bayesian networks and decision tree method use nominal or categorical values in the target variable. These values are quantitatively mentioned as present/absent, low/medium/high, yes/no, severe/very severe, etc. In pest prediction data set, to quantify the dependent variable which is pest incidence (PI) in a scientific way, it is necessary to look into economic threshold levels (ETL) for major pests on important agricultural crops. ETL is an index in integrated pest management (IPM) to make decisions for pest control measures. ETL indicates the intensity of pest population or extent of crop damage in which the cost of control measure is more than the value of the crop (Higley and Boethel 1994). Several rules have been framed for important insect pests on prime agricultural crops in different locations of India (Dhaliwal and Arora 1996). Hence, if the user likes to carry out non-linear models for pest prediction, it is necessary to follow the ETL rules for pest level classification. The appropriate class labels for pest incidence like low/medium/high can be assigned based on the index ETL.

11.4 Data Preprocessing Techniques for Independent Variables

The raw data collected from different sources for the study generally will be incomplete, contain noises and sometimes inconsistent. Preprocessing of the data is needed to avoid erroneous results. The important steps in data preprocessing are (i) cleaning the data, (ii) integration of the data, (iii) transformation of the data, (iv) reduction of the data and (v) discretization of the data.

11.4.1 Data Cleaning

Data cleaning involves handling of missing values and outliers. Missing values in the data set can be smoothened by filling of mean, mode or median values. The missing value can be replaced with global constant or with most probable value derived from Bayesian inference or EM (expectation-maximization) algorithm. Outliers can be identified and removed in the data set by using clustering, curve-fitting and hypothesis testing. Sometimes, duplicate records or inconsistent records can be deleted to converge into solution.

11.4.2 Data Integration

Data integration is required to integrate multiple databases or files related with same aspect. Import/export facilities can be used to transfer the files from one database software to another. The most compatible file formats used are .csv files (comma delimited), .xml files, .csv (tab delimited), .xls file, .json file, etc. These files are in the worksheet form and can be easily imported into databases like My-SQL, Maria-db, Mongo-db, etc.

11.4.3 Data Transformation

Data transformation is required for normalizing the data so that the model is not influenced by high or low values and falls into small specified range. It is important in artificial neural network (ANN) models. In ANN models, normalization techniques are used for scaling the variables between 0 and 1 to proportionate with the boundary of the activation functions present in the output layer (Minns and Hall 1996; Obach et al. 2001). There are several normalization techniques available, and few of them are given below:

(a) Min-max normalization
(b) Zero-mean normalization
(c) Normalization by decimal scaling

11.4.3.1 Min-Max Normalization

The formula for min-max normalization is given in Eq. (11.1).

$$V_i' = \frac{v_i - \text{Min}_A}{\text{Max}_A - \text{Min}_A}(\text{new\_Max}_A - \text{new\_Min}_A) + \text{new\_Min}_A \tag{11.1}$$

Min_A and Max_A denote minimum and maximum values in the input array 'A', and 'V_i' denotes the input value in the *ith* instance. 'new_Max$_A$' and 'new_Min$_A$' are 1 and 0 since the mapping of original values is being to the range between 0 and 1. Min-max normalization helps to maintain the relationships among the original data values.

11.4.3.2 Zero-Mean (Z-Score) Normalization

The formula for zero-mean or z-score normalization is given in Eq. (11.2).

$$Z = \frac{x - \mu}{\sigma} \tag{11.2}$$

where 'x' is a value in an array, 'μ' refers to the mean of the array and 'σ' refers to the standard deviation of the array. The attribute array has been normalized on the basis of mean and standard deviation of that array. When the situation of more outliers arises, this method is suitable to derive more appropriate results.

11.4.3.3 Normalization by Decimal Scaling

Decimal scaling is the method which provides the range between −1 and 1 (Sanjaya and Prasanta 2015). The formula for this technique is given in Eq. (11.3).

$$v^i = \frac{v}{10^j} \tag{11.3}$$

where v^i is the scale values, 'V' is a value in an array in ith instance and j is the smallest integer as Max ($|v_i|$) < 1.

11.4.4 Data Reduction

11.4.4.1 Division of Data for Training Set and Testing Set in ANN Models

In pest management programme, the study of population dynamics of target pest is necessary due to the different roles of each pest on different crops in different agroecological regions. Most of the pest prediction models are not in use, or the prediction goes wrong due to the role of pest dynamics based on several factors. Training a model is important to make the prediction correctly or accurately. The data set taken for training set is important especially in pest prediction models. There are several standard methods for partitioning of data set for training and testing phases. The proportion of dividing training set and testing set may be 1:1,

2:1, 3:1, etc. (Utans and Moody 1991; Efron and Tibshirani 1995) for cross validation. In ecological models leave-one-out cross validation method was used (Brosse et al. 2001; Guegan et al. 1998). Generally, in agriculture for prediction of crop yield or pest and disease incidences, the past years of data have been considered for training set to train the model, and the validation will be done in the forthcoming year. But, in case of classification models for pest prediction, the data set for training set is to be considered as per the target class values. This method is called soft splitting, and in this method, there is a possibility of overlapping the data set, and the net forewarning output will have better accuracy than the regular methods (Zhang and Govindaraju 2000; Shrestha and Solomatine 2006; Wu et al. 2008).

When data set contains too many attributes, data reduction helps to reduce the volume of data but produces the similar analytical results. Dimensionality reduction refers to the removal of the unimportant attributes. There are direct and indirect methods available in data reduction. In direct method, sampling techniques are used. The aim of the direct method is to choose a required attributes or features set which is adequate for data mining. Aggregation or summarizing the attribute values helps to minimize the number of attributes in the experimental data.

11.4.4.2 Sampling Techniques

Sampling techniques are used for selecting the representative subset data from the whole data set. These techniques are based on probability and also not based on probability. The probability or direct sampling method contains (i) random sampling, (ii) systematic sampling and (iii) stratified sampling. Random sampling can be used when there is very large population. Systematic sampling can be used when the given population is logically homogenous. In this method, the sample size has to be decided and then arrange the elements of the population in some order and select the samples at regular intervals from the list. In stratified sampling, the sampling population has been divided into groups based on criteria to be measured. The divided groups are called as strata and the single group is called as stratum. Then random sampling can be carried out for each group. This sampling method is suitable when the population is heterogeneous and can be split into homogeneous groups to arrive accurate results.

Indirect sampling method contains (i) cluster sampling and (ii) generalized weight share method (GWSM). In cluster sampling, the data has been divided into different groups based on the similarity, and these groups are called as clusters. Random sample can be selected from each cluster for further analysis. GWSM method is useful for finding the probability from the rare populations from the known groups, and it is based on weight values (Pierre 2007). In cluster sampling, the cluster itself is a sampling unit, and in stratified sampling, the elements in each strata are considered for analysis.

11.4.4.3 Statistical Methods

Variable or feature selection is an important task in prediction models. It is used for the selection of best subsets in predictor variables so that redundant variables can be avoided which gives clarity in results and reduce the time in computation. There are several statistical methods like principal component analysis (PCA), artificial neural networks (ANN) based PCA methods, rough set theory, regression analysis, step forward selection, step backward elimination and heuristic methods available for feature selection in modelling.

11.4.4.4 Data Discretization for Numerical Attributes

Data discretization techniques are required to transform the numerical values into nominal or categorical values since some of the classification algorithms in data mining techniques accepts only the categorical values instead of the numerical values. Data discretization is part of data reduction technique and helps to improve the accuracy by reducing the noise or non-linearity in the predictive models. In pest prediction models, data discretization is required for abiotic factors/variables like maximum temperature (MaxT), minimum temperature (MinT), rainfall (RF), number of rainy days in a week, sunshine hours, morning relative humidity, evening relative humidity, wind speed, evaporation, etc. for decision tree analysis, rule induction approach, etc. Data discretization can be done by using binning methods, and in that there are two types, one is supervised and another one is unsupervised. In the supervised binning, the model refers target or class variable while transforming the values of numerical variables into categorical values, but in unsupervised binning method, transformation of numerical variables into categorical counterparts takes place without referring the target or class variable.

Unsupervised Binning Methods

(i) Equal-width binning
(ii) Equal-frequency binning
(iii) Max-diff method

Equal-Width Binning

In equal-width binning, the algorithm partitions the data in 'k' intervals, and each interval has the same size. Let 'k' be the number of bins required and the width of bin is calculated as

$$W = (\max - \min)/k$$

where 'max' and 'min' refers to maximum and minimum values in an array. The bin range starts from the 'min' value and the bin boundaries are min + w, min + 2w, min + 3w, . . ., min + (k−1)w, min + kw + or max+. Here, the bin size is equal but the number of elements in each bin will be unequal. In a numeric array, the minimum value can be considered as zero, and the same procedure can be followed instead of taking 'min' value in an array. In that case, sometimes, there will not be any elements in some of the bins or intervals.

Equal-Frequency Binning Method

In equal-frequency binning method, the algorithm divides the data into 'k' groups where each group contains approximately same number of elements.

The steps involved in equal-frequency method for assigning label as A1, A2, etc. for the attribute MinT are given below:

(i) For M = 1 to TOTA // TOTA denotes total number of attributes to be categorized.
(ii) Input total number of bins needed and denoted as 'K'.
(iii) NB = N/K // 'N' indicates total number of tuples; NB is bin frequency.
(iv) Sort in ascending order for all the attributes to be categorized.
(v) S = 1.
(vi) For I = 1 to K.
(vii) For J = S to NB // Fixing the bin boundary.
(viii) MinT[J] = strcat('A' + str(I)) // Assigning label as A1, A2, etc.
(ix) Next J.
(x) S = NB + 1.
(xi) NB = I*NB.
(xii) Next I.
(xiii) T1 = K * NB.
(xiv) T2 = N – T1.
(xv) K1 = K + 1.
(xvi) FOR I = T1 + 1 to T2.
(xvii) MinT[I] = strcat('A' + str(K1)).
(xviii) NEXT I.
(xix) NEXT M.

The histogram for the total records of 103 by using equal-frequency binning method for the user input number of bins to be created for 5 creates 5 bins (A1,A2, . . ., A5) of each contains 20 records, and for the remaining 3 records, 6th bin (A6) will be created. But the user can modify the algorithm in such a way that the last bin label (A5) may be assigned to the remaining records (Fig. 11.2).

The example for abiotic factor minimum temperature (MinT) array has been divided into 5 bins my using equal-frequency binning method, and the minimum and maximum value of MinT are 9 °C and 23 °C (Table 11.3).

Max-Diff Method

The data array has to be sorted in ascending order, and the bin boundaries have been defined at the points where the adjoining values have the maximum difference. If the requirement is 'n' bins, n-1 maximum difference values have to be considered and n-1 cut points to be given in an array. The example of an array with three cut points will contain four bins by using max-diff method (Fig. 11.3).

The example for abiotic factor minimum temperature (MinT) array has been divided into five bins by using max-diff method containing four cut points (Table 11.4), and there was no repetition of range values in bins as in equal-frequency binning method (Table 11.3).

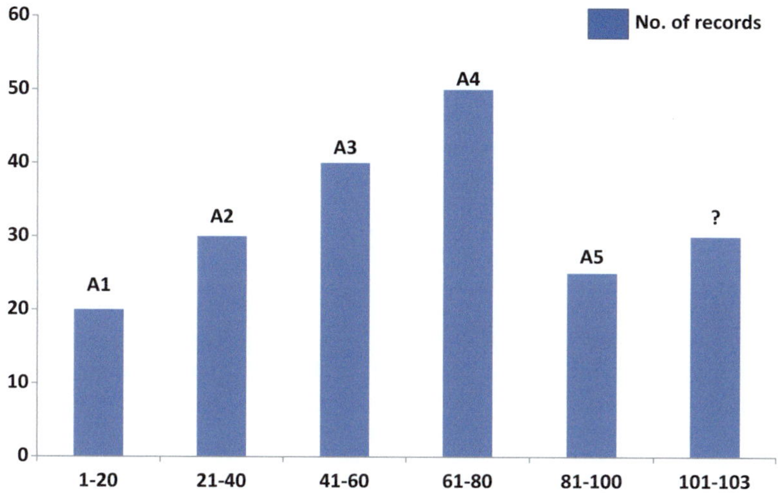

Fig. 11.2 Histogram of equal-frequency binning method

Table 11.3 Bin range values of MinT by using equal-frequency binning method

Bin label	Range values
A1	9–15.17
A2	15.8–17.4
A3	17.5–19.5
A4	19.6–20.7
A5	20.7–23

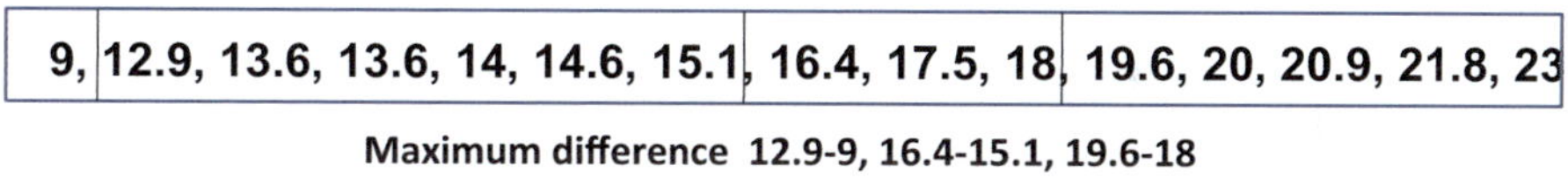

Fig. 11.3 Explanation of max-diff method

Table 11.4 Bin range values of 'MinT' by using max-diff binning method

Bin label	Range values
A1	9
A2	12.9
A3	13.6–14
A4	14.6–22.3
A5	23

11.4.4.5 Selection of Binning Method

In pest prediction data set, while considering the numerical continuous attributes for abiotic factors and for the natural enemy population, the best way for determining the number of bins 'n' is by considering the histogram. Bin optimization techniques can be used for the more accurate results.

Supervised Binning Method

Entropy-Based Binning

In information theory, entropy is called as expected information for uncertainty or choice of a variable based on probability (Shannon 1948). The equation for the calculation of entropy E is given in Eq. (11.4).

$$\text{Entropy (E)} = -\sum_{i=1}^{m} P_i log_2(P_i) \tag{11.4}$$

where '*m*' denotes the number of classifier. This is a split approach, and binning is to be done with target variable which is pest incidence (PI) in pest prediction model. The maximum gain value for the corresponding 'n' intervals or bins will be considered. In this case, the target variable PI is to be considered as numerical continuous values rather than the categorical values like high, medium, low, etc.

11.5 Conclusion

Good data preparation is an important key to produce valid and reliable models. Research findings are based on the data, and hence the appropriate preprocessing techniques have to be applied to make the data more valid. Many preprocessing techniques have been developed nevertheless vital area of research on agricultural data for the development of robust pest prediction models. Pest prediction models help the farmers in forewarning the pest attack so that control measures can be applied in advance so that crop production may be increased.

Acknowledgements The authors wish to thank the director of ICAR-National Bureau of Agricultural Insect Resources, Bengaluru, 560 024, India, for providing necessary facilities and encouragement.

References

Brosse S, Lek S, Townsend CR (2001) Abundance, diversity and structure of freshwater invertebrates and fish communities: an artificial neural network approach. New Zeal J Mar Fresh 35(1):135–145

Chakraborty P, Chakrabarti DK (2008) A brief survey of computerized expert systems for crop protection being used in India. Prog Nat Sci 18(4):469–473. https://doi.org/10.1016/j.pnsc.2008.01.001

Dhaliwal GS, Arora R (eds) (1996) Integrated pest management: achievements and challenges. In Principles of insect pest management. NATIC, Ludhiana

Dhaliwal GS, Jindal V, Mohindru B (2015) Crop losses due to insect pests: global and Indian scenario. Indian J Entomol 77(2):165–168. https://doi.org/10.5958/0974-8172.2015.00033.4

Efron B, Tibshirani RJ (1995) Cross-validation and the bootstrap: estimating the error rate of the prediction rule, Tech. Rep. No. 477. Stanford University, Stanford. https://statistics.stanford.edu/sites/default/files/EFS%20NSF%20477.pdf

Guegan JF, Lek S, Oberdorff T (1998) Energy availability and habitat heterogeneity predict global riverine fish diversity. Nature 391:382–384. https://doi.org/10.1038/34899

Henry CS, Brookes SJ, Johnson B, Venkatesan T, Duelli P (2010) The most important lacewing species in Indian agricultural crops, *Chrysoperla* sillemi (Esben-Peterson), is a subspecies of *Chrysoperla zastrowi* (Esben-Petersen) (Neuroptera:*Chrysopidae*). J Nat Hist 44 (41):2543–2555

Higley LG, Boethel DJ (eds) (1994) Handbook of soybean insect pests. Entomological Society of America, Lanham

Kumar KS, Parikh J (1998) Climate change impacts on Indian agriculture: the Ricardian approach. In: Dinar A, Mendelsohn R, Evenson R, Parikh J, Sangi A, Kumar K, Mckinse J, Lonergan S (eds) Measuring the impact of climate change on Indian agriculture, World Bank Technical Paper, 402. World Bank, Washington, DC. http://documents.worldbank.org/curated/en/793381468756570727/Measuring-the-impact-of-climate-change-on-Indian-agriculture

Limbore NV, Khillare SK (2015) An analytical study of Indian agriculture crop production and export with reference to wheat. Rev of Res 4(6):1–8. ISSN:-2249-894X ,Available online at www.ror.isrj.org

Mansingh G, Reichgelt H, Osei Bryson KM (2007) CPEST: an expert system for the management of pests and diseases in the Jamaican coffee industry. Expert Syst Appl 32(1):184–192. https://doi.org/10.1016/j.eswa.2005.11.025

Minns AW, Hall MJ (1996) Artificial neural networks as rainfall-runoff models. Hydrolog Sci J 41 (3):399–417. https://doi.org/10.1080/02626669609491511

Obach M, Wagner R, Werner H, Schmidt HH (2001) Modelling population dynamics of aquatic insects with artificial neural networks. Ecol Model 146:207–217. http://www.ephemeroptera-galactica.com/pubs/pub_o/pubobachm2001p207.pdf

Pierre L (2007) Indirect sampling, Advisors: Bickel P, Diggle P, Fienberg S, Gather U, Olkin I, Zeger S, Springer series in statistics

Pratheepa M, Verghese A, Bheemanna H (2016) Shannon information theory a useful tool for detecting significant abiotic factors influencing the population dynamics of *Helicoverpa armigera* (Hübner) on cotton crop. Ecol Model 337:25–28. https://doi.org/10.1016/j.ecolmodel.2016.06.003

Sanjaya KP, Prasanta KJ (2015) A multi-objective task scheduling algorithm for heterogeneous multi-cloud environment presented in international conference on electronic design. Computer Networks and Automated Verification (EDCAV), IEEE, Meghalaya, 29–30 January. https://doi.org/10.1109/EDCAV.2015.706054

Shannon CE (1948) A mathematical theory of communication. Reprinted with corrections from the Bell Sys Tech J 27(379–423):623–656. http://math.harvard.edu/~ctm/home/text/others/shannon/entropy/entropy.pdf

Shrestha DL, Solomatine DP (2006) Machine learning approaches for estimation of prediction interval for the model output. Neural Netw 19(2):225–235. https://doi.org/10.1016/j.neunet.2006.01.012

Singh NB (2005) *Helicoverpa* menace in the Indian subcontinent. In *Heliothis/Helicoverpa* management – emerging trends and strategies for future research. pp 39–43

Southwood TRE (1977) The relevance of population dynamics theory of pest status. In: Cherret JM, Sagar GR (eds) Origin of pest, parasite, disease and weed problems. Blackwell Scientific Publications, Oxford, pp 35–54

Trivedi TP, Yadav CP, Vishwadhar Srivastava CP, Dhandapani A, Das DK, Singh J (2005) Monitoring and forecasting of *Heliothis/Helicoverpa* population. In: Hari CS (ed) *Heliothis/Helicoverpa* management – emerging trends and strategies for future research. Oxford & IBH Publishing Co Pvt Ltd, New Delhi, pp 119–140

Utans J, Moody JE (1991) Selecting neural network architectures via the prediction risk: application to corporate bond rating predictions. In: Proceedings of the first international conference on artificial intelligence applications on wall street. IEEE Computer Society Press, Los Alamitos
Wu CL, Chau KW, Li YS (2008) River stage prediction based on a distributed support vector regression. J Hydrol 358:96–111. https://doi.org/10.1016/j.jhydrol.2008.05.028
Zhang B, Govindaraju RS (2000) Prediction of watershed runoff using Bayesian concepts and modular neural networks. Water Resour Res 36(3):753–762. https://doi.org/10.1029/1999WR900264

Genomics for Oral Cancer Biomarker Research

12

Kavitha Prasad, Roopa S. Rao, and Rupali C. Mane

Abstract

Oral cancer is one of the most common malignancies worldwide with aggressive behavior. Despite the advancements in preventive measures, diagnosis, and management of oral cancer, the 5-year survival rate has been low. For the last few decades, basic and advanced molecular techniques have been used to understand the molecular process involved in transformation of normal oral epithelium into cancer. Accumulation of genetic changes due to extrinsic or intrinsic factors results in the initiation, progression, and recurrence of oral cancer. However, the recent focus has shifted to understanding the tumor microenvironment and cancer stem cells. The common genetic alterations include mutations, amplifications, silencing, and epigenetic changes. This review elaborates the transcriptional biomarkers which are expressed in the process of carcinogenesis. The use of molecular techniques for detection of these biomarkers can aid in early diagnosis and better prognosis. Brief descriptions of relevant computational techniques are given, and databases are indicated.

Keywords

Oral squamous cell carcinomas · Oral cancer · Biomarker · Genomics · Diagnosis · Carcinogenesis

Oral cancer (OC) and oral squamous cell carcinoma (OSCC) are used interchangeably.

K. Prasad · R. S. Rao (✉) · R. C. Mane
Department of Oral and Maxillofacial Surgery, Faculty of Dental Sciences, M.S. Ramaiah University of Applied Sciences, Bangalore, Karnataka, India
e-mail: kavithaprasad.os.ds@msruas.ac.in; drroopasrao1971@gmail.com; rupalirmore77@gmail.com

H. J. Purohit et al. (eds.), *Soft Computing for Biological Systems*,
https://doi.org/10.1007/978-981-10-7455-4_12

12.1 Introduction

Oral cancer (OC), a malignancy that arises from the oral cavity, is recognized as a subtype of head and neck cancer (HNC) (Chang et al. 2013). Ninety percent of them are oral squamous cell carcinomas (OSCC) (Feller and Lemmer 2012) which originate from the epithelial lining of the oral cavity. OSCC is the sixth most common cancer with high mortality and morbidity (Macey et al. 2015; Rashid and Warnakulasuriya 2015) due to its invasive growth pattern, lymph node metastasis, and high recurrence rate (Yong-Deok et al. 2015). It is one of the common malignancies noted in males worldwide (Ferlay et al. 2015). A similar scenario has been observed in India; it is a common malignancy in males and the third most common in females (M:F = 1.5:1) due to indulgence in high-risk habits (Cancer and Consortium 2013). Individuals in the fifth decade of life are the most affected. The probability of OSCC development increases with the duration of exposure to risk factors and age-related mutagenic and epigenetic changes (Feller and Lemmer 2012).

12.1.1 Risk Factors

The etiology of OSCC is multifactorial. Tobacco, alcohol, betel nut, betel quid, genetic factors, infection with high-risk viruses like Epstein-Barr virus (EBV) and human papillomavirus (HPV), and diet deficient in fresh fruits and vegetables are a few to list (Scully 2011; Polz-Gruszka et al. 2015). The chief risk factors in many countries include tobacco, smoking, and alcohol. However, types and forms of tobacco consumption vary across the world. In the Indian subcontinent, smokeless tobacco, betel quid, and betel nut are most commonly used (Scully 2011). The ingredients of the common etiological agents, carcinogen, and their role are listed in Table 12.1. Moreover, many studies have observed that tobacco and alcohol cause DNA damage and reduce its efficiency to repair (Zedan et al. 2015). The carcinogenic effect of tobacco, betel quid, and areca nut depends upon the dose, duration, periodicity of use, or the combined use of two or more agents (Petti 2009).

12.1.2 Lesional Site and Size

OSCC can arise at any anatomical site in the mouth (Feller and Lemmer 2012). In the Western countries, the tongue (20%–40%), lips, and floor of the mouth (15%–20%) are common sites, while the retromolar area, gingivae, palate, labial, and buccal mucosa are less commonly affected (Bagan et al. 2010). In the Indian subcontinent, the most affected oral site is the buccal mucosa (52.74%) followed by the lateral border of the tongue (23.17%). The less affected sites include the floor of the mouth (6.09%), palate (5.79), lips (5.48%), retromolar trigone (2.43%), and gingiva (0.3%) (Syam sundar et al. 2012). The clinical presentation of OSCC is

Table 12.1 Risk factors and their role in carcinogenesis

Risk factors	Ingredients	Carcinogens	Role in carcinogenesis
Betel quid	Mixture of areca nut, slaked lime, and catechu, wrapped in a betel leaf	Arecoline	Arecoline induces epigenetic changes (Lin et al. 2011)
		MNPN	Causes genetic damage and tumorigenic activity (Prokopczyk et al. 1991)
		ROS	Causes oxidative DNA damage (Waris and Ahsan 2006)
Tobacco	Dried tobacco leaves	NNN, NNK	Receptor mediated tumor growth due DNA mutation and adduction (Xue et al. 2014)
Areca nut	Dried seeds of the areca palm	Arecoline	Arecoline induces epigenetic changes (Lin et al. 2011)
		MNPN	Genetic damage and tumorigenic activity (Prokopczyk et al. 1991)
Alcohol	Saturated carbon atom with functional hydroxyl group	Acetaldehyde-metabolic product of alcohol	Acetaldehyde binds to DNA and forms carcinogenic adduct (Seitz and Stickel 2010).

3-(methylnitrosamino) propionitrile (MNPN); NNK, 4-(methylnitrosamino)-1-(3-pyridyl)-1-butanone; NNN, N′-nitrosonornicotine; ROS, reactive oxygen species

varied. They are either preceded by oral potentially malignant disorders like a white lesion, mixed white and red lesion, and a red lesion or may present as a swelling, growth, an irregular necrotic ulcer, ulceroproliferative lesion, or an enlarged cervical lymph node (Markopoulos 2012; Feller and Lemmer 2012).

12.1.3 Diagnosis

The early detection of OC is important to increase the survival rate. Conventional oral examination (COE) is the standard method for screening the oral cavity. Suspected lesions are often biopsied and confirmed by histopathological examination as it is the gold standard for diagnosis (Masthan et al. 2012).

Some of the noninvasive techniques developed for early OC detection include ViziLite Plus, salivary diagnostic methods, oral brush biopsy kits, multispectral optical imaging systems, VELscope, etc. (Shah et al. 2011). Although useful for routine screening, confirmation has to be done by biopsy (Macey et al. 2015). Approximately 70% of OCs exhibit an advanced stage diagnosis due to a long asymptomatic period, similar to inflammatory diseases (Yong-Deok et al. 2015). About 21% of head and neck cancers which were reported as negative at the surgical margins under the microscope turned to be positive in highly sensitive molecular techniques like PCR. Hence, it is suggested that the advanced molecular techniques can assist in early diagnosis and prevention, more so in high-risk individuals to improve the prognosis in OC (Brennan et al. 1995).

Molecular techniques capture the genetic changes at an early stage that are not detected under the microscope (Brennan et al. 1995). Genomic techniques can measure the majority of mRNA, proteins, metabolites, protein-protein interactions, genetic mutations, amplifications, epigenetic changes, and microRNAs. It can be accessed by PCR, microarray (Macgregor and Squire 2002), RNA-seq, exome sequencing, and whole-genome sequencing (Meldrum et al. 2011). These techniques could be beneficial to detect early diagnostic biomarkers.

Proteomic analysis is equally important in understanding the physiologic status of the cell. Recent studies have used 2D gel electrophoresis, Western blotting, and LC-MS/MS for OC salivary protein biomarker detection (Hu et al. 2010). Immunohistochemistry (IHC) is a widely used technique to identify the specific proteins overexpressed in certain cancers (Duraiyan et al. 2012). Garewal et al. (2014) employed IHC to evaluate Bcl-2 and MIB-1 as a protein biomarker for OSCC. A detailed list of diagnostic methods and their merits is collated in Table 12.2.

Table 12.2 Diagnostic aids with merits in detection of oral precancer and cancer

Diagnostic method	Merits
Routine method	
Vital staining	Determines the suspected site and margins of the lesions (Shah et al. 2011)
Oral brush biopsy	Easy, noninvasive, and painless (Shah et al. 2011)
Histopathology	Reliable and inexpensive (Shah et al. 2011)
Liquid-based cytology (LBC)	Simple and sensitive technique (Sigurdsson 2013)
Laser-induced fluorescence (LIF)	Simple, sensitive technique, and noninvasive (Wei et al. 2013)
Immunohistochemistry (IHC)	Sensitive and specific (Duraiyan et al. 2012)
Molecular technique	
Polymerase chain reaction (PCR)	Detects genetic alteration and molecular markers (Macgregor and Squire 2002)
Microarray technology	Parallel nucleic acid quantification (Macgregor and Squire 2002)
Next-generation sequencing (NGS)	Detects genetic and epigenetic changes (Meldrum et al. 2011)
In situ hybridization (ISH)	Study of cell development, human gene mapping, and cytogenetics (Lichter et al. 1988)
Optical techniques/imaging technique	
Optical coherence tomography (OCT)	Yields in situ imaging, without the excision of specimen (Lee et al. 2012b)
Flow cytometry	Detects DNA aneuploidy and loss of heterozygosity (Adan et al. 2016)

12.1.4 Survival

The 5-year survival rate of OSCC is 50%, which has remained unchanged since the last few decades, despite advancements in the treatment modalities (surgery, radiation, and chemotherapy) (Shah et al. 2011). An 80% recurrence-free 5-year survival has been observed in stage I OSCC; however, it is reduced to 20% for stage IV cancer (Shin et al. 2010). Survival is not often related to age, gender, socioeconomic status, or habits (Feller and Lemmer 2012), but it is significantly related to early detection and prevention of cancer (Shin et al. 2010). Improving early diagnosis is one of the best ways to increase the survival rate, improve the quality of life, and reduce health-care costs (Shah et al. 2011).

12.2 Carcinogenesis

Carcinogenesis is a complicated, multistep process that alters normal cellular physiology quantitatively and qualitatively (Wong and Todd 1996; Khan et al. 2012). Repeated exposure of the oral mucosa to carcinogenic insult could result in accumulation of genetic alterations and development of premalignant and malignant changes (Slaughter et al. 1953). A eukaryotic cell has genetic material (genes) composed of deoxyribonucleic acid (DNA), incorporated in the form of a chromosome, located in the nucleus. The DNA sequence is responsible for the formation of proteins (Albarts et al. 1994). Altered DNA sequence due to mutations, polymorphisms, etc. can result in the formation of abnormal proteins which disrupt normal cellular functions (Fig. 12.1).

The epithelial cells are constantly renewing cells, wherein the rate of production of new cells and loss of old cells is regulated to be constant. This process is maintained by two mechanisms: proliferation and apoptosis. Tumor suppressor genes, oncogenes, and growth factors control this critical balance between apoptosis and proliferation. Any imbalance in this process can lead to cancer (Bertram 2000; Khan et al. 2012).

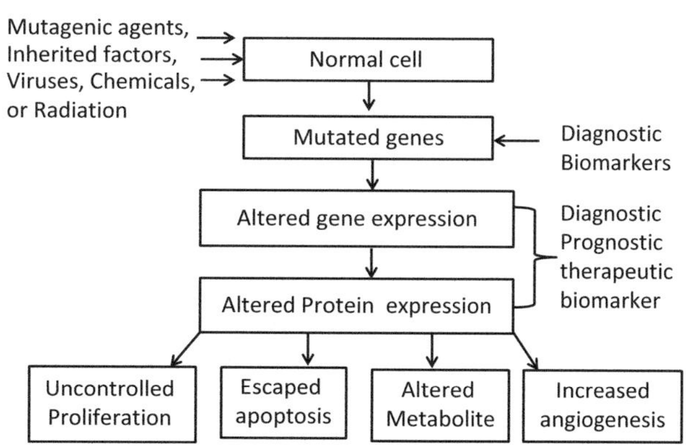

Fig. 12.1 Schematic representation of steps in carcinogenesis

Cancer is the accumulation of genetic and epigenetic changes caused by the mutation of cancer-related genes, such as tumor suppressor genes or oncogenes, as well as genes involved in cell cycle control, cell adhesion, apoptosis, DNA repair, and angiogenesis (Macgregor and Squire 2002).

12.2.1 Cell Proliferation

Proliferation is an important cellular process in carcinogenesis. It has a crucial role in the normal process of the cell cycle. Uncontrolled cell proliferation leads to the development of many different types of carcinoma (Ramires et al. 1997). Proliferation abnormalities can be evaluated using molecular techniques like IHC (Stankiewicz et al. 2009). Ki67 is a nuclear protein, which is most commonly used as a proliferative marker. Its expression is seen in all active phases of the cell cycle (G1, S, G2, and M phases) but is absent in G0 phase. Ki67 could be a very helpful biomarker in OSCC, to detect the aggressiveness of the tumor and prognosis (Tumuluri et al. 2002; Kurokawa et al. 2005).

In precancerous lesions, there is an increased expression of TGF-α when compared to EGFR in the proliferative pool of the oral epithelium. This suggests that an initial upregulation of TGF-α was likely to have a paracrine effect on the adjacent nonproliferative cells, thereby increasing the expression of the cell surface receptor, EGFR (Mendes 2012).

12.2.2 Apoptosis

Apoptosis is a strongly regulated physiologic cellular mechanism. This is a programmed process of elimination of useless, mutated, or harmful cells (Sen 1992). Apoptosis can be initiated by intrinsic or extrinsic cellular pathways. Tumor cells escape this process by an increased resistance and survival (Manning and Patierno 1996). Apoptosis is controlled by many anti-apoptotic (Bcl-2, Bcl-XL) and pro-apoptotic (Bax, Bak) regulators. P53 is one of the more frequently studied apoptotic biomarkers (Wilson et al. 2001).

12.3 Invasion and Metastasis

The propensity for invasion into foreign tissues and metastatic activity at a distant location is a distinct characteristic of a cancer cell. The invading cells reach the blood/lymphatic vessels and can spread to other organs in the body resulting in metastasis. Metastasis is a complicated process involving sequential steps and multiple factors (Khan et al. 2012). Invasion and increased cell motility are the first step in metastasis (Thiery 2002). It starts with the degradation of extracellular matrix (ECM) and basement membrane (BM). The matrix metalloproteinases (MMPs) are of the protease family; they degrade ECM and BM in several conditions including cellular development, tissue repair, tumor invasion, and metastasis (Shah et al. 2011). OSCC is highly invasive and metastatic in nature. There is a correlation between higher expression of MMP-2 and MMP-9 with poorer prognosis in OSCC (Singh et al. 2010).

Many studies have shown that epithelial-mesenchymal transition (EMT) is responsible for malignant tumor cell migration and may play a key role in tumor invasion and metastasis (Zhou et al. 2015). It has been assumed that metastasis may be connected with EMT. Areca nut exposure causes increased vimentin expression, which has an important role in EMT through PI3K/AKT pathway (Lee et al. 2012a). Similarly, cadherins, key molecules in EMT, play a major role in the maintenance of cell-cell adhesion of normal cells and are responsible for epithelial integrity (Thiery 2002). Integrins are protein molecules that maintain the tissue integrity and regulate cell proliferation, differentiation, and migration (Thomas and Speight 2001). The overexpression of cyclin D1 protein is significantly related to tumor cell differentiation, stage, and lymph node metastasis (Jones and Walker 1997).

12.4 Tumor Microenvironment

The tumor microenvironment (TME) is the environment in which the tumor exists; it is composed of cancerous cells, adjacent blood vessels and epithelial cells, immune cells, and surrounding matrix (Curry et al. 2014). Its dynamic function is to fulfill tumor cell requirements to survive and grow (Bellone et al. 2013).

12.4.1 Angiogenesis

Angiogenesis (formation of new blood vessels) is one of the most important hallmarks of cancer development. Increased blood supply is an important requirement for tumor growth and metastasis (Hanahan and Weinberg 2000).

Vascular endothelial growth factor (VEGF) is a well-known agent which boosts angiogenesis. Analysis of VEGF gene in OSCC revealed an overexpression of mitogen that is related to tumor size (Christopoulos et al. 2011).

12.4.2 Hypoxia

Hypoxia is associated with increased invasion and metastasis in several cancers. In tumor tissue, the hypoxic condition is responsible for increased angiogenesis with the rapid development of new blood vessels. These rapidly synthesized blood vessels have an insufficient blood supply. Hypoxia is harmful to both normal cells and cancer cells; however, molecular changes during carcinogenesis support the cancerous cells to survive and proliferate in a hypoxic condition. Therefore, hypoxic tumor growth could result in an aggressive malignancy (Nagaraj et al. 2004).

Hypoxia-inducible factor-1α (HIF-1α) is a significant factor for those tumor cells which are under hypoxia. HIF-1α accelerates the transcription of many hypoxic genes which are involved in angiogenesis, glucose metabolism, oxygen supply, and invasion and prevention of apoptosis (Gronroos et al. 2014).

12.4.3 Lipid Metabolism

Lipids are important biomolecules required for maintenance of various biological functions such as DNA stabilization and cell proliferation in both normal and neoplastic cells (Mehta et al. 2014). Rapidly dividing cells in a malignant condition require more amount of lipid. Increased requirement of lipids is fulfilled either through exogenous uptake (cholesterol in the proliferating tissues and in blood compartments) or by lipogenesis (endogenous lipid synthesis) (Beloribi-Djefaflia et al. 2016). Lipid biomolecule synthesis by activation of lipogenesis, especially endogenous fatty acid synthesis, is an important occurrence in the metabolic transformation of normal cells to neoplastic tumor cells (Menendez and Lupu 2007). Stearoyl-CoA desaturase (SCD) is a key enzyme for fatty acid metabolism; fatty acid is the building block of lipid. In many cancer studies, SCD was considered to be the key factor in cancer development (Ariel 2011).

12.5 Epigenetic Changes

Along with genetic changes, epigenetic changes also play a critical role in carcinogenesis. Epigenetic changes refer to any heritable genetic modification without changing DNA sequence, mediated through mechanisms like methylation, acetylation, or phosphorylation, which results in dysregulation of gene expression in a number of different ways (Egger et al. 2004). Recent studies suggest that hypermethylation, acetylation (Arif et al. 2010), and phosphorylation (Kaneko et al. 2016) play an important role in carcinogenesis (Egger G.et al., 2004). Hypermethylation, a frequent event in carcinogenesis, induces functional silencing of genes like p16, E-cadherin (Asokan et al. 2014), and phosphatase and tensin homolog (PTEN) (Sushma et al. 2016). These genes could be useful biomarkers for diagnostic purpose.

12.6 Cancer Stem Cells

Most cancers contain a subpopulation of cells which have the potential for self-renewal, differentiation, tumor invasion, metastasis, and disease recurrence called cancer stem cells (CSC). These cancer stem cells activate resistance mechanisms, like EMT, resistance to hypoxia and induction of angiogenesis, and resistance to immune escape by a reduction in tumor-specific antigens while increasing cytokines and growth factors (Albini et al. 2015). CSC research could be a useful tool for basic understanding of intrinsic and extrinsic features of the OC (Patil et al. 2013). Chen et al. (2013) have shown that CSC markers like CD44, CD133, side population cells, Bmi-1, c-Met, and ALDH1 have a positive correlation with tumor invasiveness and metastases.

12.7 Biomarkers

Hulka and colleagues defined biomarker as a cellular, biochemical, or molecular alteration that is measurable in biological media such as human tissues, cells, or fluids (Hulka and Wilcosky 1988). Biomarkers could be used for understanding the cause of malignancy, diagnosis, progression, treatment of disease, and prognosis (Mayeux 2004).They have the ability to disclose genetic and molecular changes involved in the process of oral carcinogenesis, hence helpful in the management of oral carcinomas (Shah and Kaur 2014).

There are different types of biomarkers, which can be categorized as nucleic acid (gene, microRNA, noncoding RNA), protein (enzyme, receptor), antibodies, and peptides. They could be a panel of altered genes, proteins, metabolites, etc. (Henry and Hayes 2012).

List of the biomarkers is given in Table 12.3.

12.8 Oncogenes and Tumor Suppressor Genes

Identifying genetic changes is the first step of cancer development mechanism research (Yamamoto et al. 2015). The genetic alterations in the cancer cells are of two categories: (i) gain of function by dominant damage in proto-oncogene and (ii) loss of function by recessive damage in tumor suppressor genes (Khan et al. 2012).

12.8.1 Oncogenes

Proto-oncogenes regulate the cell growth and differentiation. Precise regulation of these genes maintains the normal behavior of cells. During carcinogenesis, proto-oncogenes get altered by mechanisms like point mutations, gene amplification, and gene overexpression and become oncogenes. These abnormal genes encode modified proteins and affect the normal regulatory mechanisms (Saranath et al. 1991). Oncogenes are responsible for initiation and progression of oral neoplasia (Field 1992); many of these genes are associated with oral carcinogenesis (Wong and Todd 1996).

Various studies observed abnormal expression in members of gene families like myc, ras (H-ras, K-ras, N-ras), int-2, hst, bcl, and PRAD-1 as well as in epidermal growth factor receptor (EGFR)/c-erb 1 that are considered as contributors to OC development (Wong and Todd 1996).

Table 12.3 Biomarker: expression and role in cancer

Author	Biomarker	Expression	Role in cancer
DNA			
Sushma et al. (2016)	P16	↓	Proliferation biomarker
	Promotor methylation		Helpful for diagnosis
Sushma et al. (2016)	PTEN	↓	Proliferation biomarker
	Promotor methylation		
Rowley et al. (1998) and Hsieh et al. (2001)	P53	↑	Tumor suppression and apoptosis
Gronroos et al. (2014)	HIF-1α	↑	Transcriptional regulator in response to hypoxia
Cancer and Consortium (2013)	FAT4	↑	Cell proliferation
mRNA			
Li et al. (2004) and Panta and Venna (2014)	Spermidine N1-acetyltransferase (SAT)	↑	Catabolism of polyamines
St John et al. (2004) and Li et al. (2004)	Interleukin 8 (IL8)	↑	Angiogenesis, cell cycle arrest, and cell adhesion
Li et al. (2004) and Panta and Venna (2014)	IL-1β	↑	Important cytokine for inflammatory response
St John et al. (2004)	IL6	↑	Inflammatory cytokine, promotes tumor progression
Li et al. (2004)	DUSP	↑	Important role in MAPK pathway
Li et al. (2004)	OAZ1	↑	Intracellular polyamine levels regulator which maintains cell growth and proliferation
Protein			
Chen et al. (2014)	CD44	↑	Tumor growth and metastasis
Christopoulos et al. (2011)	VEGF	↑	Increased angiogenesis
Shpitzer et al. (2009)	Ki67	↑	Proliferation and cell cycle acceleration
	cyclinD1		
	Mmp9	↑	Metastasis
Harshani et al. (2014)	Glut-1	↑	Hypoxia
microRNAs			
Tiwari et al. (2014)	miR-125a	↓	Increased cell proliferation and decreased apoptosis
Panta and Venna (2014)	miR-200a	↓↑	Tumor suppression and early metastasis

(continued)

Table 12.3 (continued)

Author	Biomarker	Expression	Role in cancer
Hung et al. (2014)	miR31	⇧	Hypoxia pathway regulator

⇩ Decreased
⇧ Increased

Table 12.4 Proto-oncogenes: role in normal cellular function and in cancer

Proto-oncogenes	Function	Altered function	References
K-ras	Signal-transducing and cell cycle regulatory protein	Activates signaling pathway in advanced stage of OSCC	Al-Rawi et al. (2014)
		Tumor development	
H-ras	Regulation of cell growth, transduction of mitogenic cell signaling from the surface of the cell to the nucleus	Uncontrolled proliferation, survival, and apoptosis	Saranath et al. (1991), Deo MG (1991)
BCL2	Regulator of anti-apoptotic mechanism	Promotes prolonged cell survival	Teni et al. (2002)
EGFR	Transmembrane protein	Uncontrolled cell proliferation and survival	Massano et al. (2006)
c-myc	Crucial role in cell growth control, differentiation, and apoptosis	Uncontrolled proliferation, survival, and apoptosis	Krishna et al. (2015)
Cyclin D1	Cell cycle regulation	Regional nodal metastases and advanced tumor stage	Scully et al. (2000) and
			Miyamoto et al. (2003)
Cyclin A	DNA synthesis	Advanced tumor stage, larger tumor volume, lymph node metastases and recurrence	Chen et al. (2003)
	G2 phase to M phase progression		

However, the incidence of H-ras mutation in Indian OSCC (35%) is higher (Saranath et al. 1991) than that seen in Western Europe and the USA (5%). This may be due to the prevalent use of tobacco in the Indian population. Tobacco could be a possible reason for ras gene family mutation in the corresponding populations (Khan et al. 2012). List of proto-oncogenes is given in Table 12.4.

12.8.2 Tumor Suppressor Genes

Tumor suppressor genes (TSG) are regulators of fundamental cellular processes like cell division, DNA repair, and apoptosis. Genetic changes in TSG inactivate

Table 12.5 Tumor suppressor genes: role in normal cellular function and in cancer

Tumor suppressor genes	Function	Changes	References
P53	Transcription factor; regulates cell cycle and apoptosis, controls genome integrity and DNA repair	Uncontrolled proliferation of abnormal cells Tumor progression	Levine et al. (2004)
E-cadherin (epithelial cadherin)	Transmembrane glycoprotein, maintains cell polarity, and normal tissue structure	Lymph node metastasis	Chaw et al. (2012)
Adenomatous polyposis coli (APC)	Controls cytoplasmic β-catenin concentration	Invasion and metastasis	Chaw et al. (2012)
Phosphatase and tensin homolog (PTEN)	Stimulates apoptosis by inhibiting PI3K-PKB/Akt signaling pathway activity	Silencing of signal transduction from membrane growth factor receptors (EGFR, HER-2, IGFR) through the AKT pathway	Sushma et al. (2016)
p16/cyclin-dependent kinase inhibitor 2A (CDKN2A)	Cell cycle regulatory protein; inhibits the activity of cyclinD6 and prevents Rb phosphorylation	Uncontrolled cell proliferation	Pande et al. (1998)
Retinoblastoma (Rb)	Controls transition to S-phase by regulating transcription factor E2F activity, regulation of cellular proliferation	Uncontrolled cell proliferation	Pande et al. (1998)

them. These genes are frequently inhibited by point mutations, deletions, and gene rearrangements (Shah et al. 2015). P53 is a well-documented tumor suppressor gene. One of the most common genetic alterations in different types of human cancers is alteration of p53 gene (Scully et al. 2000). List of tumor suppressor genes is given in Table 12.5.

12.9 Bioinformatics Resources for Transcriptional Biomarker Research

Cancer bioinformatics is the new field which is a combination of bioinformatics, mathematics, information technology, and medical informatics (Wu et al. 2012). Genomic techniques produce a huge amount of cancer-related molecular data. Therefore, the role of computers is crucial for the structured organization of data and understanding of molecular knowledge (Luscombe et al. 2001). Finding gene expression changes between cancerous and normal specimens at different stages of cancer could be useful in identifying specific genetic signatures for diagnostic, therapeutic, and prognostic information (Kihara et al. 2006). Gene transcription is a dynamic process, allowing cells to adapt rapidly to the external, environmental, or physiological changes affecting organs, target tissues, or cells. Therefore, identification of biomarkers that describe a given physiological status, a disease, an exposure to a carcinogen, or other exogenous stimuli is possible with gene expression profiling (Riedmaier et al. 2012).

12.9.1 Methods for Transcriptome Analysis

Microarrays and high-throughput RNA sequencing, along with the development of computational tools, could be useful for biomarker identification (Riedmaier et al. 2012). These techniques allow expression profiling of huge number of genes in a given biological sample in a single experiment. Data analysis tools for microarray and RNA-seq, as well as related databases of OC, are detailed.

12.9.1.1 Microarray

Microarray is a powerful genomics tool, designed to capture the expression of thousands of genes in cells. A microarray is defined as a collection of probes arranged in an array attached to the solid surface. These specially designed probes bind to specific nucleic acids corresponding to a particular gene through the hybridization process (Jaluria et al. 2007). Microarray results in an image for each sample. Softwares are used to analyze images and to obtain the intensity at each spot, followed by data normalization, differential gene expression, and gene annotation. Free and commercial softwares are available for this purpose (Mehta and Rani 2011).

Bioconductor (http://www.bioconductor.org) is an open development initiative for computational biology. Its main focus is to provide a platform to the end user for expression analysis. Packages like affy and limma (linear models for microarray analysis) are available for raw data normalization and statistical analysis of Affymetrix and Agilent data, respectively. Three packages can be used for differentially expressed gene identification, i.e., multtest, genefilter, and edd. This is followed by genomic annotation of differentially expressed (DE) genes from databases such as GenBank, the Gene Ontology (GO) Consortium, LocusLink, UniGene, and the UCSC Human Genome Project. They can be accessed using packages like Annotate and AnnBuilder (Dudoit et al. 2003). However, programming skills and command-line interface are vital for using this computer language, which could be difficult for many biologists. Hence user-friendly graphical user interface (GUI)-based tools and software were developed (Xia et al. 2005). They are collected together in Table 12.6.

Table 12.6 Microarray and RNA-seq tools for data analysis

Software	Specification	Website/reference
Microarray		
Spotfinder	Microarray images analysis	https://omictools.com/spotfinder-tool
MultiExperiment Viewer (Mev)	Clustering, visualization, classification, and statistical analysis of normalized data files. Accepts several input file formats (.mev, .tav, .txt, .gpr)	http://mev.tm4.org/#/welcome
GeneSpring GX	Commercial software used for microarray data analysis and visualization	http://genespring-support.com/user/register
ArrayAssist	Tool for processing and visualization of expression data. It has strong support for the Affymetrix platform	http://www.biocompare.com/Product-Reviews/40281-ArrayAssist-Advanced-Software-From-Stratagene/
WebArray	It is a web platform for analysis of two-color Affymetrix microarray data	http://www.webarraydb.org/webarray/index.html
DAVID	Functional gene annotation of genes	http://david.abcc.ncifcrf.gov/
Gene Ontology	The Gene Ontology Project provides a controlled vocabulary to describe gene and gene product attributes in any organism	http://www.geneontology.org/
RNA-seq		
FastQC	Raw data preprocessing	http://www.bioinformatics.babraham.ac.uk/projects/fastqc/
HTQC	Raw data preprocessing	Yang et al. (2013)
FLEXBAR	Read trimming	Dodt et al. (2012)
FASTX-Toolkit	Read trimming	from: http://hannonlab.cshl.edu/fastx_toolkit/
Bowtie	Unspliced read aligner	Langmead et al. (2009)
TopHat	Spliced read aligner	Trapnell et al. (2009)
	Call variants	
	Detect gene fusions	
MapSplice	Spliced read aligner	Wang et al. (2010)
Cufflinks	Reference-based transcriptome assembly and isoform-level expression quantification	Trapnell et al. (2009)
Scripture	Reference-based transcriptome assembly	Guttman et al. (2010)
Trinity	Reference-independent transcriptome assembly	Grabberr MG (2011)
Trans-ABySS	Reference-independent transcriptome assembly	Robertson et al. (2010)
ALEXA-seq	Gene-level expression quantification	Griffith et al. (2010)

(continued)

Table 12.6 (continued)

Software	Specification	Website/reference
edgeR	Gene-level differential expression	Robinson et al. (2010)
DESeq	Gene-level differential expression	Anders and Huber (2010)
SAMseq	Isoform-level differential expression	Li and Tibshirani (2013)
Cuffdiff	Isoform-level differential expression	Trapnell et al. (2009)

12.9.1.2 RNA-Seq

RNA-seq is a powerful tool for analyzing changes across the entire transcriptome during cancer development. This technique is capable of detecting a wide range of transcripts compared to microarrays. RNA-seq reveals information about distinct transcript isoforms and their abundance and can be used to detect mutations in more abundantly expressed transcripts and to analyze allele-specific expression. Altogether, this technology provides a detailed genomic characterization that was previously not possible (Trapnell et al. 2009; Ozsolak and Milos 2011).

RNA-seq experiment starts with the extraction of RNA from biological samples followed by quality check and cDNA library preparation. All fragments in cDNA library are parallely sequenced using high-throughput sequencing technology (Riedmaier et al. 2012). It yields a massive amount of raw sequencing reads (Trapnell et al. 2009). Analysis of this data is a complex process comprising of multiple steps. There is no single best pipeline; however, it is dependent upon experimental design, the organism being studied, and the research goals. The routine RNA-seq workflow consists of preprocessing of raw data, mapping, quantification of expressed genes, differential expression analysis, and gene annotation (Dai et al. 2010; Conesa et al. 2016).

12.9.2 Databases

The completion of the Human Genome Project and the development of large-scale molecular techniques have led to a massive accumulation of biological data. As a result, biological databases have been developed for the systematic organization of fast-growing data. It allows the users to access existing information and to submit the new entries produced (Zou et al. 2015).

12.9.2.1 OrCGDB

Oral cancer gene database (OrCGDB) is a collection of tumor-related genes, available at http://www.tumor-gene.org/Oral/oral.html. At present, this database has 300 genes. The user can search for oral cancer-specific genes in this database (Levine and Steffen 2001).

Table 12.7 List of database

Database	Database details	Website
OMIM	Collection of human genes and genetic disorders	https://www.omim.org/
Gene Expression Omnibus (GEO)	Collection of functional genomics experimental data	https://www.ncbi.nlm.nih.gov/geo/
ArrayExpress	Collection of functional genomics experimental data	https://www.ebi.ac.uk/arrayexpress/
KEGG pathway	Collection of manually drawn pathway maps representing molecular interaction and reaction networks knowledge	http://www.genome.jp/kegg/pathway.html
Gene Ontology	Collection of structured ontologies	http://www.geneontology.org/
The Cancer Genome Atlas	Database of cancer genomic data	https://cancergenome.nih.gov/
International Cancer Genome Consortium	A comprehensive description of genomic, transcriptomic, and epigenomic changes in 50 different tumor types and/or subtypes including oral cancer	http://icgc.org/
Sequence Read Archive (SRA)	Sequencing data from high-throughput sequencing platforms	https://www.ncbi.nlm.nih.gov/sra

12.9.2.2 OCDB

Oral Cancer Database (OCDB) is user friendly and is freely available at http://www.actrec.gov.in/OCDB/index.htm that provides information and external links for genes involved in oral cancer. It also furnishes information about 374 genes involved in oral cancer, interactions between them, and their role in oral cancer along with clinical relevance. This database can be queried by keyword search that will give the gene name and chromosomal region. Hence, it can act as a complete web resource (Gadewal and Zingde 2011).

12.9.2.3 HNOCDB

Head and neck and oral cancer database (HNOCDB) is a repository of information for genes and miRNAs involved in the head and neck/oral cancer. This information is linked to chromosomal map. Information about causes of oncogenic activation, genetic mutation, and chromosomal localization of the gene/miRNA is also available in this database (http://gyanxet.com/hno.html). Other databases for getting OC-related information are mentioned in Table 12.7.

12.10 Perspectives

Early diagnosis of oral cancer remains a challenge in spite of technical advancements. The 5-year survival rate following treatment for oral cancer is still low with the current treatment protocols. Molecular methods offer promise in improving this scenario. If the stage and behavior of the disease can be correctly gauged with the help of biomarkers, targeted therapy and personalized treatment can improve the treatment outcome. Therefore, knowledge of recent advancements and their judicious use is recommended for improved cancer care.

References

Adan A, Alizada G, Kiraz Y, Baran Y, Nalbant A (2016) Flow cytometry: basic principles and applications. Crit Rev Biotechnol 37:163–176. https://doi.org/10.3109/07388551.2015.1128876

Albarts B, Bray D, Lewis J (1994) Molecular biology of the cell. Garland Publishing Inc, London

Albini A, Bruno A, Gallo C, Pajardi G, Noonan DM, Dallaglio K (2015) Cancer stem cells and the tumor microenvironment: interplay in tumor heterogeneity. Connect Tissue Res 56:414–425. https://doi.org/10.3109/03008207.2015.1066780

Al-Rawi N, Ghazi A, Merza M (2014) PIK3CB and K-ras in oral squamous cell carcinoma. A possible cross-talk! J Orofac Sci 6:99. https://doi.org/10.4103/0975-8844.143049

Anders S, Huber W (2010) Differential expression analysis for sequence count data. Genome Biol 11:R106. https://doi.org/10.1186/gb-2010-11-10-r106

Ariel IR (2011) Roles of stearoylCoA desaturase-1 in the regulation of cancer cell growth, survival and tumorigenesis. Cancers 3:2462–2477. https://doi.org/10.3390/cancers3022462

Arif M, Vedamurthy BM, Choudhari R, Ostwal YB, Mantelingu K, Kodaganur GS, Kundu TK (2010) Nitric oxide-mediated histone hyperacetylation in oral cancer: target for a water-soluble HAT inhibitor, CTK7A. Chem Biol 17:903–913. https://doi.org/10.1016/j.chembiol.2010.06.014

Asokan GS, Jeelani S, Gnanasundaram N (2014) Promoter hypermethylation profile of tumour suppressor genes in oral leukoplakia and oral squamous cell carcinoma. J Clin Diagn Res 8(10):ZC09–ZC12. https://doi.org/10.7860/JCDR/2014/9251.4949. Epub 2014 Oct 20

Bagan J, Sarrion G, Jimenez Y (2010) Oral cancer: clinical features. Oral Oncol 46:414–417. https://doi.org/10.1016/j.oraloncology.2010.03.009

Bellone M, Calcinotto A, Filipazzi P, De Milito A, Fais S, Rivoltini L (2013) The acidity of the tumor microenvironment is a mechanism of immune escape that can be overcome by proton pump inhibitors. Oncoimmunology 2:e22058. https://doi.org/10.4161/onci.22058

Beloribi-Djefaflia S, Vasseur S, Guillaumond F (2016) Lipid metabolic reprogramming in cancer cells. Oncogenesis 5:e189. https://doi.org/10.1038/oncsis.2015.49

Bertram JS (2000) The molecular biology of cancer. Mol Asp Med 21:167–223

Brennan JA, Mao L, Hruban RH, Boyle JO, Eby YJ, Koch WM, Goodman SN, Sidransky D (1995) Molecular assessment of histopathological staging in squamous-cell carcinoma of the head and neck. N Engl J Med 332:429–435. https://doi.org/10.1056/NEJM199502163320704

Cancer I, Consortium G (2013) Mutational landscape of gingivo-buccal oral squamous cell carcinoma reveals new recurrently-mutated genes and molecular subgroups. Nat Commun 4:2873. https://doi.org/10.1038/ncomms3873

Chang KY, Tsai SY, Chen SH, Tsou HH, Yen CJ, Liu KJ, Fang HL, Wu HC, Chuang BF, Chou SW, Tang CK (2013) Dissecting the EGFR-PI3K-AKT pathway in oral cancer highlights the role of the EGFR variant III and its clinical relevance. J Biomed Sci 27:20–43. https://doi.org/10.1186/1423-0127-20-43

Chaw SY, Abdul Majeed A, Dalley AJ, Chan A, Stein S, Farah CS (2012) Epithelial to mesenchymal transition (EMT) biomarkers – E-cadherin, beta-catenin, APC and Vimentin – In oral squamous cell carcinogenesis and transformation. Oral Oncol 48:997–1006. https://doi.org/10.1016/j.oraloncology.2012.05.011

Chen HM, Kuo MYP, Lin KH, Lin CY, Chiang CP (2003) Expression of cyclin A is related to progression of oral squamous cell carcinoma in Taiwan. Oral Oncol 39:476–482. https://doi.org/10.1016/S1368-8375(03)00007-1

Chen C, Zimmermann M, Tinhofer I, Kaufmann AM, Albers AE (2013) Epithelial-to-mesenchymal transition and cancer stem (−like) cells in head and neck squamous cell carcinoma. Cancer Lett 338:47–56. https://doi.org/10.1016/j.canlet.2012.06.013

Chen J, Zhou J, Lu J, Xiong H, Shi X, Gong L (2014) Significance of CD44 expression in head and neck cancer: a systemic review and meta-analysis. BMC Cancer 14:15. https://doi.org/10.1186/1471-2407-14-15

Christopoulos A, Ahn SM, Klein JD, Kim S (2011) Biology of vascular endothelial growth factor and its receptors in head and neck cancer: beyond angiogenesis. Head Neck 33:1220–1229. https://doi.org/10.1002/hed.21588

Conesa A, Madrigal P, Tarazona S, Gomez-Cabrero D, Cervera A, McPherson A, Szczesniak MW, Gaffney DJ, Elo LL, Zhang X, Mortazavi A (2016) A survey of best practices for RNA-seq data analysis. Genome Biol 17:13. https://doi.org/10.1186/s13059-016-0881-8

Curry JM, Sprandio J, Cognetti D, Luginbuhl A, Ad VB, Pribitkin E, Tuluc M (2014) Tumor microenvironment in head and neck squamous cell carcinoma. Semin Oncol 41:217–234. https://doi.org/10.1053/j.seminoncol.2014.03.003

Dai M, Thompson RC, Maher C, Contreras-Galindo R, Kaplan MH, Markovitz DM, Omenn G, Meng F (2010) NGSQC: cross-platform quality analysis pipeline for deep sequencing data. BMC Genomics 4:S7. https://doi.org/10.1186/1471-2164-11-S4-S7

Deo MG (1991) High frequency mutation in codons 12 and 61 of H-ras oncogene in chewing tobacco-related human oral carcinoma in India. Br J Cancer 63:573–578

Dodt M, Roehr JT, Ahmed R, Dieterich C (2012) FLEXBAR—flexible barcode and adapter processing for next-generation sequencing platforms. Biology 1:895–905. https://doi.org/10.3390/biology1030895

Dudoit S, Gentleman RC, Quackenbush J (2003) Microarray data. BioTechniques 34:S45–S51

Duraiyan J, Govindarajan R, Kaliyappan K, Palanisamy M (2012) Applications of immunohistochemistry. J Pharm Bioallied Sci 4:S307–S309. https://doi.org/10.4103/0975-7406.100281

Egger G, Liang G, Aparicio A, Peter AJ (2004) Epigenetics in human disease and prospects for epigenetic therapy. Nature 429:457–463. https://doi.org/10.1038/nature02625

Feller L, Lemmer J (2012) Oral squamous cell carcinoma: epidemiology, clinical presentation and treatment. J Cancer Ther 3:263–268. https://doi.org/10.4236/jct.2012.34037

Ferlay J, Soerjomataram I, Dikshit R, Eser S, Mathers C, Rebelo M, Parkin DM, Forman D, Bray F (2015) Cancer incidence and mortality worldwide: sources, methods and major patterns in GLOBOCAN 2012. Int J Cancer 136:E359–E386. https://doi.org/10.1002/ijc.29210

Field JK (1992) Oncogenes and tumour-suppressor genes in squamous cell carcinoma of the head and neck. Eur J Cancer B Oral Oncol 28:67–76. https://doi.org/10.1016/0964-1955(92)90016-T

Gadewal NS, Zingde SM (2011) Database and interaction network of genes involved in oral cancer: version II. Bioinformation 6:169–170. https://doi.org/10.6026/97320630006169

Garewal J, Garewal R, Sircar K (2014) Expression of Bcl-2 and MIB-1 markers in oral squamous cell carcinoma (OSCC)- a comparative study. J Clin Diagn Res 8:QC01–QC04. https://doi.org/10.7860/JCDR/2014/6474.4562

Grabherr MG, Haas BJ, Yassour M, Levin JZ, Thompson DA, Amit I, Adiconis X, Fan L, Raychowdhury R, Zeng Q, Chen Z, Mauceli E, Hacohen N, Gnirke A, Rhind N, Palma FD, Birren BW, Nusbaum C, Lindblad-Toh K, Friedman N, Regev A (2011) Full-length transcriptome assembly from RNA-Seq data without a reference genome. Nature Biotechnol 29:644–652. https://doi.org/10.1038/nbt.1883

Griffith M, Griffith OL, Mwenifumbo J, Goya R, Morrissy AS, Morin RD (2010) Alternative expression analysis by RNA sequencing. Nat Methods 7:843–847. https://doi.org/10.1038/nmeth.1503

Gronroos TJ, Lehtio K, Soderstrom KO, Kronqvist P, Laine J, Eskola O, Minn H (2014) Hypoxia, blood flow and metabolism in squamous-cell carcinoma of the head and neck: correlations between multiple immunohistochemical parameters and PET. BMC Cancer 14:876. https://doi.org/10.1186/1471-2407-14-876

Guttman M, Garber M, Levin JZ, Donaghey J, Robinson J, Adiconis X, Regev A (2010) Ab initio reconstruction of transcriptomes of pluripotent and lineage committed cells reveals gene structures of thousands of lincRNAs. Nat Biotechnol 28:503–510. https://doi.org/10.1038/nbt.1633

Hanahan D, Weinberg RA (2000) The hallmarks of cancer. Cell 100:57–70. https://doi.org/10.1016/S0092-8674(00)81683-9

Harshani JM, Yeluri S, Guttikonda VR (2014) Glut-1 as a prognostic biomarker in oral squamous cell carcinoma. J Oral Maxillofac Pathol 18:372–378. https://doi.org/10.4103/0973-029X.151318

Henry NL, Hayes DF (2012) Cancer biomarkers. Mol Oncol 6:140–146. https://doi.org/10.1016/j.molonc.2012.01.010

Hsieh LL, Wang PF, Chen IH, Liao CT, Wang HM, Chen MC, Chang JT, Cheng AJ (2001) Characteristics of mutations in the p53 gene in oral squamous cell carcinoma associated with betel quid chewing and cigarette smoking in Taiwanese G: C to A: T transitions were the predominant mutations still showed an independent effect on G: C to A: Oxford University Press. Carcinogenesis 22:1497–1503. https://doi.org/10.1093/carcin/22.9.1497

Hu S, Jiang J, Wong DT (2010) Proteomic analysis of saliva: 2D gel electrophoresis, LC-MS/MS, and western blotting. Methods Mol Biol 666:31–41. https://doi.org/10.1007/978-1-60761-820-1_3

Hulka BS, Wilcosky T (1988) Biological markers in epidemiologic research. Arch Environ Health 43:83–89. https://doi.org/10.1080/00039896.1988.9935831

Hung PS, Tu HF, Kao SY, Yang CC, Liu CJ, Huang TY, Chang KW, Lin SC (2014) miR-31 is upregulated in oral premalignant epithelium and contributes to the immortalization of normal oral keratinocytes. Carcinogenesis 35:1162–1171. https://doi.org/10.1093/carcin/bgu024

Jaluria P, Konstantopoulos K, Betenbaugh M, Shiloach J (2007) A perspective on microarrays: current applications, pitfalls, and potential uses. Microb Cell Fact 6:4. https://doi.org/10.1186/1475-2859-6-4

Jones JL, Walker RA (1997) Control of matrix metalloproteinase activity in cancer. J Pathol 183:377–379. https://doi.org/10.1002/(SICI)1096-9896(199712)183:4<377::AID-PATH951>3.0.CO;2-R

Kaneko T, Dehari H, Sasaki T, Igarashi T, Ogi K, Okamoto JY, Kawata M, Kobayashi JI, Miyazaki A, Nakamori K, Hiratsuka H (2016) Hypoxia-induced epithelial-mesenchymal transition is regulated by phosphorylation of GSK3-β via PI3 K/Akt signaling in oral squamous cell carcinoma. Oral Surg Oral Med Oral Pathol Oral Radiol 122:719–730. https://doi.org/10.1016/j.oooo.2016.06.008

Khan M, Mishu M, Imam S (2012) Current molecular concept of oral carcinogenesis and invasion. Med Today 22:38–42. https://doi.org/10.3329/medtoday.v22i1.5605

Kihara D, Yang YD, Hawkins T (2006) Bioinformatics resources for cancer research with an emphasis on gene function and structure prediction tools. Cancer Inform 2:25–35

Krishna A, Singh S, Kumar V, Pal US (2015) Molecular concept in human oral cancer. Natl J Maxillofac Surg 6:9–15. https://doi.org/10.4103/0975-5950.168235

Kurokawa H, Zhang M, Matsumoto S, Yamashita Y, Tanaka T, Tomoyose T, Takano H, Funaki K, Fukuyama H, Takahashi T, Sakoda S (2005) The relationship of the histologic grade at the deep invasive front and the expression of Ki-67 antigen and p53 protein in oral squamous cell carcinoma. J Oral Pathol Med 34:602–607. https://doi.org/10.1111/j.1600-0714.2005.00358.x

Langmead B, Trapnell C, Pop M, Salzberg SL (2009) Ultrafast and memory-efficient alignment of short DNA sequences to the human genome. Genome Biol 10:R25. https://doi.org/10.1186/gb-2009-10-3-r25

Lee SS, Tsai CH, Tsai LL, Chou MC, Chou MY, Chang YC (2012a) B-catenin expression in areca quid chewing-associated oral squamous cell carcinomas and upregulated by Arecoline in human oral epithelial cells. J Formos Med Assoc 111:194–200. https://doi.org/10.1016/j.jfma.2010.11.002

Lee CK, Chi TT, Wu CT, Tsai MT, Chiang CP, Yang CC (2012b) Diagnosis of oral precancer with optical coherence tomography. Biomed Opt Express 3:1632–1646. https://doi.org/10.1364/BOE.3.001632

Levine AE, Steffen DL (2001) OrCGDB: a database of genes involved in oral cancer. Nucleic Acids Res 29:300–302. https://doi.org/10.1093/nar/29.1.300

Levine AJ, Finlay CA, Hinds PW (2004) P53 is a tumor suppressor gene. Cell 116:S67–S70. https://doi.org/10.1016/S0092-8674(04)00036-4

Li J, Tibshirani R (2013) Finding consistent patterns: a nonparametric approach for identifying differential expression in RNA-Seq data. Stat Methods Med Res 22:519–536. https://doi.org/10.1177/0962280211428386

Li Y, St John MA, Zhou X, Kim Y, Sinha U, Jordan RC, Eisele D, Abemayor E, Elashoff D, Park NH, Wong DT (2004) Salivary transcriptome diagnostics for oral cancer detection salivary transcriptome diagnostics for oral cancer detection. Clin Cancer Res 10:8442–8450. doi:10/24/8442[pii]\r10.1158/1078-0432.CCR-04-1167

Lichter P, Cremer T, BordenJ ML, Ward DC (1988) Delineation of individual human chromosomes in metaphase and interphase cells by in situ suppression hybridization using recombinant DNA libraries. Hum Genet 80:224–234. https://doi.org/10.1007/BF01790090

Lin PC, Chang WH, Chen YH, Lee CC, Lin YH, Chang JG (2011) Cytotoxic effects produced by arecoline correlated to epigenetic regulation in human K-562 cells. J Toxicol Environ Health A 74:737–745. https://doi.org/10.1080/15287394.2011.539123

Luscombe NM, Greenbaum D, Gerstein M (2001) What is bioinformatics? An introduction and overview. Yearb Med Inform:83–99. https://doi.org/10.1053/j.ro.2009.03.010

Macey R, Walsh T, Brocklehurst P, Kerr AR, Liu JL, Lingen MW, Ogden GR, Warnakulasuriya S, Scully C (2015) Diagnostic tests for oral cancer and potentially malignant disorders in patients presenting with clinically evident lesions. Cochrane Database Syst Rev 29:CD010276. https://doi.org/10.1002/14651858.CD010276

Macgregor PF, Squire JA (2002) Application of microarrays to the analysis of gene expression in cancer. Clin Chem 48:1170–1177

Manning FCR, Patierno SR (1996) Apoptosis: inhibitors or instigators of carcinogenesis. Cancer Invest 14:455–465. https://doi.org/10.3109/07357909609018903

Markopoulos AK (2012) Current aspects on oral squamous cell carcinoma. Open Dent J 6:126–130. https://doi.org/10.2174/1874210601206010126

Massano J, Regateiro FS, Januario G, Ferreira A (2006) Oral squamous cell carcinoma: review of prognostic and predictive factors. Oral Surg Oral Med Oral Pathol Oral Radiol Endod 102:67–76. https://doi.org/10.1016/j.tripleo.2005.07.038

Masthan KMK, Aravindha Babu N, Dash KC, Elumalai M (2012) Advanced diagnostic aids in oral cancer. Asian Pac J Cancer Prev 13:3573–3576. https://doi.org/10.7314/APJCP.2012.13.8.3573

Mayeux R (2004) Biomarkers: potential uses and limitations. NeuroRx 1:182–188. https://doi.org/10.1602/neurorx.1.2.182

Mehta JP, Rani S (2011) Software and tools for microarray data analysis. Methods Mol Biol 784:41–53. https://doi.org/10.1007/978-1-61779-289-2_4

Mehta R, Gurudath S, Dayansoor S, Pai A, Ganapathy KS (2014) Serum lipid profile in patients with oral cancer and oral precancerous conditions. Dent Res J 11:345–350. https://doi.org/10.4103/1735-3327.135889

Meldrum C, Doyle M, Tothill RW (2011) Next-generation sequencing for cancer diagnostics: a practical perspective. Clin Biochem Rev 32:177–195. https://doi.org/10.1016/j.jmoldx.2016.08.002

Mendes RA (2012) Oncogenic pathways in the development of oral cancer. J Carcinog Mutagen 3:2–3. https://doi.org/10.4172/2157-2518.1000133

Menendez JA, Lupu R (2007) Fatty acid synthase and the lipogenic phenotype in cancer pathogenesis. Nat Rev Cancer 7:763–777. https://doi.org/10.1038/nrc2222

Miyamoto R, Uzawa N, Nagaoka S, Hirata Y, Amagasa T (2003) Prognostic significance of cyclin D1 amplification and overexpression in oral squamous cell carcinomas. Oral Oncol 39:610–618. https://doi.org/10.1016/S1368-8375(03)00048-4

Nagaraj NS, Vigneswaran N, Zacharias W (2004) Hypoxia-mediated apoptosis in oral carcinoma cells occurs via two independent pathways. Mol Cancer 3:38. https://doi.org/10.1186/1476-4598-3-38

Ozsolak F, Milos PM (2011) RNA sequencing: advances, challenges and opportunities. Nat Rev Genet 12(2):87–98. https://doi.org/10.1038/nrg2934

Pande P, Mathur M, Shukla NK, Ralhan R (1998) pRB and p16 protein alterations in human oral tumorigenesis. Oral Oncol 34:396–403. https://doi.org/10.1016/S1368-8375(98)00024-4

Panta P, Venna VR (2014) Salivary RNA signatures in oral cancer detection. Anal Cell Pathol 2014:450629. https://doi.org/10.1155/2014/450629

Patil S, Rao RS, Amrutha N, Agarwal A, Sanketh DS (2013) Cancer stem cells: a revolution in cancer research. J Dent Orofac Res 9:16–20

Petti S (2009) Lifestyle risk factors for oral cancer. Oral Oncol 45:340–350. https://doi.org/10.1016/j.oraloncology.2008.05.018

Polz-Gruszka D, Morshed K, Stec A, Polz-Dacewicz M (2015) Prevalence of human papillomavirus (HPV) and Epstein-Barr virus (EBV) in oral and oropharyngeal squamous cell carcinoma in south-eastern Poland. Infect Agent Cancer 10:37. https://doi.org/10.1186/s13027-015-0031-z

Prokopczyk B, Rivenson A, Hoffmann D (1991) A study of betel quid carcinogenesis. IX. Comparative carcinogenicity of 3-(methylnitrosamino)propionitrile and 4-(methylnitrosamino)-1-(3-pyridyl)-1-butanone upon local application to mouse skin and rat oral mucosa. Cancer Lett 60:153–157. https://doi.org/10.1016/0304-3835(91)90222-4

Ramires M, David L, Leitao D, Seixas M, Sansonetty F, Sobrinho-Simoes M (1997) Ki67 labelling index in gastric carcinomas. An immunohistochemical study using double staining for the evaluation of the proliferative activity of diffuse-type carcinomas. J Pathol 182:62–67. https://doi.org/10.1002/(SICI)1096-9896(199705)182:1<62::AID-PATH849>3.0.CO;2-2

Rashid A, Warnakulasuriya S (2015) The use of light-based (optical) detection systems as adjuncts in the detection of oral cancer and oral potentially malignant disorders: a systematic review. J Oral Pathol Med 44:307–328. https://doi.org/10.1111/jop.12218

Riedmaier I, Pfaffl MW, Meyer HH (2012) The physiological way: monitoring RNA expression changes as new approach to combat illegal growth promoter application. Drug Test Anal 1:70–74. https://doi.org/10.1002/dta.1386

Robertson G, Schein J, Chiu R, Corbett R, Field M, Jackman SD (2010) De novo assembly and analysis of RNA-seq data. Nat Methods 7:909–912. https://doi.org/10.1038/nmeth.1517

Robinson MD, McCarthy DJ, Smyth GK (2010) edgeR: a bioconductor package for differential expression analysis of digital gene expression data. Bioinformatics 26:139–140. https://doi.org/10.1093/bioinformatics/btp616

Rowley H, Sherrington P, Helliwell TR, Kinsella A, Jones AS (1998) P53 expression and P53 gene mutation in oral cancer and dysplasia. Otolaryngol Head Neck Surg 118:115–123. doi.org/S0194-5998(98)70387-0 [pii]

Saranath D, Chang SE, Bhoite LT, Panchal RG, Kerr IB, Mehta AR, Johnson NW, Deo MG (1991) High frequency mutation in codons 12 and 61 of H-ras oncogene in chewing tobacco-related human oral carcinoma in India. Br J Cancer 63:573–578

Scully C (2011) Oral cancer aetiopathogenesis; past, present and future aspects. Med Oral Patol Oral Cir Bucal 16:306–311. https://doi.org/10.4317/medoral.16.e306

Scully C, Field JK, Tanzawa H (2000) Genetic aberrations in oral or head and neck squamous cell carcinoma 2: chromosomal aberrations. Oral Oncol 36:311–327. https://doi.org/10.1016/S1368-8375(00)00021-X

Seitz HK, Stickel F (2010) Acetaldehyde as an underestimated risk factor for cancer development: role of genetics in ethanol metabolism. Genes Nutr 5:121–128. https://doi.org/10.1007/s12263-009-0154-1

Sen S (1992) Programmed cell death: concept, mechanism and control. Biol Rev Camb Philos Soc 67:287–319. https://doi.org/10.1111/j.1469-185X.1992.tb00727.x

Shah S, Kaur M (2014) Biomarkers and chemopreventives in oral carcinogenesis and its prevention. JOMFP 18:69–76. https://doi.org/10.4103/0973-029X.131914

Shah FD, Begum R, Vajaria BN, Patel KR, Patel JB, Shukla SN, Patel PS (2011) A review on salivary genomics and proteomics biomarkers in oral cancer. Indian J Clin Biochem 26:326–334. https://doi.org/10.1007/s12291-011-0149-8

Shah S, Pathak P, Gulati N (2015) Cell signaling pathways in oral cancer: a review. JOADMS 1:69–74

Shin D, Vigneswaran N, Gillenwater A, Kortum RR (2010) Advances in fluorescence imaging techniques to detect oral cancer and its precursors. Future Oncol 6:1143–1154. https://doi.org/10.2217/fon.10.79.Advances

Shpitzer T, Hamzany Y, Bahar G, Feinmesser R, Savulescu D, Borovoi I, Gavish M, Nagler RM (2009) Salivary analysis of oral cancer biomarkers. Br J Cancer 101:1194–1198. https://doi.org/10.1038/sj.bjc.6605290

Sigurdsson K (2013) Is a liquid-based cytology more sensitive than a conventional pap smear? Cytopathology 24:254–263. https://doi.org/10.1111/cyt.12037

Singh R, Srivastava P, Srivastava A, Mittal RD (2010) Matrix metalloproteinase (MMP-9 and MMP-2) gene polymorphisms influence allograft survival in renal transplant recipients. Nephrol Dial Transplant 25:3393–3401. https://doi.org/10.1093/ndt/gfq174

Slaughter DL, Southwick HW, Smejkal W (1953) Field cancerization in oral stratified squamous epithelium: clinical implication of multicentric origins. Cancer 6:963–968

St John MA, Li Y, Zhou X, Denny P, Ho CM, Montemagno C, Shi W, Qi F, Wu B, Sinha U, Jordan R, Wolinsky L, Park NH, Liu H, Abemayor E, Wong DT (2004) Interleukin 6 and interleukin 8 as potential biomarkers for oral cavity and oropharyngeal squamous cell carcinoma. Arch Otolaryngol Head Neck Surg 130:929–935. https://doi.org/10.1001/archotol.130.8.929

Stankiewicz E, Kudahetti SC, Prowse DM, Ktori E, Cuzick J, Ambroisine L, Zhang X, Watkin N, Corbishley C, Berney DM (2009) HPV infection and immunochemical detection of cell-cycle markers in verrucous carcinoma of the penis. Mod Pathol 22:1160–1168. https://doi.org/10.1038/modpathol.2009.77

Sushma PS, Jamil K, Uday Kumar P, Satyanarayana U, Ramakrishna M, Triveni B (2016) PTEN and p16 genes as epigenetic biomarkers in oral squamous cell carcinoma (OSCC): a study on south Indian population. Tumor Biol 37:7625. https://doi.org/10.1007/s13277-015-4648-8

Syam sundar B, Nageswara RR, Faheem K (2012) Epidemiological and clinicopathological study of oral cancers in a tertiary care hospital. Int J Biol Med Res 3:2376–2380

Teni T, Pawar S, Sanghvi V, Saranath D (2002) Expression of bcl-2 and bax in chewing tobacco-induced oral cancers and oral lesions from India. Pathol Oncol Res 8:109–114. doi.org/PAOR.2002.8.2.0109

Thiery JP (2002) Epithelial-mesenchymal transitions in tumour progression. Nat Rev Cancer 2:442–454. https://doi.org/10.1038/nrc822

Thomas GJ, Speight PM (2001) Cell adhesion molecules and oral cancer. Crit Rev Oral Biol Med 12:479–498

Tiwari A, Shivananda S, Gopinath KS, Kumar A (2014) MicroRNA-125a reduces proliferation and invasion of oral squamous cell carcinoma cells by targeting estrogen-related receptor α: implications for cancer therapeutics. J Biol Chem 289:32276–32290. https://doi.org/10.1074/jbc.M114.584136

Trapnell C, Pachter L, Salzberg SL (2009) TopHat: discovering splice junctions with RNA-Seq. Bioinformatics 25:1105–1111. https://doi.org/10.1093/bioinformatics/btp120

Tumuluri V, Thomas GA, Fraser IS (2002) Analysis of the Ki-67 antigen at the invasive tumour front of human oral squamous cell carcinoma. J Oral Pathol Med 31:598–604. https://doi.org/10.1034/j.1600-0714.2002.00042.x

Wang K, Singh D, Zeng Z, Coleman SJ, Huang Y, Savich GL, He X, Mieczkowski P, Grimm SA, Perou CM, MacLeod JN (2010) MapSplice: accurate mapping of RNA-seq reads for splice junction discovery. Nucleic Acids Res 38:e178. https://doi.org/10.1093/nar/gkq622

Waris G, Ahsan H (2006) Reactive oxygen species: role in the development of cancer and various chronic conditions. J Carcinog 5:14. https://doi.org/10.1186/1477-3163-5-14

Wei L, Xiaohe Z, Kunping L, Sida Z, Yixiang D (2013) Laser-induced fluorescence: progress and prospective for in vivo cancer diagnosis. Chin Sci Bull 8:2003–2016. https://doi.org/10.1007/s11434-013-5826-y

Wilson GD, Saunders M, Dische S, Richman P, Daley F, Bentzen SM (2001) Bcl-2 expression in head and neck cancer: an enigmatic prognostic marker. Int J Radiat Oncol Biol Phys 49:435–441. https://doi.org/10.1016/S0360-3016(00)01498-X

Wong D, Todd R (1996) Molecular biology of human oral cancer. Crit Rev Oral Biol Med 7:319–328. https://doi.org/10.1007/978-1-4020-3186-1

Wu D, Rice CM, Wang X (2012) Cancer bioinformatics: a new approach to systems clinical medicine. BMC Bioinformatics 13:71. https://doi.org/10.1186/1471-2105-13-71

Xia X, McClelland M, Wang Y (2005) WebArray: an online platform for microarray data analysis. BMC Bioinformatics 6:306. https://doi.org/10.1186/1471-2105-6-306

Xue J, Yang S, Seng S (2014) Mechanisms of cancer induction by tobacco-specific NNK and NNN. Cancers 6:1138–1156. https://doi.org/10.3390/cancers6021138

Yamamoto N, Onda T, Sugahara K, Nomura T, Shibahara T (2015) Molecular biological change in oral cancer, summary of our researches. Jpn Dent Sci Rev 51:25–33. https://doi.org/10.1016/j.jdsr.2014.09.003

Yang X, Liu D, Liu F, Wu J, Zou J, Xiao X, Zou F, Zhu B (2013) HTQC: a fast quality control toolkit for Illumina sequencing data. BMC Bioinformatics 14:33. https://doi.org/10.1186/1471-2105-14-33

Yong-Deok K, Eun-Hyoung J, Yeon-Sun K, Kang-Mi P, Jin-Yong L, Sung-Hwan C, Tae-Yun K, Tae-Sung P, Soung-Min K, Myung-Jin K, Jong-Ho L (2015) Molecular genetic study of novel biomarkers for early diagnosis of oral squamous cell carcinoma. Med Oral Patol Oral Cir Bucal 20:e167–e179. https://doi.org/10.4317/medoral.20229

Zedan W, Mourad MI, El-Aziz SMA, Salamaa NM, Shalaby A (2015) Cytogenetic significance of chromosome 17 aberrations and P53 gene mutations as prognostic markers in oral squamous cell carcinoma. Diagn Pathol 10:2. https://doi.org/10.1186/s13000-015-0232-1

Zhou J, Tao D, Xu Q, Gao Z, Tang D (2015) Expression of E-cadherin and vimentin in oral squamous cell carcinoma. Int J Clin Exp Pathol 8:3150–3154

Zou D, Ma L, Yu J, Zhang Z (2015) Biological databases for human research. Genomics Proteomics Bioinformatics 13:55–63. https://doi.org/10.1016/j.gpb.2015.01.006

Soft Computing Methods and Tools for Bacteria DNA Barcoding Data Analysis

13

Ravi Prabhakar More and Hemant J. Purohit

Abstract

DNA barcoding is a modern and extensively used molecular-based recognition method that aims to categorize biological specimens and to affiliate them to a given species. Newly, the progress of next-generation sequencing technology has become growingly important in the bacterial taxonomy analysis, sequence classification, and species recognition. This chapter describes the major 16S rRNA gene sequence databases and tools available for DNA barcoding studies. Here we reviewed bioinformatics, tools and methods are summarized that can support researchers who accurately prepare a database query to be capable of retrieving the most proper information required for their area of research. The aim of the present work is to draw outline of the current scenario of bacterial DNA barcoding with respect to bacterial sequence classification and species identification.

Keywords

Bacterial DNA barcoding · 16S rRNA gene · Ribosomal databases · Similarity and compositional methods · Signatures

R. P. More
ADBS, TIFR-National Centre for Biological Sciences (NCBS), Bangalore, Karnataka, India
e-mail: ravimore7@yahoo.in

H. J. Purohit (✉)
Environmental Biotechnology and Genomics Division, CSIR-National Environmental Engineering Research Institute (NEERI), Nagpur, Maharashtra, India
e-mail: hj_purohit@neeri.res.in; hemantdrd@hotmail.com

H. J. Purohit et al. (eds.), *Soft Computing for Biological Systems*,
https://doi.org/10.1007/978-981-10-7455-4_13

13.1 Introduction

16S ribosomal RNA is a part of the 30S small subunit of the prokaryotic ribosome. Sequencing the ribosomal RNA gene (rRNA) is the technique of option for nucleotide sequence-based bacterial identification to estimate microbial diversity. The function of the 16S rRNA gene over time has not changed, suggesting it has a more accurate means for the study of the evolutionary analysis (Janda and Abbott 2007). Bioinformatics is playing an important role in the analysis of DNA barcoding data using various tools and databases. Different methods have been implemented to deal with DNA barcoding data along with similarity-based methods. Recent studies suggest that alignment-free and Bayesian algorithms are used for taxonomic classification method for 16S rRNA gene sequence analysis (Gao et al. 2017; Lu et al. 2017) and has proposed pattern-based signature approach for taxa classification in Bacteria and plants (More and Purohit 2016; More et al. 2016). The procedure of DNA barcoding primarily includes the accessible data collection of the existing databases, and several databases have been available in recent years. The NCBI Taxonomy offers easy access to the Entrez search engine for users to find all the information about a particular taxon, from the species level up to genus, family, order, and higher levels of the hierarchy. The NCBI Taxonomy database (http://www.ncbi.nlm.nih.gov/taxonomy) was designed to provide nomenclature and identification of the taxonomic origin (Federhen 2012). On another side, recent advances in sequencing technologies have significantly improved our understanding of microbial diversity (Fadrosh et al. 2014). A quick fall in sequencing cost per nucleotide has enabled to generate a huge amount of data, and researcher has deposited into the various public databases. The accurate identification of species is depending upon the error-free unambiguous and overall quality of nucleotide sequences available in public databases. There are evidences available that many sequences are deposited with poor quality (Heikens 2005). Therefore, it is important to refer good-quality non-redundant database for DNA barcoding study. In this survey, we are discussing about the useful databases and methods for DNA barcoding.

13.2 16S rRNA Sequence Databases

Researchers can deposit and access 16S rRNA sequences in a number of public and commercial depositaries. The very popular public database of the National Center for Biotechnology Information (NCBI) is the GenBank database (Wheeler et al. 2007), which is having multiple record information of sequence (Clayton et al. 1995). It has many redundant sequence records. For DNA barcoding purpose, many database available, which are having quality-checked entries. There are three major databases, mainly Ribosomal Database Project (RDP) (Cole et al. 2005, 2014), SILVA (Quast et al. 2013), and Greengenes (DeSantis et al. 2006), which possibly

Table 13.1 The major online 16S ribosomal RNA (rRNA) sequence databases

Sr. no.	Database	URL	References
1	Ribosomal Database Project (RDP II)	http://rdp.cme.msu.edu/	Cole et al. (2005)
2	SILVA	http://www.arb-silva.de	Quast et al. (2013)
3	Greengenes	http://greengenes.lbl.gov	DeSantis et al. (2006)
4	EzTaxon	http://eztaxon-e.ezbiocloud.net/	Chun et al. (2014)
5	BIBI	http://pbil.univ-lyon1.fr/bibi/	Devulder et al. (2003)

comprise the high-quality datasets that have been designed for bacterial classification by using 16S rRNA gene sequence as shown in Table 13.1.

13.2.1 Ribosomal Database Project (RDP) Database

The Ribosomal Database Project (RDP) designed to analysed rRNA gene sequences of Bacteria. It includes the 3,356,809 16S rRNA and 125,525 fungal 28S rRNA sequences. RDP is a versatile database with a large number of functions and annotation tools for a wide range of analyses for Bacteria. RDP is the rRNA gene sequence database of Michigan State University (MSU) and offers a relevant information along with appropriate tools for sequence analysis. It has many useful tools and related links to supporting data (Wang et al. 2007; Cole et al. 2014). Some of them are listed below:

1. Hierarchy Browser: With this page, the user can download 16s rDNA sequences with various filters such as strain, source, size, quality, and taxonomy. It is available at https://rdp.cme.msu.edu/hierarchy/hb_intro.jsp.
2. RDP Classifier (Wang et al. 2007): This is a rapid tool for sequence classification using the naïve Bayesian algorithm. The user can submit 16S rRNA or ITS sequence and get taxonomic affiliation against RDP or UNITE database. RDP Classifier takes input in the form of the FASTA, GenBank, or EMBL format and searches against target training database to get respective taxonomy of query sequence. At present, the user can submit 100,000 sequences at a time to perform the analysis. This tool can be obtained at https://rdp.cme.msu.edu/classifier/classifier.jsp.
3. RDP Pipeline: RDP's Pipeline offers the handling of vast rRNA arrangement libraries, which contain paired- and single-end sequences obtained through high-throughput sequencing techniques. This pipeline performs assembly of sequences, quality trimming, and taxonomic classification. The output of this tool can be used for different statistical software packages. It is available at http://pyro.cme.msu.edu/.

4. ProbeMatch: This tool can be used for searching best-fit sequences to query probe in the RDP's database. It is available at https://rdp.cme.msu.edu/probematch/search.jsp.
5. FunGene: This tool offers an interactive front end of sequence search results for those concerned about specific gene family (Fish et al. 2013). It is very useful for functional genomics and related studies. It is available at http://fungene.cme.msu.edu/.

13.2.2 SILVA Database

SILVA database offers detailed, quality-checked, and recently restructured entries of aligned ribosomal RNA (rRNA) sequences (large subunit (23S/28S, LSU) and small (16S/18S, SSU)) for Archaea, Bacteria, and Eukarya (Quast et al. 2013). All 16S rRNA sequences were downloaded from SILVA database version (SSURef_NR99_115_tax_silva) located at https://www.arb-silva.de/. SILVA database is a high-quality ribosomal RNA database for the retrieval of the 16S rRNA sequences of the culturable microorganisms. The importance of the database is that it provides the reference datasets of high-quality, full-length sequences capable of in-depth phylogenetic analysis and probe design (Pruesse et al. 2012). SSU Parc and LSU Parc are a small subunit rRNA database, which comprises all aligned entries with an alignment identity score equal and above 50 and 40, respectively.

13.2.3 Greengenes Database

Greengenes (DeSantis et al. 2006) is a chimera-checked 16S rRNA gene database which is famous for 16S rRNA sequence collection. Greengenes is the publicly available database (http://greengenes.lbl.gov/cgi-bin/nph-index.cgi.) that provides access of 16S rRNA gene sequences and downloading sequence entries, similarity search using BLAST, and probing. It also offers tools for probes, microarray data, and annotation of sequences. It has the following various tools to analyze 16S rRNA gene sequences:

1. Trim (http://greengenes.lbl.gov/cgi-bin/nph-trim_fasta_by_qual.cgi): This tool can be used for trim input fast sequences based on their quality scores.
2. Export (http://greengenes.lbl.gov/cgi-bin/nph-export_records.cgi): User can input a list of NCBI accession numbers separated by spaces/tabs/newlines. Also, the user can apply filters like minimum nucleotide count and maximum non-ACGT character count; sequences must have a prokMSAname assigned, and sequences must have a chimera test result.
3. SimRank (http://greengenes.lbl.gov/cgi-bin/nph-compare_choices.cgi): This is a sequence search tool that is useful for similarity searching by comparison of k-mers against Greengenes sequences. The user can perform a search for a batch of aligned sequences.

13.3 Approaches Used for DNA Barcoding

In Bacteria, current methods to identify unknown taxa sequences frequently depend on algorithm recall and precision, and many identification methods do not offer a degree of confidence for all the taxa. In silico predictions showed that precise taxonomic identification is highly reliant on 16S rRNA sequence quality, sequencing technology, and computational algorithm (Golob et al. 2017; Ramiro-Garcia et al. 2016; Edgar 2016; Chen et al. 2013, 2016). So far, methodological papers available on DNA barcoding have typically focused on the three types of approaches that are utilized for the taxonomic identification using 16S rRNA sequences: composition-based (word-based), similarity-based, and phylogenetic-based approaches as shown in Table 13.2. First is a composition-based method, which is based on oligonucleotide features that can be considered in primary sequence data and shows a direct association from the sequences (e.g., k-mer frequencies and Markov models). Second, similarity-based (alignment or homology) methods rely on a sequence similarity with reference library sequences. Similarity-based approach primarily utilized pairwise or multiple alignments to assign the taxonomic nodes of query 16S rRNA sequences (e.g., BLAST and

Table 13.2 Overview of the 16S rRNA sequence mining tools and their characteristics

Methods	Program	Algorithm used	Interface	URL	References
Composition-based (alignment-free)	RDP naïve Bayesian classifier	Supervised	Command line/Web-based	http://rdp.cme.msu.edu/	Wang et al. (2007)
	k-means/Knn	Supervised	Command line	http://www.mothur.org/wiki/Classify.seqs	
Similarity-based (alignment-based)	MEGAN	BLAST	GUI	http://ab.inf.uni-tuebingen.de/software/megan/	Huson et al. (2007)
	SILVA Incremental Aligner (SINA)	MSA	Web-based/command line	http://www.arb-silva.de/aligner/	Pruesse et al. (2012)
	MG-RAST	BLAST	Web-based	http://metagenomics.anl.gov/	Meyer et al. (2008)
	MARTA	BLAST	Command line	http://bergelson.uchicago.edu/software/marta	Horton et al. (2010)
	TUIT	BLAST	Web-based	http://sourceforge.net/projects/tuit	Tuzhikov et al. (2014)
Phylogeny-based	Greengenes (NAST, SimRank)	Other	Web-based command	http://greengenes.lbl.gov/cgi-bin/nph-classify.cgi	DeSantis et al. (2006)

MSA). Third is the phylogenetic methods, in which defined query best "fits" in the phylogeny by applying an evolutionary algorithm to the close relative sequences on a phylogenetic tree; and the method uses the following algorithms: neighbor joining (NJ), Bayesian methods, and maximum likelihood (ML) (Munch et al. 2008).

13.3.1 Similarity-Based Methods

There are many types of approaches reported to deal with 16S rRNA sequences. BLAST hosted by NCBI is the most common for similarity search (homology) against reference databases (nucleotides and proteins). However, the BLAST does not automatically offer an exact taxonomic identification for the user query sequence. In fact, BLAST examinations give an extensive list of homolog hits and need to describe the taxonomy of the query at the deepest taxonomic rank possible. In order to obtain taxonomic node of query sequence, generally the researcher does the BLAST homology search and then performs multiple sequence alignment (MSA) using tools like ClustalW (Thompson et al. 1994), MEGA7 (Kumar et al. 2016), T-Coffee (Notredame et al. 2000), etc. followed by the use of phylogenetic methods such as neighbor joining, parsimony, Bayesian (Munch et al. 2008), and nearest neighbor to detect taxonomic nodes using phylogeny packages like PHYLIP (Felsenstein 2002). However, this approach has certain limitations such as difficult to align in large sequence sets and computational task processing time (Mohammed et al. 2011).

In the literature, many factors have been mentioned that accounts for the misleading identification, such as a partial alignment process; since no close homologs are available in the database which can be used as a reference (Chu et al. 2006), Van Velzen et al. (2012) have introduced the performance of methods including Fitch and Margoliash, neighbor joining, parsimony, and nearest neighbor. They mainly utilize algorithms and databases that also require domain knowledge and computational time to perform the analysis. However, MSA approach has certain limitations such as difficult to align in large sequence datasets and processing time (Cameron et al. 2006; Nielsen and Matz 2006). Recently, alignment-based efficient method SINA (SILVA Incremental Aligner) has been proposed for taxonomic classification, which utilizes a blend of k-mer searching and partial order alignment (POA) to keep very great alignment accuracy. It has implemented an option to categorize sequences with the least common ancestor (LCA) method. SINA can accurately align hundred thousand of sequences on the basis of reference of curated SEED alignment. In the analysis, the first step requires the aligner to define the next associated sequences by an optimized suffix tree server (Quast et al. 2013). Another Greengenes classifier correctly aligned query sequence with the prokMSA to discover near neighbors using SimRank, and then sequence deviation from near neighbors will be considered using the DNAML selection of DNADIST (PHYLIP package) (DeSantis et al. 2006). Alignment-based programs, like MEGAN (MEtaGenome ANalyzer) (Huson et al. 2007) and MG-RAST (the Metagenomics RAST) (Meyer et al. 2008), compare 16S rRNA sequences against sequences in public databases (e.g., NCBI nr database, RDP, or SILVA) using BLASTn and then allocate them according to their most recent common ancestor (LCA) algorithm such as of source organisms. Basically, the similarity-based (BLASTn) identification assumed to obtain alignment strategy between the query and reference sequences. On the other hand, MG-RAST automatic annotation server has been included in the databases such as Greengenes, RDP II, and European ribosomal RNA as the reference information to perform 16s rRNA classification of sequences. In the case of TUIT (Taxonomic Unit Identification Tool), it relies on standard BLAST results and a taxonomic database search engine for effective taxonomic identification of nucleotide sequences (Tuzhikov et al. 2014). Similarly, the MARTA tool is utilized on NCBI BLAST software and taxonomy database with different inbuilt parameter options to predict taxonomic ranks (phylum to genus) of the query sequence (Horton et al. 2010). Also, DNA QR Code Web Server is developed to identify plant species by using BLAST (Liu et al. 2012).

13.3.2 Composition-Based Methods

Recently, methods using sequence composition-based features are widely used in the analysis. Perticularly, k-mer frequencies have commonly been utilized since they carry phylogenetic information (Liu et al. 2013; Fan et al. 2014; Raje et al. 2010) which showed that k-mers in terms of pentanucleotide frequencies were highly significant within and between bacterial sequences. Supervised composition-based methods need a reference library sequences with identified taxonomic source. It referred the reference information to find out sequence features of each taxonomic rank during a training stage. Accordingly, the trained classifier is applied to detect the taxa of nucleotide string of unknown source (Bazinet and Cummings 2012).

The RDP II Classifier has implemented a naïve Bayesian algorithm that reaches its level of efficacy by using a library of known bacterial 16S rRNA gene sequences and algorithm that does not need sequence alignment scheming. It gives the taxonomic assignment for a query sequence with a bootstrap confidence score for each taxonomic rank according to the taxonomic hierarchy. It is based on Bayes' theorem by observing overall probability of k-mer (8 bp) composition in sequence, and it is faster than the BLAST-based methods (Porter et al. 2014). Recently, it has been proposed that the RDP Classifier (i.e., naïve Bayesian) is one of the most efficient tools to classify 16S rRNA sequences (Lan et al. 2012). There is another well-known similar classifier based on a k-nearest neighbor (k-NN) algorithm that uses a character-matching scheme that determines the percentage of heptamer frequencies between a query and members of a database of sequences (Cole et al. 2005). Both RDP and k-NN classifiers can offer the facility to select RDP or SILVA database as their reference for taxonomic identification.

Recent studies have pointed out that RDP Classifier usually results in higher prediction accuracy in most 16S rRNA sequence dataset with optimal sensitivity and specificity (Porter et al. 2014). Even though the RDP Classifier is usually considered superior, other computational approaches, such as similarity-based approaches, have shown a comparable level of accuracy detection rate (Liu et al. 2008). Also, the RDP training library sequences comprise only a limited number of well-categorized sequences at different taxonomic levels. Databases from the NCBI, on the other hand, are regularly updated and contain the latest deposited sequence information. There is a chance a BLAST search against databases can give additional information that somehow detects sequences that RDP Classifier failed to classify; thus, these two approaches are complementary to each other and have their own strength and limitations (Tuzhikov et al. 2014).

This can be summarized as composition- and similarity-based methods can proficiently and with higher accuracy detect specimens as mentioned in Table 13.3 (Chan and Ragan 2013). There are however many challenges are still to be addressed for accurate analyses that are needed for DNA barcoding including the specificity of identification as well as the efficiency and scalability of computational methods. Although multiple sequence alignment is routinely referred by researchers, it is observed that MSA has computationally time-consuming procedure with large number of sequences. Hence, it is one of the possible limitations of MSA. Therefore, alignment-free methods have been focused by researchers to overcome the boundaries of alignment-based methods (Van Velzen et al. 2012; Kuksa and Pavlovic 2009). The composition vector (CV) method comes under the alignment-free method, which utilized the frequencies of nucleotide or amino acid patterns to signify sequence identity and showed good results in comparative

Table 13.3 Comparison of key features between multiple sequence alignment and alignment-free approaches

S. N.	Multiple sequence alignment	Alignment-free methods
1	Assumes contiguity (with gaps) of homologous regions	Does not assume contiguity of homologous regions
2	Based on all possible pairwise comparisons of whole sequences; computationally expensive	Based on occurrences of subsequences, computationally inexpensive, can be memory- intensive
3	Well-established and well-studied approach in phylogenomics	Application in phylogenomics limited; requires further testing for robustness and scalability
4	More dependent on substitution/ evolutionary models	Less dependent on substitution/ evolutionary models
5	More sensitive to stochastic sequence variation, recombination, lateral genetic transfer, rate heterogeneity, and sequences of varied lengths	Less sensitive to stochastic sequence variation, recombination, lateral genetic transfer, rate heterogeneity, and sequences of varied lengths
6	Best practice uses inference algorithms with complexity at least O(n2); less time-efficient	Inference algorithms typically O(n2) or less; more time-efficient
7	Heuristic solutions; statistical significance of how alignment scores relate to homology is difficult to assess	Exact solutions; statistical significance of the sequence distances (and degree of similarity) can be readily assessed

genomics of prokaryotes (Liu et al. 2013; Chan et al. 2010). Composition-based method is comprised of three main steps: first is a fixed integer k, to count the number of overlapping k-mers in one nucleotide sequence and create a vector based on calculated frequency or probability of k-mers. Usually, the difference among two k-mer frequency vectors is used to calculate the distances among two taxa based on distances among all taxa (Yu et al. 2013).

The generation of signatures based on k-mer frequencies by utilizing nucleotide sequences (Bacteria or plant) can offer valued indications to taxonomic affiliation (More and Purohit 2016; More et al. 2016). On a similar line, the TETRA program has been reported for microorganism identification based on comparative tetranucleotide frequency analysis (Teeling et al. 2004). Moreover, it referred to pre-computed tetranucleotide usage patterns for 166 prokaryote genomes as a reference dataset, indicating that k-mer frequencies play an important role for genome discrimination. Sequence classification based on CV analysis could have other applications in DNA barcoding purpose. The determination of oligonucleotide frequencies of DNA fragments would facilitate easy classification of taxon-specific motifs that can be used as taxon-specific motifs for taxon classification (Summerbell et al. 2005). It was reported that classification of the large dsDNA viruses on the basis of the molecular composition vector method is reliable than with those based on the conventional analysis (Yu et al. 2010). Qi et al. (2004) reported the composition vector (CV) method for the whole-genome-based prokaryotic phylogeny analysis. Due to its achievement toward sequence analysis, quite a few more approaches have been developed along this direction.

The CV method comprises of the following four features: (i) construct the frequency vectors based on different oligonucleotide frequencies from sequences; (ii) for each species sequence, compute signal-to-noise ratio; (iii) calculate the distance between every pair of composition vectors; and (iv) build the phylogenetic trees based on the distance matrix (Chan et al. 2010). A combination of both phylogeny-based and composition-based methods—Metagenome Composition Vector (MetaCV)—for recognition and taxonomic origin of sequenced environmental reads has been recently described (Liu et al. 2013). On similar lines, Chu et al. (2009) confirmed that the analysis of composite vector based on COI, 18S, and 16S rRNA sequences is a trustworthy clustering approach for DNA barcoding purpose. Interestingly, the capability of unaligned rRNA gene sequences as DNA barcodes using composition vectors was tested on datasets from Archaea to tetrapods at taxonomic ranks (class to species); this has indicated that clustering without alignment is always reliable with the phylogenetic trees created by conventional methods (Chu et al. 2006). Moreover, the use of such k-mer composition vector signatures could provide as a taxon-specific signature (Qi et al. 2004). Our earlier study by Raje et al. (2010) reported that tetra- and pentanucleotide features in self-organizing maps (SOM) could discriminate the 16S rRNA sequence with more than 90% accuracy. They have also demonstrated that five most closely associated bacterial genera could be differentiated using the dimer nucleotide frequencies of 16S rRNA genes and suggested the use of k-mer features for creating signatures in the bacterial taxonomy identification process. To decrease the computational task processing time of the similarity search tool like BLAST step, the analysis can be limited to a specific signature (Segata et al. 2012) or marker (Liu et al. 2011) genes. One of the major advantages of this composition-based approach is their capability to rapidly provide accurate taxonomy identification of a large number of query sequences without doing conventional methods mainly BLAST (Altschul et al. 1990), MSA, and phylogeny (Swoford 2002). Therefore, we believe that the CV approach-based regular expression can also sidestep the problems associated with sequence alignment in analyzing large datasets of 16S rRNA.

13.3.3 Gene-Specific Signature (i.e., Regular Expression)-Based Method

Gene-specific signature method deals with taxonomic classification of microorganisms using regular expression (signature), which comprises of discriminating patterns from 16S rRNA marker genes. In the present context, a pattern or motif (e.g., GCCCA) is a term represented by a subsequence that is highly conserved in nucleotide region within an entire sequence.

The term "regular expression" (now commonly abbreviated to "RegExp" or even "RE") simply refers to a set of patterns that follows the rules of syntax mainly the distance between two consecutive patterns. A regular expression is comprised of pattern and word boundary that can be searched against a fragment of text. Detection of regular expression against a fragment of text either succeeds or fails. In other words, any pattern (e.g., set of nucleotides) in a regular expression requires a corresponding pattern in the given string (Schwartz et al. 1997).

Here is an example of signature (i.e., regular expression), considered as a DNA barcode that is specific to 16S rRNA gene.

Example: GCCSR\w{19,63}AKHAKGG\w{117,150}BRGCWWAMTWC\w{372,408}GDVWHTYHHD\w{92,170}HSWWRWD\w{60,141}RKWWKD\w{15,80}SYYYHTDWK

- Where GCCSR pattern considers degeneracy that could be presented as one of the following ways: GCCCA, GCCCG, GCCGA, or GCCGG due to S(C/G) and R (A/G) base-pair IUPAC code scheme.
- \w represents the word characters.
- {*x*,*y*}—in notation *x* represented as minimum base-pair boundary and *y* as maximum base-pair boundary between the two consecutive patterns.
- All patterns are represented in regular expression as per IUPAC code system.

To demonstrate signature as a DNA barcode concept, the pattern-based gene signature is depicted in Fig. 13.1 by referring 16S rRNA sequence of *Alicyclobacillus acidiphilus* TA-67 [Ribosomal Database Project (RDP) Accession No. AB076660]. The pattern-based gene-specific signature possesses two important characteristics: first, it is a unique combination of taxa-specific nucleotide patterns and, second, the separating distance between consecutive patterns. These

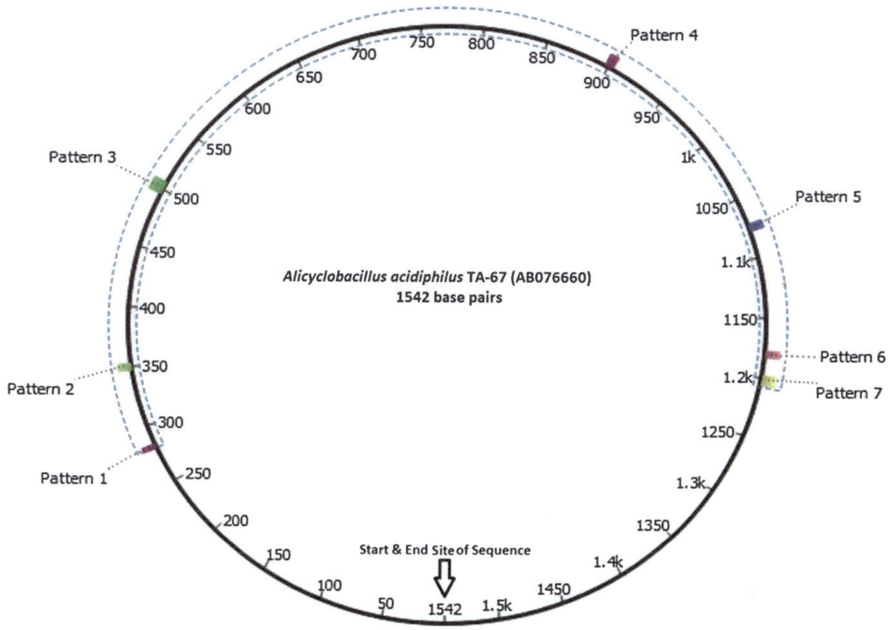

Fig. 13.1 The signature as a DNA barcode for sample sequence of *Alicyclobacillus acidiphilus* TA-67 [RDP Accession No. AB076660]. The dotted blue line represents the signature region covered as DNA barcode by comprising patterns 1 to 7 in regular expression

characteristics collectively provide an identity to particular taxa. The conceptual signature as a DNA barcode is represented in Fig. 13.1 for *Bacilli* class using *Alicyclobacillus acidiphilus* TA-67 as a representative sequence.

The signature subsequence of 925 base pairs (bp) was obtained from a total of 1542 bp. It spans a region from 281 bp to 1205 bp, having unique discriminating patterns at 281, 348, 497, 897, 1071, 1177, and 1197 start positions. All these patterns, their sizes, and the separating distance between consecutive patterns are shown in Table 13.4. The region span of signature at each taxonomic level (class to species) could be different from each other. Table 13.4 showed the minimum and maximum distance between two patterns and the overall span of signature at each taxon.

In previous study, it was demonstrated that based on 16S rRNA gene, the set of features (*k*-mers refer to a specific n-tuple of nucleic acid) discriminate particular organisms with higher precision (Raje et al. 2010). This is helpful in data mining research area taxa-specific (different taxonomic level, class to species) features which could be recognized with respect to their frequencies and applicable to develop regular expressions as shown in Fig. 13.2. In case of class, signature holds highly variable C_1 to C_5 patterns with certain fix distance between two consecutive patterns, and both characteristics are responsible for discrimination at class level. In order level, O_4 and O_5 patterns are discriminated from other classes,

Table 13.4 Summary of nuclutide patterns participated into the molocular signature

Pattern no.	Pattern	Size	Position		Distance between consecutive patterns
			Start	End	
Pattern 1	GCCSR	5	281	285	64
Pattern 2	AKHAKGG	7	348	354	142
Pattern 3	BRGCWWAMTWC	11	497	507	390
Pattern 4	GDVWHTYHHD	10	897	903	168
Pattern 5	HSWWRWD	7	1071	1077	100
Pattern 6	RKWWKD	6	1177	1182	15
Pattern 7	SYYYHTDWK	9	1197	1205	

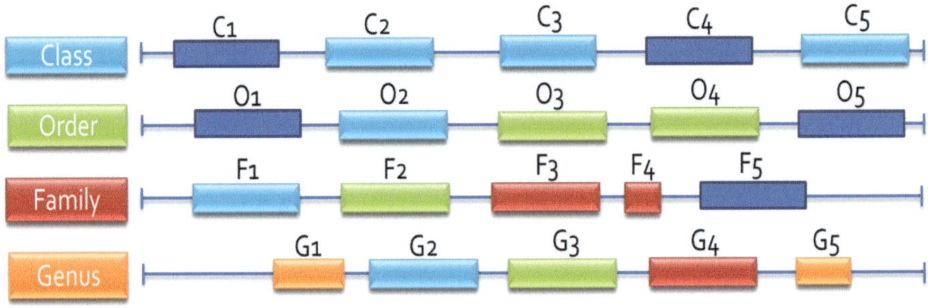

Fig. 13.2 The schematic representation of signatures according to the taxonomic hierarchical level (from class to genus)

and O_3, O_4, and O_5 patterns are responsible for order-level discrimination in the same class. The logical patterns of discrimination are considered at family and genus level. Patterns related to typical biological motifs may be likely to raise due to the statistical nature of large sequence datasets. In other words, motifs with strength similar to real discriminating motifs begin to occur by chance (Zia and Moses 2012). On similar line, it reported the use of pattern-based DNA barcodes in 16S rRNA gene, 26 base pair (bp) for phyla *Firmicutes*, and 12 bp for *Bacteriodetes*, which is a rapid method for taxonomy identification (Armougom and Raoult 2008). Conceptually, it is globally accepted that unique oligonucleotide frequencies exist in marker gene that make them generally discriminated from other sequences. However, to our knowledge, there is no methodical study that explores the use of patterns (oligonucleotide frequencies) with ordered distance arrangement to generate signature-based approach to assign the taxonomy to unknown sequence.

13.4 Perspectives

DNA barcoding is widely used worldwide. The present chapter covers tools and databases to bacterial DNA barcoding. Research on 16S rRNA gene sequences is yet very limited, and still, it is in early stages. So, DNA barcoding research has to progress for future studies. Bacterial DNA barcoding will be valuable for recognition of bacterial species. Using the available resources, the detection of unspecified or unidentified bacterial sequences will be easier to classify in respective taxonomy in the future. So, these types of database and tools are needful for the DNA barcoding studies to classify bacterial sequences with high precision.

Acknowledgments The authors wish to thank the director of CSIR-NEERI for providing facility to do this work. RPM is thankful to SRTM University, Nanded, for Ph.D. work.

References

Altschul SF, Gish W, Miller W, Myers EW, Lipman DJ (1990) Basic local alignment search tool. J Mol Biol 215:403–410. doi.org/10.1016/S0022-2836(05)80360-2

Armougom F, Raoult D (2008) Use of pyrosequencing and DNA barcodes to monitor variations in Firmicutes and Bacteroidetes communities in the gut microbiota of obese humans. BMC Genomics 9:576. https://doi.org/10.1186/1471-2164-9-576

Bazinet AL, Cummings MP (2012) A comparative evaluation of sequence classification programs. BMC Bioinformatics 13:92. doi.org/10.1186/1471-2105-13-92

Cameron S, Rubinoff D, Will K (2006) Who will actually use DNA barcoding and what will it cost? Syst Biol 55:844–847. doi.org/10.1080/10635150600960079

Chan CX, Ragan MA (2013) Next-generation phylogenomics. Biol Direct 8:3. doi.org/10.1186/1745-6150-8-3

Chan RH, Wang RW, Yeung H M (2010) Composition vector method for phylogenetics-a review. In Proc. 9th Int. Symp. Operations Research and Its Applications (ORSC & APORC, Chengdu, China, 2010). p 13

Chen W, Zhang CK, Cheng Y, Zhang S, Zhao H (2013) A comparison of methods for clustering 16S rRNA sequences into OTUs. PLoS One 8:e70837. doi.org/10.1371/journal.pone.0070837

Chen SY, Deng F, Huang Y, Jia X, Liu YP, Lai SJ (2016) bioOTU: an improved method for simultaneous taxonomic assignments and operational taxonomic units clustering of 16s rRNA gene sequences. J Comp Biol 23:229–238. doi.org/10.1089/cmb.2015.0214

Chu KH, Li CP, Qi J (2006) Sequence analysis Ribosomal RNA as molecular barcodes: a simple correlation analysis without sequence alignment. Bioinformatics 22:1690–1701. doi.org/10.1093/bioinformatics/btl146

Chu KH, Xu M, Li CP (2009) Rapid DNA barcoding analysis of large datasets using the composition vector method. BMC Bioinformatics 9:1–9. doi.org/10.1186/1471-2105-10-S14-S8

Chun J, Lee J, Jung Y, Kim M, Kim S, Kim BK, Lim Y (2014) EzTaxon: a web-based tool for the identification of prokaryotes based on 16S ribosomal RNA gene sequences. Int J Syst Evol Microbiol 57:2259–2226. doi.org/10.1099/ijs.0.64915-0

Clayton RA, Sutton G, Hinkle PS Jr, Bult C, Fields C (1995) Intraspecific variation in small-subunit rRNA sequences in GenBank: why single sequences may not adequately represent prokaryotic taxa. Int J Syst Bacteriol 45:595–599. doi.org/10.1099/00207713-45-3-595

Cole JR, Chai B, Farris RJ, Wang Q, Kulam SA, McGarrell DM, Garrity GM, Tiedje JM (2005) The ribosomal database project (RDP-II): sequences and tools for high-throughput rRNA analysis. Nucleic Acids Res 33:D294–D296. doi.org/10.1093/nar/gki038

Cole JR, Wang Q, Fish JA, Chai B, McGarrell DM, Sun YC, Brown T, Porras-Alfaro A, Kuske CR, Tiedje JM (2014) Ribosomal database project: data and tools for high throughput rRNA analysis. Nucleic Acids Res 42:D633–D642. doi.org/10.1093/nar/gkt1244

DeSantis TZ, Hugenholtz P, Larsen N, Rojas M, Brodie EL, Keller K, Huber T, Dalevi D, Hu P, Andersen GL (2006) Greengenes, a chimera-checked 16S rRNA gene database and workbench compatible with ARB. Appl Environ Microbiol 72:5069–5072. doi.org/10.1128/AEM.03006-05

Devulder G, Perriere G, Bernard C, Flandrois J, Devulder G, Perrie G (2003) BIBI, a bioinformatics bacterial identification tool. J Clin Microbiol 41:1785–1787. doi.org/10.1128/JCM.41.4.1785

Edgar R (2016) SINTAX: a simple non-Bayesian taxonomy classifier for 16S and ITS sequences. bioRxiv 074161. https://doi.org/10.1101/074161

Fadrosh DW, Ma B, Gajer P, Sengamalay N, Ott S, Brotman RM, Ravel J (2014) An improved dual-indexing approach for multiplexed 16S rRNA gene sequencing on the Illumina MiSeq platform. Microbiome 2:6. doi.org/10.1186/2049-2618-2-6

Fan L, Hui JH, Yu ZG, Chu KH (2014) VIP Barcoding: composition vector-based software for rapid species identification based on DNA barcoding. Mol Ecol Resour 14:871–881. doi.org/10.1111/1755-0998.12235

Federhen S (2012) The NCBI taxonomy database. Nucleic Acids Res 40:D136–D143. doi.org/10.1093/nar/gkr1178

Felsenstein J (2002) {PHYLIP}(Phylogeny Inference Package) Version 3.6 a3

Fish JA, Chai B, Wang Q, Sun Y, Brown CT, Tiedje JM, Cole JR (2013) FunGene: the functional gene pipeline and repository. Front Microbiol 4:291. doi.org/10.3389/fmicb.2013.00291

Gao X, Lin H, Revanna K, Dong Q (2017) A Bayesian taxonomic classification method for 16S rRNA gene sequences with improved species-level accuracy. BMC Bioinformatics 18:247. https://doi.org/10.1186/s12859-017-1670-4

Golob JL, Margolis E, Hoffman NG, Fredricks DN (2017) Evaluating the accuracy of amplicon-based microbiome computational pipelines on simulated human gut microbial communities. BMC Bioinformatics 18:283. https://doi.org/10.1186/s12859-017-1690-0

Heikens EA (2005) Comparison of genotypic and phenotypic methods for species-level identification of clinical isolates of coagulase-negative staphylococci. J Clin Microbiol 43:2286–2290. https://doi.org/10.1128/JCM.43.5.2286-2290.2005

Horton M, Bodenhausen N, Bergelson J (2010) MARTA: a suite of Java-based tools for assigning taxonomic status to DNA sequences. Bionforma Appl Note 26:568–569. doi.org/10.1093/bioinformatics/btp682

Huson DH, Auch AF, Qi J, Schuster SC (2007) MEGAN analysis of metagenomic data. Genome Res 17:377–386. doi.org/10.1101/gr.5969107

Janda JM, Abbott SL (2007) 16S rRNA gene sequencing for bacterial identification in the diagnostic laboratory: pluses, perils, and pitfalls. J Clin Microbiol 45:2761–2764. doi.org/10.1128/JCM.01228-07

Kuksa P, Pavlovic V (2009) Efficient alignment-free DNA barcode analytics. BMC Bioinformatics 18:1–18. doi.org/10.1186/1471-2105-10-S14-S9

Kumar S, Stecher G, Tamura K (2016) MEGA7: molecular evolutionary genetics analysis version 7.0 for bigger datasets. Mol Biol Evol 22:msw054. doi.org/10.1093/molbev/msw054

Lan Y, Wang Q, Cole JR, Rosen GL (2012) Using the RDP classifier to predict taxonomic novelty and reduce the search space for finding novel organisms. PLoS One 7:e32491. doi.org/10.1371/journal.pone.0032491

Liu Z, DeSantis TZ, Andersen GL, Knight R (2008) Accurate taxonomy assignments from 16S rRNA sequences produced by highly parallel pyrosequencers. Nucleic Acids Res 36:e120–e120. doi.org/10.1093/nar/gkn491

Liu B, Gibbons T, Ghodsi M, Treangen T, Pop M (2011) Accurate and fast estimation of taxonomic profiles from metagenomic shotgun sequences. BMC Genomics 12:S4. doi.org/10.1186/1471-2164-12-S2-S4

Liu C, Shi L, Xu X, Li H, Xing H, Liang D, Jiang K, Pang X, Song J, Chen S (2012) DNA barcode goes two-dimensions: DNA QR code web server. PLoS One 7:e35146. doi.org/10.1371/journal.pone.0035146

Liu J, Wang H, Yang H, Zhang Y, Wang J, Zhao F, Qi J (2013) Composition-based classification of short metagenomic sequences elucidates the landscapes of taxonomic and functional enrichment of microorganisms. Nucleic Acids Res 41:e3. doi.org/10.1093/nar/gks828

Lu YY, Tang K, Ren J, Fuhrman JA, Waterman MS, Sun F (2017) CAFE: aCcelerated Alignment-FrEe sequence analysis. Nucleic Acids Res. doi.org/10.1093/nar/gkx351

Meyer F, Paarmann D, D'Souza M, Olson R, Glass EM, Kubal M, Paczian T, Rodriguez A, Stevens R, Wilke A, Wilkening J (2008) The metagenomics RAST server – a public resource for the automatic phylogenetic and functional analysis of metagenomes. BMC Bioinformatics 9:386. doi.org/10.1186/1471-2105-9-386

Mohammed MH, Ghosh TS, Singh NK, Mande SS (2011) SPHINX – an algorithm for taxonomic binning of metagenomic sequences. Bioinformatics 27:22–30. doi.org/10.1093/bioinformatics/btq608

More RP, Purohit HJ (2016) The identification of discriminating patterns from 16S rRNA gene to generate signature for bacillus genus. J Comput Biol 23:651–661. https://doi.org/10.1089/cmb.2016.0002

More RP, Mane RC, Purohit HJ (2016) matK-QR classifier: a patterns based approach for plant species identification. BioData Min 9:39. https://doi.org/10.1186/s13040-016-0120-6

Munch K, Boomsma W, Huelsenbeck JP, Willerslev E, Nielsen R (2008) Statistical assignment of DNA sequences using Bayesian phylogenetics. Syst Biol 57:750–757. doi.org/10.1080/10635150802422316

Nielsen R, Matz M (2006) Statistical approaches for DNA barcoding. Syst Biol 55:162–169. doi.org/10.1080/10635150500431239

Notredame C, Higgins DG, Heringa J (2000) T-Coffee: a novel method for fast and accurate multiple sequence alignment. J Mol Biol 302:205–217. doi.org/10.1006/jmbi.2000.4042

Porter TM, Gibson JF, Shokralla S, Baird DJ, Golding GB, Hajibabaei M (2014) Rapid and accurate taxonomic classification of insect (class Insecta) cytochrome c oxidase subunit 1 (COI) DNA barcode sequences using a naïve Bayesian classifier. Mol Ecol Resour 14:929–942. doi.org/10.1111/1755-0998.12240

Pruesse E, Peplies J, Glöckner FO (2012) SINA: accurate high-throughput multiple sequence alignment of ribosomal RNA genes. Bioinformatics 28:1823–1829. doi.org/10.1093/bioinformatics/bts252

Qi J, Wang B, Hao BI (2004) Whole proteome prokaryote phylogeny without sequence alignment: a K-string composition approach. J Mol Evol 58:1–11. doi.org/10.1007/s00239-003-2493-7

Quast C, Pruesse E, Yilmaz P, Gerken J, Schweer T, Yarza P, Peplies J, Glöckner FO (2013) The SILVA ribosomal RNA gene database project: improved data processing and web-based tools. Nucleic Acids Res 41:D590–D596. doi.org/10.1093/nar/gks1219

Raje DV, Purohit HJ, Bandhe YP, Tambe SS, Kulkarni BD (2010) Self-organizing maps: a tool to ascertain taxonomic relatedness based on features derived from 16S rDNA sequence. J Biosci 35:617–627. doi.org/10.1007/s12038-010-0070-y

Ramiro-Garcia J, Hermes GD, Giatsis C, Sipkema D, Zoetendal EG, Schaap PJ, Smidt H (2016) NG-Tax, a highly accurate and validated pipeline for analysis of 16S rRNA amplicons from complex biomes. F1000Research 5. 10.12688/f1000research.9227.1

Schwartz R et al (1997) Learning Perl, ISBN 1-56592-284-0, 302 pages. 2nd edition

Segata N, Waldron L, Ballarini A, Narasimhan V, Jousson O, Huttenhower C (2012) Metagenomic microbial community profiling using unique clade-specific marker genes. Nat Methods 9:811–814. doi.org/10.1038/nmeth.2066

Summerbell RC, Lévesque CA, Seifert KA, Bovers M, Fell JW, Diaz MR, Boekhout T, De Hoog GS, Stalpers J, Crous PW (2005) Microcoding: the second step in DNA barcoding. Philos Trans R Soc Lond Ser B Biol Sci 360:1897–1903. doi.org/10.1098/rstb.2005.1721

Swoford DL (2002) Phylogenetic analysis using parsimony (*and other methods) PAUP, 4:b10

Teeling H, Waldmann J, Lombardot T, Bauer M, Glockner FO (2004) TETRA: a web-service and a stand-alone program for the analysis and comparison of tetranucleotide usage patterns in DNA sequences. BMC Bioinforma 5:163. doi.org/10.1186/1471-2105-5-163

Thompson JD, Higgins DG, Gibson TJ (1994) CLUSTAL W: improving the sensitivity of progressive multiple sequence alignment through sequence weighting, position-specific gap penalties and weight matrix choice. Nucleic Acids Res 22:4673–4680

Tuzhikov A, Panchin A, Shestopalov VI (2014) TUIT, a BLAST-Based tool for taxonomic classification of nucleotide sequences. BioTechniques 56:78–84. doi.org/10.2144/000114135

Van Velzen R, Weitschek E, Felici G, Bakker FT (2012) DNA barcoding of recently diverged species: relative performance of matching methods. PLoS One 7:e30490. doi.org/10.1371/journal.pone.0030490

Wang Q, Garrity GM, Tiedje JM, Cole JR (2007) Naive Bayesian classifier for rapid assignment of rRNA sequences into the new bacterial taxonomy. Appl Environ Microbiol 73:5261–5267. https://doi.org/10.1128/AEM.00062-07

Wheeler DL, Barrett T, Benson DA, Bryant SH, Canese K, Chetvernin V, Church DM, DiCuccio M, Edgar R, Federhen S, Geer LY (2007) Database resources of the National Center for Biotechnology Information. Nucleic Acids Res 35:D5–D12. doi.org/10.1093/nar/gkl1031

Yu ZG, Chu KH, Li CP, Anh V, Zhou LQ, Wang RW (2010) Whole-proteome phylogeny of large dsDNA viruses and parvoviruses through a composition vector method related to dynamical language model. BMC Evol Biol 10:192. doi.org/10.1186/1471-2148-10-192

Yu WB, Huang PH, Li DZ, Wang H (2013) Incongruence between nuclear and chloroplast DNA phylogenies in Pedicularis section Cyathophora (Orobanchaceae). PLoS One 8:e74828. doi.org/10.1371/journal.pone.0074828

Zia A, Moses AM (2012) Towards a theoretical understanding of false positives in DNA motif finding. BMC Bioinformatics 13:151. https://doi.org/10.1186/1471-2105-13-151

Fish DNA Barcoding: A Comprehensive Survey of Bioinformatics Tools and Databases

14

Rupali C. Mane, Ganesh Hegde, Ravi Prabhakar More, Rajesh Ramavadh Pal, and Hemant J. Purohit

Abstract

A paradigm shift took place with the advent of molecular taxonomy, which is a combinatorial approach utilizing both computational and molecular biology. DNA barcoding is a reliable, cost-effective method that uses the cytochrome *c* oxidase I (COI) mitochondrial gene to recognize animal species. This gene has a short subsequence 658 bp region that is used for species discrimination. The availability of amplification standard operation protocols and sequence databases for barcoding enables the use of COI sequences for studying taxonomic aspects, particularly in phylogeny, phylogeography, and population genetics studies. The overall process of DNA barcoding in fish is widely performed under the umbrella of molecular and computational methods. In this chapter, we report the current status of fish DNA. barcoding with respect to the databases and software tools available in the public domain.

R. C. Mane
Department of Oral and Maxillofacial Surgery, Faculty of Dental Sciences, M.S. Ramaiah University of Applied Sciences, Bangalore, Karnataka, India

G. Hegde
Central Institute of Freshwater Aquaculture-ICAR (CIFA), Regional Research Center, Bangalore, Karnataka, India

R. P. More (✉)
ADBS, TIFR-National Centre for Biological Sciences (NCBS), Bangalore, Karnataka, India
e-mail: ravimore7@yahoo.in

R. R. Pal
Nagarjuna Fertilizers and Chemicals Limited, Hyderabad, Telangana, India

H. J. Purohit (✉)
Environmental Biotechnology and Genomics Division, CSIR-National Environmental Engineering Research Institute (NEERI), Nagpur, Maharashtra, India
e-mail: hj_purohit@neeri.res.in; hemantdrd@hotmail.com

H. J. Purohit et al. (eds.), *Soft Computing for Biological Systems*,
https://doi.org/10.1007/978-981-10-7455-4_14

Keywords
Fish DNA barcoding · Cytochrome *c* oxidase I (COI) mitochondrial gene · Barcode of Life Data System (BOLD) · FISH-BOL · DNA barcoding tools and databases

14.1 Introduction

Bioinformatics has emerged into a fully fledged multidisciplinary field that integrates statistics and informatics for the analysis of biological data. Due to the advancement in next-generation sequencing (NGS) technology, there has been a dramatic growth in studies of fish genomics (Kumar and Kocour 2017). Public databases now host a catalogue of complete genomes of biological species (mainly fish), which contain protein sequences, protein three-dimensional structures, metabolic pathways, and biodiversity-related information (Vera-Escalona et al. 2017; Adrian-Kalchhauser et al. 2017). Bioinformatics is helping to solve biological problems using software and databases in areas such as functional genomics, bimolecular structure, proteome analysis, taxonomy, and pesticide molecule design (Cambiaghi et al. 2016).

Our earth harbors approximately 8.7 million species, of which around 2.2 million are marine (Mora et al. 2011). IUCN Red List version 2016–3 estimates that the number of described fish species is 33,400. The challenging part was to identify and classify this many species. Earlier methods employed to identify species relied mainly on morphology, protein electrophoresis, and chromatography (Yilmaz et al. 2007; Strauss and Bond 1990; Viswanathan and Pillai 1956). The barcoding technique is effectively utilized in fisheries and has been used to identify recently radiated megadiverse fauna from neotropical areas. The mitochondrial gene encoding cytochrome *c* oxidase subunit I (COI) is used as a marker in phylogeny, phylogeography, and population genetics studies (Pereira et al. 2012; Sbordoni 2010). It has been used for systematic study of native freshwater fish, to monitor the geographic distribution of species (Hubert et al. 2008), and to monitor threatened shark species (Velez-zuazo et al. 2015). These applications facilitate authentication of commercially important species and thereby enhance transparency and fair trade in the domestic fisheries market (Cawthorn et al. 2012). Recent developments include meta-barcoding, in which DNA released by organisms into the environment (eDNA) via cells, excreta, gametes, and decaying materials can effectively be used for species identification. A study conducted in the English Lake District described fish communities in large lakes, both quantitatively and qualitatively (Hanfling et al. 2016). The DNA meta-barcoding approach is considered a next-generation tool for biodiversity monitoring in aquatic ecosystems (Valentini et al. 2016). Mini-barcode primer pairs of length 127–314 bp were developed for authentication of fish food products (Shokralla et al. 2015).

In 2004, an international initiative by the Consortium for the Barcode of Life (CBOL) was taken to make DNA barcoding a standard method or tool for identification of species (http://www.barcodeoflife.org/content/about/what-cbol) (Group et al. 2009). The Barcode of Life Data System (BOLD) is the central informatics platform for DNA barcoding (ibol.org). The Fish Barcode of Life (FISH-BOL) and Shark Barcode of Life (Shark-BOL) initiatives are two important fish barcoding projects at the global level. In India, the Fish Barcode Information System (FBIS), a DNA barcode database on fish, was developed by the National Bureau of Fish Genetic Resources (NBFGR). The overall process of DNA barcoding in fish exploits both molecular and computational methods. A unique region of the specimen is considered as a barcoding marker. In the case of fish, the marker is the gene encoding cytochrome *c* oxidase I (COI) (Hebert et al. 2003).

The general strategy of barcoding involves DNA extraction from the specimen, amplification of a unique marker region using the polymerase chain reaction (PCR), and sequencing. Computational aspects such as editing and aligning sequences is carried out using software such as BOLD v 3.0 (Pereira et al. 2012), TaxI (Steinke et al. 2005), MEGA (Kumar et al. 2008), MEGA 5.05 (Landi et al. 2014), CodonCode Aligner 3.7.1.1(Shokralla et al. 2015), and GENIOUS PRO 5.4.2, (Henriques et al. 2015). Results are later submitted to GenBank or BOLD databases. Hence, once sequencing is completed, the computational aspect plays a key role not only in identification but also in addressing questions related to evolution, diversity (Shen et al. 2016), and taxonomy (Hebert and Gregory 2005).

14.2 Molecular and Computational Approaches for Fish DNA Barcoding

The tissue sample collected from the fish specimen is subjected to DNA extraction. PCR amplifies the target COI gene using a universal primer cocktail (Ivanova et al. 2007). Sequencing of amplified PCR products by BigDye Terminator v.3.1 Cycle Sequencing Kit (Cawthorn et al. 2012) gives both forward and reverse strand sequences. Subsequent important steps are editing, alignment, and sequence submission.

A full-length sequence is made up of aligned reverse and forward strand sequences for all samples of a species (http://mail.nbfgr.res.in/fbis/protocol.php). All the aligned sequences are translated into amino acids to approve the efficiency of the sequence and to identify the presence of nuclear DNA pseudogenes, insertions, deletions, or stop codons (Shen et al. 2016). Edited sequences are placed into the BLAST tool of the National Center for Biotechnology Information (NCBI) to obtain the nearest similar sequence matches and are later submitted to GenBank or BOLD. (http://mail.nbfgr.res.in/fbis/protocol.php). Available editing packages are DNASTAR multiple packages (Chen et al. 2015), Sequencer 4.8 (Gene Codes) (Velez-zuazo et al. 2015), GAP 4 (Shirak et al. 2016; Baxevanis and Ouellette 2004), MEGA version 4.1 (Costa et al. 2012), and MEGA 5.05 (Landi et al. 2014). Useful software packages, alignment tools, databases, and web pages pertaining to barcoding and other related analysis are listed in Tables 14.1, 14.2.

Table 14.1 Fish DNA barcoding databases

Database for barcoding	Website
BOLD	http://www.boldsystems.org/
FBIS	http://mail.nbfgr.res.in/fbis/protocol.php
FISH-BOL	http://www.fishbol.org/
iBOL	http://www.barcodeoflife.org/
NCBI GenBank	https://www.ncbi.nlm.nih.gov/genbank/barcode/

Table 14.2 Software used for DNA barcoding

Software	Type	Website
BioEdit	Alignment	www.mbio.ncsu.edu/bioedit/bioedit.html
MUSCLE	Alignment	www.ebi.ac.uk/Tools/msa/muscle/
CLUSTULW2	Alignment	www.ebi.ac.uk/Tools/msa/clustalw2/
GENEIOUS	Alignment	www.geneious.com/
CLC Genomics	Alignment	https://www.qiagenbioinformatics.com/products/clc-genomics-workbench/
PHYLIP	Phylogenetic	evolution.genetics.washington.edu/phylip.html
MrBayes	Phylogenetic	mrbayes.sourceforge.net/
DNASTAR	Alignment/ phylogenetic	https://www.dnastar.com/
MEGA	Alignment/ phylogenetic	www.megasoftware.net/

Sequence alignment is a method for finding commonality and conserved sequence regions between two or more sequences using a statistical algorithm. It is an important step in identifying the functional, structural, and evolutionary roles of a molecular sequence. A number of sequence alignment packages are available, among which BLAST (Altschul et al. 1990; Madden 2013), MUSCLE (Henriques et al. 2015), CLUSTULX 2.0 (Chen et al. 2015), ClustalW (Velez-zuazo et al. 2015), SeqScape v. 2.1.1 (Applied Biosystems. Inc.) (Zhang and Hanner 2012), BOLD v.3.0 (Pereira et al. 2012), and CodonCode Aligner v 3.7.1.1 (CodonCode Corp., Dedham, MA, USA) (Shokralla et al. 2015) are routinely used.

The usefulness of DNA barcode data in deciphering the phylogenetic relationship between and within species is well studied and involves a series of steps such as alignment, determination of substitution model, and tree building. The latter includes either distance-based tree building or character-based tree building. The distance-based method utilizes the distance between two aligned sequences to generate phylogenetic trees, whereas character-based methods use the composition of oligonucleotide frequencies (e.g., di-, tri-, tera-, penta-, hexa-, heptanucleotides) in the sequences (Baxevanis and Ouellette 2004; Higgs and Manchester 2001). The most commonly employed distance-based methods are neighbor-joining (Saitou and Nei 1987), the Fitch–Margoliash method, the unweighted pair group method with arithmetic mean (UPGMA), and minimum evolution (ME). Maximum parsimony (MP) and maximum likelihood (ML) are two major character-based methods used for phylogentics (Felsenstein 1981). In addition, Bayesian analysis has been proposed for phylogeny (Huelsenbeck and Ronquist 2001). Tests for evaluating constructed trees include the skewness test, permutation test, and bootstrapping, which can be parametric or nonparametric, and the likelihood ratio test . Software packages for phylogenetic analysis include PHYLIP, PAUP, PUZZLE, FastDNAml, MACCLADE, and MOLPHY, along with internet-accessible phylogenetic software such as WEBPHYLIP, PhyloBLAST, BLAST 2, and Orthologue Search Server (Baxevanis and Ouellette 2004).

Noncoding internal transcribed spacer genes have also been suggested as candidate barcodes, along with the COI gene for animal and plant DNA barcoding (Gao et al. 2017; Yang et al. 2017). Two new approaches (DV-RBF and FJ-RBF) have been used to align the noncoding regions for DNA barcoding and showed 100% success rate in identifying marine fish species. (Zhang et al. 2012). On other hand, alignment-free methods such as normalized compression distance (NCD) and information-based distance (IBD) have been utilized for taxonomic analysis of barcode sequences (La Rosa et al. 2013). Taxonomic classification methods are mainly categorized into (1) tree-based approaches, (2) composition-based approaches, (3) similarity-based approaches, and (4) hybrids. These methods required reference databases to predict the taxonomy (Tanabe and Toju 2013).

In a recent study, similarity-based methods such as nearest-neighbor, centric auto-k-NN (NN Cauto), and query-centric auto-k-NN (Q Cauto) were proposed for barcoding studies (Tanabe and Toju 2013). A method of string kernel-based sequence analysis of barcode data sets was proposed that considerably improves species identification accuracy compared with traditional approaches (Kuksa and Pavlovic 2007). The few sequence identification methods that use pairwise alignment (e.g., BLAST) are not able to discriminate species that have highly similar sequences, because only very few base pairs are different between the sequences. To address this issue, alignment-free methods (e.g., BRONX) were developed to identify species sequences (Little 2011). BRONX detects short subsequence regions and matching regions in reference sequences. Based on these regions, the algorithm generates a score without use of multiple sequence alignment to identify sequences at the genus level (Little 2011).

14.3 Public Domain Databases

Recent progress in next-generation sequencing (NGS) platforms has led to advancement of the discipline of bioinformatics for the annotation of genome data. Public databases contain huge amounts of accessible data on whole genome sequences, which have improved research in applied fish science. There are some very popular primary, secondary, and specialized databases available from BOLD, FISH-BOL, GenBank, and FBIS.

14.3.1 Barcode of Life Data System

The Barcode of Life Data System (BOLD) (http://www.barcodinglife.org) facilitates a detailed collection of specimens deposited by researchers from different barcoding studies. This database holds three main categories of information. The first category is basic information on the specimen and sequence entries. The second maintains quality assurance and manages barcode data with all related information. The third category facilitates a detailed catalogue of specimen data entries from geographically different researchers. A user can store specimen information in the following sections:

- Species name
- Voucher data, institution storing, and catalogue number
- Collection record, which includes collector name, location with GPS coordinates, and data of collection
- Identifier of the specimen
- COI sequence with minimum 500 bp
- PCR primers referred for amplicon capture of trace files

BOLD is an informatics workbench used for collection, storing, scrutiny, and publication of DNA barcode entries and is freely accessible. It involves more than 65,000 lines of combined code written in Java, C++, and PHP. To gain formal barcode status, certain criteria must be satisfied, including species name, voucher data, and collection record. BOLD employs many tools to identify data anomalies or low-quality records. All acquiesced sequences are translated into amino acids and are matched against a hidden Markov model (HMM) of COI protein to confirm that they essentially originate from the COI sequence. Later sequences are checked for stop codons, and also against a small set of possible contaminants. If any errors are detected, the submitter is informed and the sequence is flagged. After providing a trace file, BOLD further determines a PHRED score for each nucleotide position and a mean value for the full sequence based on these results. Next, it manages each sequence entry into one of four classes: failed (no sequence), low quality (mean PHRED < 30), medium quality (mean PHRED $= 30$–40), and high quality (mean PHRED > 40). The data stored in BOLD can be readily exported in FASTA format for use in other analytical packages. BOLD provides an examination utility that permits users to determine sequence coverage for a specific taxonomic or geographic region. It includes an integrated analytic system (MAS), which provides data analysis tools such as the taxon identification (ID) tree. Unknown sequences are identified by pasting their sequence record into the input box on the ID form. Core data element records in BOLD consist of a specimen page and a sequence page. Barcodes in the search archives are grouped into two categories. Species are considered with three representatives and maximum divergence of 2%, A HMM method is used to align the query sequence with archive sequences. The HMM method is faster than BLAST because of its efficient data processing capability.

BOLD detects species if the query sequence displays a close match with at least <1% divergence against the archive sequences (Ratnasingham and Hebert 2007).

14.3.2 Fish Barcode of Life Campaign and Fish Barcode Information System

The campaign FISH-BOL was started in 2004 with the aim of generating tools for identifying all types of fish species. Its primary goal was to gather barcodes for all of the world's fish. FISH-BOL comprises sequences, geographical information, and images for examined specimens, thereby creating a valuable public resource. Information organized and analyzed through the BOLD database is later delivered via a data feed to the FISH-BOL web portal. This depository utilizes taxonomic information resulting from FishBase and maintains a catalogue of fish (Ward et al. 2008). The International Nucleotide Sequence Database Collaboration (INSDC) archives DNA sequences from the FISH-BOL campaign and annotates each sequence with the key word "barcode" when it meets the barcode data standards. It requires the bidirectionally sequenced 5′-end of the COI gene sequence, valid species name, details concerning voucher specimens, coordinates of the collection locality, collection date, collecter, and identifier. Also required are a list of the primers used to generate reference sequences and archiving of the underlying electropherogram trace files in a publically accessible NCBI trace archive. All this information is useful for using barcodes in molecular diagnostics applications. BOLD provides an online workbench to FISH-BOL (Ward et al. 2008).

The FBIS web-based tool is designed for the fish of India. The database has a total of 2334 COI gene sequences belong to 472 aquatic species. It works both as a local DNA barcode library and as an analysis system and contains valuable data regarding the phenotype, distribution, and IUCN Red List status of fish (Nagpure et al. 2012). This database enables saving and extracting data in an easy way with simple steps. A user can submit species sequences through a submission protocol. Species identification is performed using similarity search programs; it finds homologues with almost 99% similarity to the query sequence, which accurately assigns the species (Nagpure et al. 2012).

14.3.3 NCBI GenBank

GenBank is a comprehensive database that contains nucleotide sequences for more than 250,000 species (Benson et al. 2013). NCBI offers an online/offline sequence submission platform to deposit sets of barcode sequences to the GenBank database. Along with the barcode data, the submission platform collects other annotations such as specimen voucher, geographical information, sample collection date, primer data, and raw files to help recognize the sequence's source organism and to maintain the accuracy of the sequence. The GenBank file structure format is easy to understand for users. It contains sequence data along with the accession numbers and gene names, taxonomy, references to published literature, and other meaningful information. The GenBank format comprises the locus, definition, accession, keywords, source, reference, and features fields for the gene. The user can download the FASTA format nucleotide or amino acid sequence from the FASTA link given on files or send to menu option (https://www.ncbi.nlm.nih.gov/genbank/barcode/). It is important to give the publication details related to barcodes and sequences in FASTA format with reverse and forward primers. Protein sequence submission is optional.

14.4 DNA Barcoding Repositories and Their Associated Tools

It is difficult to preserve the data integrity, interoperability, and utility of information generated relating to the "what", "where", and "when" of biodiversity data. Furthermore, DNA barcoding and other biodiversity information systems must maintain data standards so that appropriate metadata is efficiently included. Three main organizations (the International Barcode of Life Project (iBOL), CBOL, and BOLD), promote barcoding research with the aim of generating reference barcodes (Group et al. 2009; Ratnasingham and Hebert 2007). These organisations are focused toward development of barcoding as a universal standard and offer an online workbench for collection, management, analysis, and use of DNA barcodes.

iBOL (http://ibol.org) has network of collaborators from about 150 countries, includes more than 190,000 marine species, and has identified 6000 potentially new species (flowering plants, ants, birds, butterflies, ants, mammals, bees, fish, and fungi). It has collections in the form of ecosystems such as rain forests, kelp forests, poles, seas, and coral reefs. CBOL generated the BOLD system as a catalogue of living beings and has collections covering more than 790,000 sequences, conforming to more than 67,000 correctly called "species." The BOLD database entries contain barcode sequences and specimen information such as images, morphology, collection date, and geographical site. To provide practical utility for BOLD data, the mobile-based software DNA Barcoding Assistant efficiently maintains metadata for the gathering and management of specimen data for BOLD and other biodiversity information databases.

The DNA Barcoding Assistant (http://www.dnabarcodingassistant.org/) enables users to store and retrieve data such as provisional user-allocated taxonomic classification, geospatial data, digital images, and collection event information for specimens found in the field. Another web-based data-processing system tool, BioBarcode (http://www.asianbarcode.org), focuses on the collection of Asiatic organisms and encompasses about 11,300 specimen entries (Lim et al. 2009). On similar lines, a field information management system (FIMS) has been developed that provides information associated with tissues, collecting events, and specimens (Deck et al. 2012). Similarly, the Quick Response (QR) barcode system could be efficiently implemented to identify and track samples, together with relevant information such as site details, time of collection, and taxonomic identity (Diazgranados and Funk 2013). These indicate that continuous progress is being made in DNA barcoding.

Acknowledgements Authors are thankful to the Director of CSIR-NEERI, Nagpur, Maharashtra, India for this work.

References

Adrian-Kalchhauser I, Svensson O, Kutschera VE, Rosenblad MA, Pippel M, Winkler S, Schloissnig S, Blomberg A, Burkhardt-Holm P (2017) The mitochondrial genome sequences of the round goby and the sand goby reveal patterns of recent evolution in gobiid fish. BMC Genomics 18:177. https://doi.org/10.1186/s12864-017-3550-8

Altschul SF, Gish W, Miller W, Myers EW, Lipman DJ (1990) Basic local alignment search tool. J Mol Biol 215:403–410. https://doi.org/10.1016/S0022-2836(05)80360-2

Baxevanis AD, Ouellette BF (2004) Bioinformatics: a practical guide to the analysis of genes and proteins, vol 43. Wiley, Hoboken

Benson DA, Cavanaugh M, Clark K, Karsch-Mizrachi I, Lipman DJ, Ostell J, Sayers EW (2013) GenBank. Nucleic Acids Res 41:36–42. https://doi.org/10.1093/nar/gks1195

Cambiaghi A, Ferrario M, Masseroli M (2016) Analysis of metabolomic data: tools, current strategies and future challenges for omics data integration. Brief Bioinform 12:bbw031. https://doi.org/10.1093/bib/bbw031

Cawthorn DM, Steinman HA, Corli witthuhn R (2012) DNA barcoding reveals a high incidence of fish species misrepresentation and substitution on the South African market. Food Res Int 46:30–40. https://doi.org/10.1016/j.foodres.2011.11.011

Chen W, Ma X, Shen Y, Mao Y, He S (2015) The fish diversity in the upper reaches of the Salween River Nujiang River revealed by DNA barcoding. Sci Rep 5:17437. https://doi.org/10.1038/srep17437

Costa FO, Landi M, Martins R, Costa MH, Costa ME, Carneiro M et al (2012) A ranking system for reference libraries of dna barcodes: application to marine fish species from Portugal. PLoS One 7:e35858. https://doi.org/10.1371/journal.pone.0035858

Deck J, Gross J, Stones-Havas S, Davies N, Shapley R, Meyer C (2012) Field information management systems for DNA barcoding. Methods Mol Biol 858:255–267. https://doi.org/10.1007/978-1-61779-591-6_12

Diazgranados M, Funk VA (2013) Utility of QR codes in biological collections. PhytoKeys 25:21–34. https://doi.org/10.3897/phytokeys.25.5175

Felsenstein J (1981) Evolutionary trees from DNA sequences: a maximum likelihood approach. J Mol Evol 17:368–376

Gao LM, Li Y, Phan LK, Yan LJ, Thomas P, Phan LK, Möller M, Li DZ (2017) DNA barcoding of east Asian Amentotaxus (Taxaceae): potential new species and implications for conservation. J Syst Evol 55:16–24. https://doi.org/10.1111/jse.12207

Group CP, Hollingsworth PM, Forrest LL, Spouge JL, Hajibabaei M, Ratnasingham S, van der Bank M, Chase MW, Cowan RS, Erickson DL, Fazekas AJ (2009) A DNA barcode for land plants. Proc Natl Acad Sci U S A 106:12794–12797. https://doi.org/10.1073/pnas.0905845106

Hanfling B, Lawson HL, Read DS, Hahn C, Li J, Nichols P, Winfield IJ (2016) Environmental DNA metabarcoding of lake fish communities reflects long-term data from established survey methods. Mol Ecol 25:3101–3119. https://doi.org/10.1111/mec.13660

Hebert PD, Gregory TR (2005) The promise of DNA barcoding for taxonomy. Syst Biol 54:852–859. https://doi.org/10.1080/10635150500354886

Hebert PDN, Cywinska A, Ball SL, deWaard JR (2003) Biological identifications through DNA barcodes. Proc Biol Sci 270:313–332. https://doi.org/10.1098/rspb.2002.2218

Henriques JM, da Costa Silva GJ, Ashikaga FY, Hanner R, Foresti F, Oliveira C (2015) Use of DNA barcode in the identification of fish species from Ribeira de Iguape Basin and coastal rivers from São Paulo state (Brazil). DNA 3:118–128. https://doi.org/10.1515/dna-2015-0015
Higgs P, Manchester U (2001) Introduction to phylogenetics methods (ITP series on-line seminars). http://online.kitp.ucsb.edu/online/infobio01/higgs/
Hubert N, Hanner R, Holm E, Mandrak NE, Taylor E, Burridge M, Bernatchez L (2008) Identifying Canadian freshwater fishes through DNA barcodes. PLoS One 3:e2490. https://doi.org/10.1371/journal.pone.0002490
Huelsenbeck JP, Ronquist F (2001) MRBAYES: Bayesian inference of phylogenetic trees. Bioinformatics 17:754–755
Ivanova NV, Zemlak TS, Hanner RH, Hebert PD (2007) Universal primer cocktails for fish DNA barcoding. Mol Ecol Notes 7:544–548. https://doi.org/10.1111/j.1471-8286.2007.01748.x
Kuksa P, Pavlovic V (2007) Fast kernel methods for SVM sequence classifiers. In: Giancarlo R, Hannenhalli S (eds) Algorithms in bioinformatics, Lecture Notes in Computer Science, vol 4645. Springer, Berlin/Heidelberg, pp 228–239
Kumar G, Kocour M (2017) Applications of next-generation sequencing in fisheries research: a review. Fish Res 186:11–22. https://doi.org/10.1016/j.fishres.2016.07.021
Kumar S, Nei M, Dudley J, Tamura K (2008) MEGA: a biologist-centric software for evolutionary analysis of DNA and protein sequences. Brief Bioinform 9:299–306. https://doi.org/10.1093/bib/bbn017
La Rosa M, Fiannaca A, Rizzo R, Urso A (2013) Alignment-free analysis of barcode sequences by means of compression-based methods. BMC Bioinforma 14:S4. https://doi.org/10.1186/1471-2105-14-S7-S4
Landi M, Dimech M, Arculeo M, Biondo G, Martins R, Carneiro M, Carvalho GR, Brutto SL, Costa FO (2014) DNA barcoding for species assignment: the case of mediterranean marine fishes. PLoS One 9:e106135. https://doi.org/10.1371/journal.pone.0106135
Lim J, Kim SY, Kim S, Eo HS, Kim CB, Paek WK, Bhak J (2009) BioBarcode: a general DNA barcoding database and server platform for Asian biodiversity resources. BMC Genomics 10:1. https://doi.org/10.1186/1471-2164-10-S3-S8
Little DP (2011) DNA barcode sequence identification incorporating taxonomic hierarchy and within taxon variability. PLoS One 6:e20552. https://doi.org/10.1371/journal.pone.0020552
Madden T (2013) The BLAST sequence analysis tool. In: The NCBI handbook. NCBI, Bethesda. https://unmc.edu/bsbc/docs/NCBI_blast.pdf
Mora C, Tittensor DP, Adl S, Simpson AGB, Worm B (2011) How many species are there on earth and in the ocean? PLoS Biol 9:e1001127. https://doi.org/10.1371/journal.pbio.1001127
Nagpure NS, Rashid I, Pathak AK, Singh M, Singh SP, Sarkar UK (2012) FBIS: a regional DNA barcode archival analysis system for Indian fishes. Bioinformation 8:483–488. https://doi.org/10.6026/97320630008483
Pereira LH, Hanner R, Foresti F, Oliveira C (2012) Can DNA barcoding accurately discriminate megadiverse Neotropical freshwater fish fauna? BMC Genet 14:20–20. https://doi.org/10.1186/1471-2156-14-20
Ratnasingham S, Hebert PD (2007) BOLD: the barcode of life data system (wwwbarcodinglifeorg). Mol Ecol Notes 7:355–364. https://doi.org/10.1111/j.1471-8286.2007.01678.x
Saitou N, Nei M (1987) The neighbor-joining method: a new method for reconstructing phylogenetic trees. Mol Biol Evol 4:406–425. https://doi.org/10.1093/oxfordjournals.molbev.a040454
Sbordoni V (2010) Strength and limitations of DNA barcode under the multidimensional species perspective. Prog Probl:271–276. ISBN 978-88-8303-295-0
Shen Y, Guan L, Wang D, Gan X (2016) DNA barcoding and evaluation of genetic diversity in Cyprinidae fish in the midstream of the Yangtze River. Ecol Evol 6:2702. https://doi.org/10.1002/ece3.2060

Shirak A, Dor L, Seroussi E, Ron M, Hulata G, Golani D (2016) DNA barcoding of fish species from the mediterranean coast of Israel. Mediterr Mar Sci 17:459–466. https://doi.org/10.12681/mms.1384

Shokralla S, Hellberg RS, Handy SM, King I, Hajibabaei M (2015) A DNA mini-barcoding system for authentication of processed fish products. Sci Rep 5:15894. https://doi.org/10.1038/srep15894

Steinke D, Vences M, Salzburger W, Meyer A (2005) TaxI: a software tool for DNA barcoding using distance methods. Philos Trans R Soc Lond Ser B Biol Sci 360:1975–1980. https://doi.org/10.1098/rstb.2005.1729

Strauss R, Bond C (1990) Taxonomic methods: morphology. In: Schreck CB, Moyle PB (eds) Methods for fish biology. American Fisheries Society, Bethesda, pp 109–140

Tanabe AS, Toju H (2013) Two new computational methods for universal DNA barcoding: a benchmark using barcode sequences of bacteria archaea animals fungi and land plants. PLoS One 8:e76910. https://doi.org/10.1371/journal.pone.0076910

Valentini A, Taberlet P, Miaud C, Civade R, Herder J, Thomsen PF, .. Gaboriaud C (2016) Next-generation monitoring of aquatic biodiversity using environmental DNA metabarcoding. Mol Ecol 25:929–942. https://doi.org/10.1111/mec.13428

Velez-zuazo X, Alfaro-shigueto J, Mangel J, Papa R, Agnarsson I (2015) What barcode sequencing reveals about the shark fishery in Peru. Fish Res 161:34–41. https://doi.org/10.1016/j.fishres.2014.06.005

Vera-Escalona I, Habit E, Ruzzante DE (2017) The complete mitochondrial genome of the freshwater fish Galaxias Platei and a comparison with other species of the genus galaxias (faraway, so close?). Mitochondrial DNA 28:176–177. https://doi.org/10.3109/19401736.2015.1115497

Viswanathan R, Pillai VK (1956) Paper chromatography in fish taxonomy. Proc Indian Acad Sci 43:334–339. https://doi.org/10.1007/BF03050245

Ward R, Hanner R, Hebert P (2008) The campaign to DNA barcode all fishes FISH-BOL. J Fish Biol 74:329–356. https://doi.org/10.1111/j.1095-8649.2008.02080.x

Yang J, Vázquez L, Chen X, Li H, Zhang H, Liu Z, Zhao G (2017) Development of chloroplast and nuclear DNA markers for Chinese oaks (Quercus subgenus Quercus) and assessment of their utility as DNA barcodes. Front Plant Sci 8:816. https://doi.org/10.3389/fpls.2017.00816

Yilmaz M, Yilmaz HR, Alas A (2007) An electrophoretic taxonomic study on serum proteins of Acanthobrama Marmid Leuciscus Cephalus and Chondrostoma Regium. Eurasia J Biosci 3:22–27

Zhang J, Hanner R (2012) Molecular approach to the identification of fish in the South China Sea. PLoS One 7:e30621. https://doi.org/10.1371/journal.pone.0030621

Zhang AB, Feng J, Ward RD, Wan P, Gao Q, Wu J, Zhao WZ (2012) A new method for species identification via protein-coding and non-coding DNA barcodes by combining machine learning with bioinformatic methods. PLoS One 7:e30986. https://doi.org/10.1371/journal.pone.0030986

Microbes and Mountains: The Mid-Domain Effect on Mt. Fuji, Japan

15

Dharmesh Singh

Abstract

Dispersal of biodiversity – a major cause of variation – leads to speciation, extinction, and dispersal. The most prominent and well-researched pattern in biogeography is the ubiquitous elevational gradient. Most studies have focused on macroorganisms. However, with but the advent of molecular biology tools such as the next-generation sequencing (NGS), these studies incorporate microorganisms into their horizon. The basic objective is to understand and broaden our perspective for other significant microbial groups and their ecology.

Keywords

Microbial community · Archaea · Evolution · Diversity · Biogeography

15.1 Introduction

Biogeography studies the dispersal of biodiversity over space and time with a goal to understand factors that are directly responsible for causing variation in diversity, such as speciation, extinction, and dispersal (Brown and Lomolino 2005). Questions like how and why the species varies geographically have been actively motivating biogeographical research over the past many decades. The most prominent and well-researched pattern in biogeography after the latitudinal gradient is the ubiquitous elevational gradient. Also, properties like having many biological replicates, ease to carry out controlled experiments, and the absence of covarying variables like area, climate, and history along elevational gradients perhaps make them more suitable for studying the underlying causes of spatial variation in

D. Singh (✉)
Environmental Biotechnology and Genomics Division (EBGD), CSIR-National Environmental Engineering Research Institute (NEERI), Nagpur, India
e-mail: singhdharmesh24@gmail.com

H. J. Purohit et al. (eds.), *Soft Computing for Biological Systems*,
https://doi.org/10.1007/978-981-10-7455-4_15

biodiversity. Although most of these elevational studies have focused on macroorganisms but with the advent of the next-generation sequencing (NGS) in the last decades, these studies have now started to incorporate microorganisms into their horizon.

What constitutes a microorganism? Generally, the term represents bacteria and archaeal domain members, as well as the microscopic candidates from domain Eukarya such as some fungi and protists. Why there's a need to study their biogeography? Although they are perhaps the most abundant (Whitman et al. 1998) and diverse (Torsvik et al. 2002; Venter et al. 2004) organisms on Earth, their distribution and the factors controlling their distribution at all scales are still poorly understood. They practically inhabit all environments and can be found in soil, sediments, water (marine and aquatic), air, gastrointestinal tracts, geysers, and even deep underground and high up in the atmosphere. Especially bacteria and archaea have important roles in biogeochemical cycles by which an element or a compound cycles through abiotic (lithosphere, atmosphere, and hydrosphere) and biotic (biosphere) components of Earth. C, H, N, O, S, and P are the main constituents for all organic macromolecules (Schlesinger 1997), and out of these, the first five elements' biological fluxes are largely catalyzed under thermodynamically constrained redox reactions by prokaryotic drivers (Falkowski et al. 2008).

Even after such importance, only a few studies have tried to study microbial elevational gradients (Bryant et al. 2008; Fierer et al. 2011; Wang et al. 2012; Singh et al. 2014). Elevational gradients for microbes in general, relative to latitudinal gradients, have a varied range of benefits, which makes them a useful tool in understanding the fundamental basis of diversity gradients. I) Unlike two replicates of latitudinal gradients, there are unlimited replicates available for elevational diversity gradients – fundamentally each mountain or mountain range is a replicate, and each mountain range can be covered repetitively in many transects. This helps us to examine the generality of elevational diversity gradients, to check whether species occurring along the gradient originated from the same regional species pool and shared similar evolutionary history or vice versa. II) Latitudinal studies for eukaryotes generally cover a given taxon in a particular mountain range which could be overcome in the case of prokaryotic elevational study, where due to NGS technologies, we can study a whole domain like bacteria or a functional group like ammonia-oxidizing archaea, at once in a single study. III) It is comparatively much more feasible to perform controlled experiments along elevational gradients. IV) Data collection is easier along elevational gradients. V) A lot of the fundamental factors (like area, history, climate, etc.) that might shape the community covary on a latitudinal gradient that do not behave so along elevational gradients (Korner 2007). Because of these reasons, elevational gradients are now being seen as prized tools to expose the mechanisms that shape both the biodiversity patterns and ecosystem functioning (Fukami and Wardle 2005; Nogues-Bravo et al. 2008).

Given that there are only a few elevational studies to date on microbes, it becomes vital to review and summarize their results into a well-documented result which might help us in understanding and broadening our perspective for other significant microbial groups and their ecology. Especially studies from mountain systems with comparatively uniform geology and simpler climatic gradients like Mt. Fuji are promising in terms of identifying the factors which delimit the prokaryotic community composition and diversity.

This chapter discusses the microbial diversity patterns observed in the Mt. Fuji, central Japan which has a temperate climate populated by temperate mixed deciduous forest at lower elevations and subalpine/alpine vegetation at higher elevations. In brief, the following points will be discussed throughout the chapter.

1. Dominant microbial phyla in Mt. Fuji soil and which compositional variations are observed with elevation?
2. Variation in microbial diversity along Mt. Fuji elevational gradient and which environmental variables control the overall community?
3. Finally, what underlying mechanisms predict soil microbial community structure?

15.2 Mt. Fuji

Mt. Fuji, the highest peak in Japan, is a stratovolcano cone covered by a fairly uniform basaltic composition that started to grow about 8000–11,000 years ago. Hoei crater, on the east slope of the Fuji mountain, saw the most recent volcanic eruption in 1707 with uniform ash deposition on the east side of the mountain across the coastal plain [http://www.bousai.go.jp/fujisan-kyougikai/(in Japanese)]. The sampling sites discussed in this chapter are far from this side of the mountain, and hence, any part of transects discussed here hasn't been affected by this recent volcanic activity. Human disturbance in the forest (almost 300 years old) below the tree line is almost negligible as the vegetation and wildlife of Mt. Fuji are sheltered as a national park. Soil sampling for the archaeal and bacterial samples discussed in this chapter spans from 1000 masl, at the base of the mountain, to the 3760 masl summit point, whereas for the ectomycorrhizal fungi (EM fungi), sampling was limited just below the tree line ranging from 1100 masl to 2250 masl.

Vegetation on Mt. Fuji could be broadly distributed into temperate mixed deciduous forest starting at 700 masl to around 1600 masl and into subalpine forest ranging from 1600 masl to 2500 masl. Above 2500 masl, a prominent decrease in tree species richness can be seen with the emergence of an alpine zone composed of scattered shrubs. Alpine zone finally gives way to a no vascular plant zone, with only a sparse cover of lichens and moss (Ohsawa 1984). Temperate mixed deciduous forests on Mt. Fuji are composed mainly of *Fagus crenata*, *Quercus crispula*, and various *Acer* species. Subalpine forest belt is dominated by *Abies veitchii*, *Tsuga diversifolia*, *Larix leptolepis*, and *Betula ermanii*. Alpine zone is populated by shrubs such as *Polygonum cuspidatum*, *Salix reinii*, and *Alnus maximowiczii* (Ohsawa 1984).

15.3 Microbial Community on Mt. Fuji

To study the archaeal/bacterial community, the 16S rRNA gene (V1–V3 region) and, for EM fungal community, ITS regions (ITS1, 5.8S, ITS2) were used (Singh et al. 2012a, b; Miyamoto et al. 2014). To study the community, DNA is generally isolated from the processed samples and amplified using target specific primers, which are then collected and purified before being sequenced using NGS techniques. Sequencing yields a huge amount of data which is generally processed using online bioinformatics platforms like Mothur, QIIME, and UPARSE or against in-house software, which then processes the data into a comprehensible format to be understood by the scientific community. This process also takes help of several freely online available curated databases like SILVA, Greengenes, RDP, and/or EzTaxon for alignment and classification purposes.

15.4 Microbial Community Composition on Mt. Fuji

In general, a unimodal pattern could be seen for all three group of organisms discussed below with distinctive differences (Fig. 15.1).

15.4.1 Bacteria

For Mt. Fuji bacterial population, rarefaction curve ($\geq$97% sequence identity) reveals samples from the mid-elevations as the richest ones while never appearing to reach an asymptote, whereas the samples from the summit and base appear to be least diverse.

Proteobacteria (38.5%) and *Acidobacteria* (20.6%) are the two most abundant phyla on Mt. Fuji, followed by *Actinobacteria* (11.7%), *Chloroflexi* (5.2%), and *Bacteroidetes* (5.1%) (Fig. 15.2). *Alphaproteobacteria* (18.5%) is in similar abundance with the second most abundant phylum *Acidobacteria* (20.6%). *Afipia* sp. from *Alphaproteobacteria* is the most abundant species (2.19%), present across the Mt. Fuji

15.4.2 Archaea

Lack of asymptotes in the rarefaction curve for archaea suggests that a considerable aspect of archaeal diversity still rests un-sampled. Only two archaeal phyla could be found on Mt. Fuji – Thaumarchaeota dominating with 96.4% reads and lagging behind is Euryarchaeota with around 3.9% of total reads (Fig. 15.3). A single OTU or operational taxonomic unit (Nitrososphaerales (soil cluster I.1b)) emerges as the most abundant OTU across the entire site, accounting for 46.5% of total reads.

Fig. 15.1 A general unimodal pattern observed for richness/diversity on Mt. Fuji for bacteria, archaea, and EM fungi. Observed maxima for the bacterial richness/diversity is comparatively higher than archaea and EM fungi. The figure also depicts the vegetation observed at different elevations of Mt. Fuji

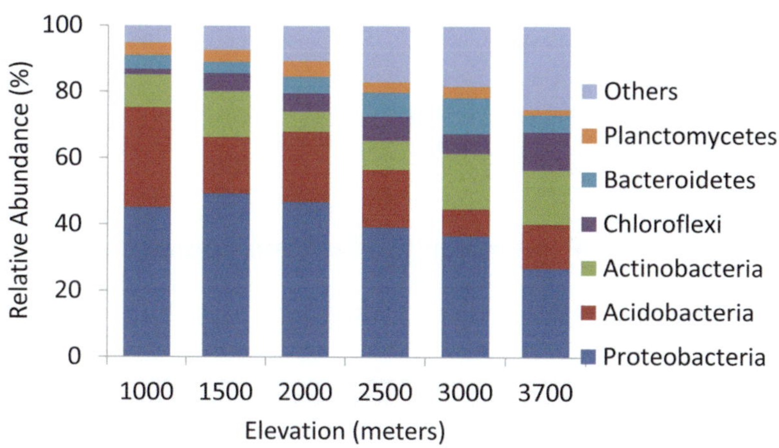

Fig. 15.2 Taxonomic breakdown of bacteria on Mt. Fuji

15.4.3 EM Fungi

EM fungi are present throughout the elevational gradient (present in 197/200 soil core) samples, taken before the tree line (2250 masl). A rarefaction curves drawn for observed richness did not reach an asymptote for any of the sites, revealing that much EM fungal diversity remains un-sampled.

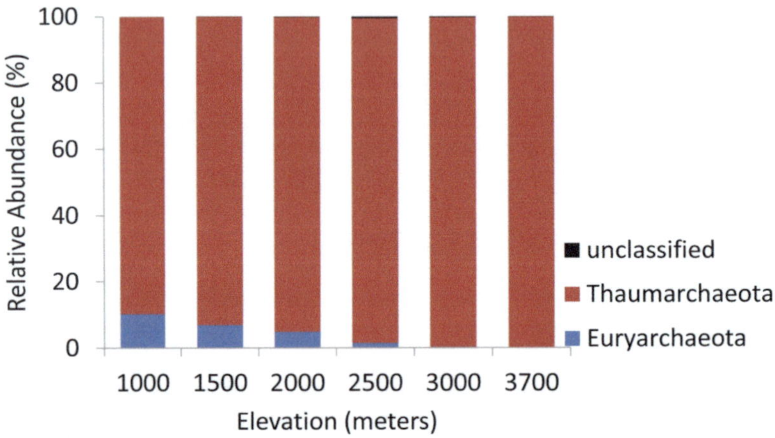

Fig. 15.3 Taxonomic breakdown of archaea on Mt. Fuji

The EM fungal community on Mt. Fuji mostly belongs to *Basidiomycetes* (96.77%) with *Ascomycetes* (3.226%) playing a much smaller role. The most abundant lineages recorded are *Tomentella-Thelephora* (18.4%), followed by *Cortinarius* (16.4%), *Sebacina* (10.55%), *Russula-Lactarius* (10.02%), and *Inocybe* (8.97%) (Fig. 15.4). 24.2% of species occur in two or more sites with *Piloderma fallax* and *Sebacina* sp. prevailing in three sites (1550–2250 m). The site just below the tree line (2250 masl) and the base site (1100 masl) does not have any common species among them. The number of shared species among adjacent site pairs is significantly higher than among nonadjacent site pairs ($P<0.001$, χ^2-test).

15.5 Elevational Patterns on Mt. Fuji and Climatic Variables

Bacterial diversity plotted against elevation follows a "unimodal pattern" with maxima at around 2500 masl. Again, richness shows a similar trend with minimum richness in samples from the summit which contains only about 79% of the OTUs present at the 2500 masl (richness maxima) (Fig. 15.5). *Proteobacteria* and *Bacteroidetes* follow a unimodal distribution ($P<0.05$), while *Acidobacteria* shows a remarkable decrease in richness/diversity with elevation. The most abundant OTU from the genus *Afipia* is found throughout the elevational gradient with maxima at mid-elevations. Elevation only, among all of the edaphic variables, significantly correlates ($P<0.05$) with both richness ($R^2=0.33$) and diversity (Shannon index, $R^2=0.18$). Principal coordinate analysis or PCoA (unweighted UniFrac distance) for the total community shows significant variability among different elevation's bacterial community and high affinity within samples taken from the same elevational zones.

Like bacteria, elevation significantly controls archaeal community diversity and richness (Fig. 15.5). A unimodal pattern with a diversity/richness "peak" at around

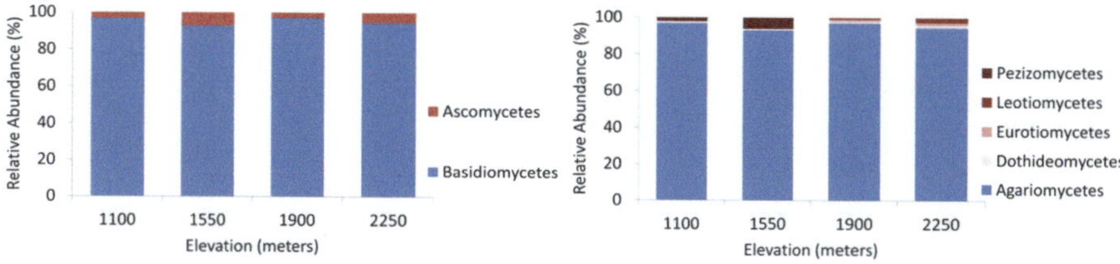

Fig. 15.4 Taxonomic breakdown of EM fungi (left, phylum; right class) on Mt. Fuji

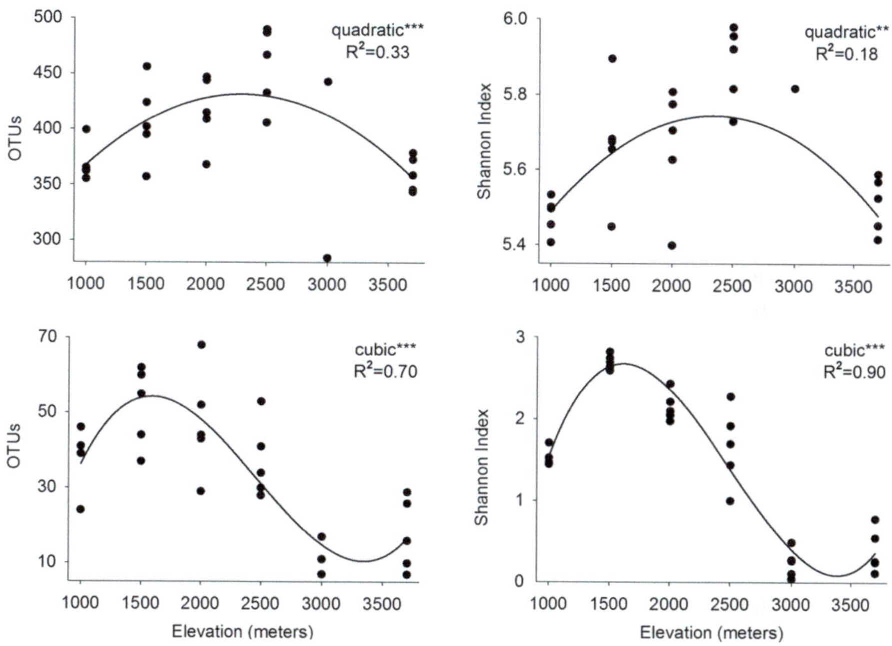

Fig. 15.5 Effect of elevation on richness (left) and diversity (right) for the bacterial (above row) and archaeal communities (below row)

1500 masl can be seen. Around 79% of the total OTUs are present at 1500 masl with least richness at 3000 masl, just above the tree line. Thaumarchael richness/diversity, when plotted against elevation, reveals the same unimodal pattern as observed for the whole community. For the total archaeal population on Mt. Fuji (unweighted UniFrac index/Bray-Curtis similarity), elevation was able to explain >38% of the total variability [MRM (multiple regression on matrices) results]. Reiterating these results, a nonmetric multidimensional scaling (NMDS) plot on Bray-Curtis similarity index matrix shows a community separation according to elevation with samples belonging to similar elevational sites harboring related communities.

The EM fungal richness also follows a unimodal pattern, with maxima at 1550 masl (Fig. 15.6). Observed fungal richness does not show any significant

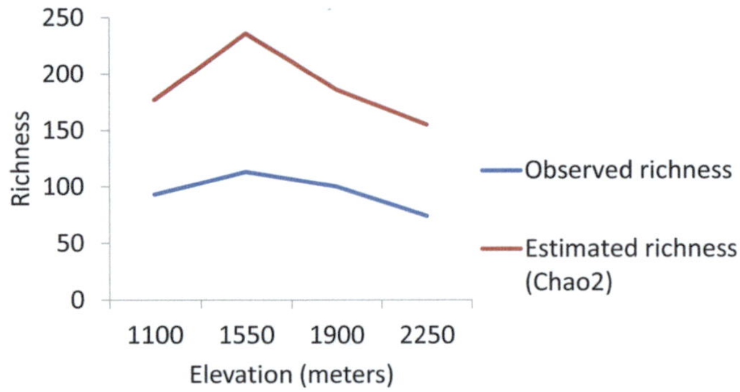

Fig. 15.6 Relationship between elevation and observed richness and estimated richness for the EM fungal community

correlation with climate or soil factors (P>0.1) but correlated with elevation ($F_{1,2}$=14.2, R^2=0.82, P=0.063), precipitation ($F_{1,2}$=103.6, R^2=0.97, P=0.009), soil C/N ($F_{1,2}$=69.9, R^2=0.96, P=0.014), and richness of host genera ($F_{1,2}$=19.0, R^2=0.86, P=0.049).

Higher elevations, starting from 2250 masl, are associated with higher relative occurrence (relative occurrence = number of cores with the presence of a given lineage/total number of soil cores containing EM-colonized roots) of lineages *Cortinarius* [18.7% (2250 masl), 20.8% (1900 masl), 7.6% (1550 masl), and 6.9% (1100 masl), respectively] and *Russula-Lactarius* (24.2%, 20.0%, 6.9%, and 15.4%). Lower-elevation sites show a higher relative occurrence of lineages *Sebacina* (2.3%, 2.0%, 11.5%, and 12.8%) and *Inocybe* (2.7%, 2.3%, 7.6%, and 8.5%). However, the relative occurrence of lineage *Tomentella-Thelephora* (5.0%, 14.6%, 18.0%, and 16.0%) shows a unimodal pattern, with maxima at 1550 masl. Relative occurrence of all these lineages mentioned above significantly varied between sites (P<0.002, Fisher's exact test). NMDS plot (Bray-Curtis distance) reveals a distribution of EM fungi on Mt. Fuji guided mostly by site and host.

15.6 Trends and the Underlying Mechanisms

This microbial mid-elevation peak in species richness/diversity diverges from the steady decline observed in richness pattern observed for the vascular plants on Mt. Fuji (Ohsawa 1984). This observed trend for microbes on Mt. Fuji still coincides with many terrestrial macroorganism patterns (McCain 2005; Rahbek 2005). One might expect to observe a decrease in microbial richness/diversity as we move up the mountain as both vascular plant population and area are decreasing, but this pattern seen on Mt. Fuji is evidently in conflict with the hypothesis that habitat area or the ultimate food supply (plants) is key to deciphering patterns in species richness with elevation (Lomolino 2001). One possible reason behind this could be the size of the microbes which allows them to maintain a vast and successful population within small niches with minute sources of nutrients and thus invalidates the habitat area hypothesis here.

So, how does this peak in mid-elevations could be explained? It could be perhaps attributed to the mid-domain effect or MDE which predicts that within a confined geographical area (such as Mt Fuji here), an increase in range overlap near the center of the area (mid-elevations) could result in an increase in phylotypic richness/diversity in that area with time. The MDE model assumes that species which have continuous dispersal ranges (Colwell et al. 2004; Currie and Kerr 2008) are possibly related to environmental gradients present in that area. Mt. Fuji in itself possesses an environmental gradient where almost all of the environmental variables like total carbon and nitrogen and extractable phosphorus and potassium linearly decrease with an increase in elevation, in a sense creating an optimal zone at mid-elevations which could be possibly suitable for most of the microbes. Also, near neutral pH range (optimal pH for most of the microbes) could be found near mid-elevations on Mt. Fuji. Thus, this range overlap perhaps provides an "optimal" environment of the two extremes demanding less physiological specialization, and species accumulate into it with time.

"Intermediate disturbance" hypothesis could also provide likely answers for this observed pattern (Huston 1994). We can observe this specifically for bacteria that just beyond the vegetation line (above 2500 masl on Mt. Fuji), the extreme fluctuations in temperature, harsher UV rays, dearth of nutrients, and a recurrent soil erosion of the loose substrate lead to lower competition rates and a subsequent increase in diversity due to "lottery" recruitment. However, at the summit and higher elevation, the environmental conditions may reach such extremes that only vastly adapted bacterial species for such environments can survive. In addition, this relatively restricted and unstable environment at highest elevations may not have allowed accumulation of substantial species through evolutionary time (especially in case of vascular plants (Grime 2001). Similarly, for archaea and EM fungi, we can say that it is possible that a more stable environment of the lowermost forest zones (1000 masl) begets out-competition between species with overlapping niches, resulting in a decrease in richness/diversity. Again, the higher elevations (3000 masl and above) might be reverse of an optimal environment with bare/sparsely vegetated by shrubs only fields, subjected to frequent frost heave and landslides. This may result in viable habitation by only a few species (or availability of a few viable niches which are highly prone to frequent disturbances and hence to recurrent drops in populations) – hence the lower diversity at summit and higher elevations (Huston 1994).

Potentially, the hypothesis of "everything is everywhere, and the environment selects" also seems to play an important role, which needs us to observe which environmental parameters are delimiting the community composition and diversity. Extractable potassium, nitrate, and ammonium concentrations significantly vary with diversity on Mt. Fuji for both archaea and bacteria, although not as strong as elevation itself. But elevation in itself is strongly covarying with different soil edaphic variables like temperature, total carbon, and nitrogen and might act as a proxy for these variables. The relative importance of each measured variable is difficult to determine at this level as they probably are correlated with some unknown and unmeasured variables such as disturbance or history, which could turn out to be the parameter actually controlling the diversity trend. The area is still open for further studies which can elucidate a better explanation for the trends observed and underlying mechanisms.

15.7 Perspectives

We could say that with studies like this, we have started to understand the elevational patterns of richness and diversity for microbes, but still further work is required to comprehend the underlying causes of these observed patterns, including both further observational studies and manipulative experiments along the gradients.

References

Brown JH, Lomolino MV (2005) Biogeography (Sinauer, Sunderland, 1998). A definitive textbook on the biogeography of macroorganisms

Bryant JA, Lamanna C, Morlon H, Kerkhoff AJ, Enquist BJ, Green JL (2008) Microbes on mountainsides: contrasting elevational patterns of bacterial and plant diversity. Proc Natl Acad Sci U S A 105:11505–11511. https://doi.org/10.1073/pnas.0801920105

Colwell RK, Rahbek C, Gotelli NJ (2004) The mid-domain effect and species richness patterns: what have we learned so far? Am Nat 163:E1–E23. https://doi.org/10.1086/382056

Currie DJ, Kerr JT (2008) Tests of the mid-domain hypothesis: a review of the evidence. Ecol Monogr 78:3–18. https://doi.org/10.1890/06-1302.1

Falkowski PG, Fenchel T, Delong EF (2008) The microbial engines that drive Earth's biogeochemical cycles. Science 320:1034–1039. https://doi.org/10.1126/science.1153213

Fierer N, McCain CM, Meir P, Zimmermann M, Rapp JM, Silaman MR, Knight R (2011) Microbes do not follow the elevational diversity patterns of plants and animals. Ecology 92:797–804. https://doi.org/10.1890/10-1170.1

Fukami T, Wardle DA (2005) Long-term ecological dynamics: reciprocal insights from natural and anthropogenic gradients. Proc R Soc B Biol Sci 272:2105–2115. https://doi.org/10.1098/rspb.2005.3277

Grime JP (2001) Plant functional types, communities and ecosystems. Ecol Achiev Chall 127:161–181

Huston MA (1994) Biological diversity: the coexistence of species on changing landscapes. Cambridge University Press, Cambridge

Korner C (2007) The use of 'altitude' in ecological research. Trends Ecol Evol 22:569–574. https://doi.org/10.1016/j.tree.2007.09.006

Lomolino MV (2001) Elevation gradients of species-density: historical and prospective views. Glob Ecol Biogeogr 10:3–13. https://doi.org/10.1046/j.1466-822x.2001.00229.x

McCain CM (2005) Elevational gradients in diversity of small mammals. Ecology 86:366–372. https://doi.org/10.1890/03-3147

Miyamoto Y, Nakano T, Hattori M, Nara K (2014) The mid-domain effect in ectomycorrhizal fungi: range overlap along an elevation gradient on Mount Fuji, Japan. ISME J 8:1739–1746. https://doi.org/10.1038/ismej.2014.34

Nogues-Bravo D, Araujo MB, Romdal T, Rahbek C (2008) Scale effects and human impact on the elevational species richness gradients. Nature 453:216–U218. https://doi.org/10.1038/nature06812

Ohsawa M (1984) Differentiation of vegetation zones and species strategies in the Subalpine region of Mt Fuji. Vegetatio 57:15–52. https://doi.org/10.1007/BF00031929

Rahbek C (2005) The role of spatial scale and the perception of large-scale species-richness patterns. Ecol Lett 8:224–239. https://doi.org/10.1111/j.1461-0248.2004.00701.x

Schlesinger WH (1997) Biogeochemistry. Geotimes 42:44–44

Singh D, Takahashi K, Adams JM (2012a) Elevational patterns in archaeal diversity on Mt. Fuji. PLoS One 7. https://doi.org/10.1371/journal.pone.0044494

Singh D, Takahashi K, Kim M, Chun J, Adams JM (2012b) A hump-backed trend in bacterial diversity with elevation on Mount Fuji, Japan. Microb Ecol 63:429–437. https://doi.org/10.1007/s00248-011-9900-1

Singh D, Lee-Cruz L, Kim WS, Kerfahi D, Chun JH, Adams JM (2014) Strong elevational trends in soil bacterial community composition on Mt. Ha lla, South Korea. Soil Biol Biochem 68:140–149. https://doi.org/10.1016/j.soilbio.2013.09.027

Torsvik V, Ovreas L, Thingstad TF (2002) Prokaryotic diversity – magnitude, dynamics, and controlling factors. Science 296:1064–1066. https://doi.org/10.1126/science.1071698

Venter JC, Remington K, Heidelberg JF, Halpern AL, Rusch D, Eisen JA, Wu DY, Paulsen I, Nelson KE, Nelson W, Fouts DE, Levy S, Knap AH, Lomas MW, Nealson K, White O, Peterson J, Hoffman J, Parsons R, Baden-Tillson H, Pfannkoch C, Rogers YH, Smith HO (2004) Environmental genome shotgun sequencing of the Sargasso Sea. Science 304:66–74. https://doi.org/10.1126/science.1093857

Wang JJ, Soininen J, He JZ, Shen J (2012) Phylogenetic clustering increases with elevation for microbes. Environ Microbiol Rep 4:217–226. https://doi.org/10.1111/j.1758-2229.2011.00324.x

Whitman WB, Coleman DC, Wiebe WJ (1998) Prokaryotes: the unseen majority. Proc Natl Acad Sci U S A 95:6578–6583. https://doi.org/10.1073/pnas.95.12.6578

Integration of Soft Computing Approach in Plant Biology and Its Applications in Agriculture

16

Archana Kumari, Minu Kesheri, Rajeshwar P. Sinha, and Swarna Kanchan

Abstract

Soft computing is a modern approach for analysis of complex problems. In agricultural field for complex problems, we require conventional methods which can give cost-effective, analytical and complete solutions. The past few decades have witnessed extensive research in the field of soft computing. In retrospect to development in agricultural sector, various analytical methods like artificial neural networks, support vector machines, fuzzy logic, decision trees and many more have been designed. These methods help to analyze soil and water regimes which are directly involved in crop growth, food processing and also help in precision farming. This review will provide an overview of the integration of soft computing approach in various fields of biology. Moreover, an extensive review of future prospects of soft computing in agriculture in particular and plant biology in general. In this book chapter, co-relation between soil and water as well as crop management has been discussed. The book chapter has been made more reader friendly and easily understandable by incorporation of appropriate diagrams providing detailed study on integration of soft computing approach in plant biology and its applications in agriculture in an easy and illustrative manner.

A. Kumari
Department of Biotechnology, Bodoland University, Kokrajhar, Assam, India

M. Kesheri
Amity School of Engineering and Technology, Amity University Jharkhand, Ranchi, India

R. P. Sinha
Centre of Advanced Study in Botany, Banaras Hindu University, Varanasi, India

S. Kanchan (✉)
Department of Biology, School of Engineering, Presidency University, Bengaluru, India
e-mail: swarnabioinfo@gmail.com

H. J. Purohit et al. (eds.), *Soft Computing for Biological Systems*,
https://doi.org/10.1007/978-981-10-7455-4_16

Keywords
Agriculture · Plant biology · Artificial neural network · Genetic algorithms · Soft computing

16.1 Introduction

Soft computing is basically a study of single or unique combination of various machine learning methods like genetic algorithm, fuzzy logic, artificial neural network, etc. All of which serves to aid in providing solution for very complex problems. Actually, it is a mixture of various biological data which includes gene and protein sequences, domains, secondary and tertiary protein structures and computing techniques. Different techniques provide varied ways of solution for a particular problem. For example, fuzzy logic (FL) generally gives a multivalued, nonnumeric variable, artificial neural networks (ANNs) provide the solution based on the connection of a number of artificial neurons (ANN mimics biological neurons and their connections), and genetic algorithms (GAs) provide solution based on the same process which nature uses, i.e. best selection for survival, recombination and mutation. Soft computing is a modern approach to get a cost-effective as well as less time-consuming solution in a precise way. Due to these features, soft computing techniques become the most conventional method providing solution in analytical way. Among all machine learning techniques, FL is the first soft computing technique which laid the foundation of machine learning (Zadeh 1965, 1981, 1973) techniques. The idea of ANN was given by Rumelhart and McClelland (1986), whereas idea of GA was given by John Holland in 1975, and later this idea was promoted by his student, David Goldberg.

In the present scenario, these three methods are considered as the heart of soft computing. In modern sciences, some other techniques are also included in soft computing/machine learning tools like support vector machine, probabilistic reasoning, chaos theory, etc. From the last few years, soft computing emerged as the hot area for various scientific researches. Although techniques used in soft computing are providing new approaches for efficient and reliable solutions for complex biological problems yet, support vector machines (SVMs) which provide the higher accuracies as compared to ANN and is based on supervised linear classifiers (Burges 1998), have emerged as one of the major areas of research in the last few years. Another approach of modern soft computing is the fusion of two or more soft computing techniques. Actually, fusion is a cascade or combination of advanced soft computing techniques for the best system performance. The best example for this type of system is neuro-fuzzy (Simpson and Jahns 1993; Takagi and Hayashi 1991; Horikawa et al. 1992; Nie and Linkens 1992). It is quite easy to predict the stability and behaviour of hard computing, whereas the burden of algorithms in hard computing is typically very low or moderate. The combination of soft computing and hard computing has great potential. Due to its good compatibility, it is easy to develop high enactment, cost-effective and trustworthy computing arrangements which ultimately provide innovative solutions (Ovaska et al. 2002).

In plant biology and agriculture, soft computing emerged very slowly as compared to other branches of sciences (Whittaker et al. 1991; Eerikäinen et al. 1993; Zhang and Litchfield 1992); however, it suddenly increased from the last few decades. In plant biology and agriculture, scientists have developed novel approaches for decision tree (DT), FL, ANN, GA, Bayesian inference (BI) and SVM. In plant biology these soft computing/machine learning methods are used to explore the 3D protein structures (Priya et al. 2016; Kumari et al. 2016; Kesheri et al. 2015a, b). These soft computing methods are used in evolutionary study of various proteins having plant and animal origin (Kanchan et al. 2014, 2015). These approaches are mostly used for the study of soil as well as water in relation to growth of agricultural crop. It also helps to improve the process of food processing as well as helps to take good decisions in precision farming. They also used combination techniques in solving problems related to plant biology and agriculture.

Still, we have not found any proper application which can establish a strong link between soft computing and hard computing. While elucidating the phenomena of stress tolerance in *Nostoc commune*, the use of computational biology/soft computing revealed the evolutionary relationship through phylogenetic tree (Kesheri et al. 2014). The significance of soft computing techniques is also visualized while exploring the potential of antioxidants in retarding ageing (Kesheri et al. 2017). This could be a unique research topic which could also have a great potential. In this chapter, we briefly reviewed the various machine learning/soft computing tools and their applications in plant biology and agriculture. This chapter will explain a basic concept of soft computing techniques in a very precise way with reference to crop improvement and management.

16.2 Methods of Soft Computing

Generally, soft computing/machine learning methods include FL, ANN, GA, probabilistic computing, chaotic systems, etc. These methods are considered as primary methods particularly for research and development in plant biology and agricultural fields. In other words, ANNs are based on the idea of the presence of a number of neurons in the human brain. GAs is known as genetic algorithm, whereas BI deals with probabilistic computing. For huge data management, DT is one of the interesting and unique ways of learning as well as organizing the data sets in soft computing. Hence all these methods are applied in plant biology and agriculture, significantly since the last few decades. A list of soft computing methods is mentioned in (Fig. 16.1).

16.2.1 Fuzzy Logic

Fussy logic implements fuzzy set theory which consists of multivalued logic. Fussy logic displays 0 and 1 as outputs, which represent fuzzy (degree of truth). In FL, the

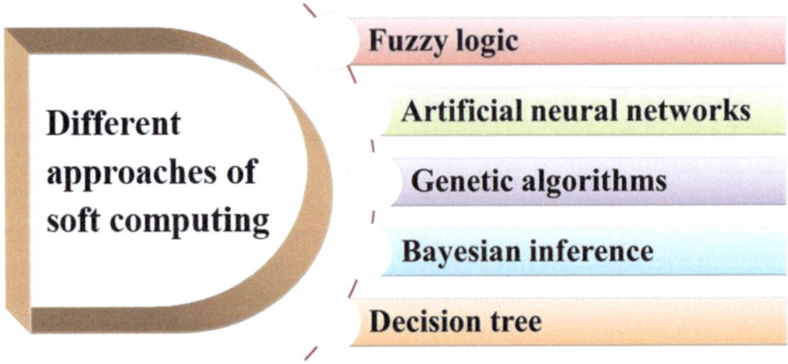

Fig. 16.1 Different approaches of soft computing

fuzzy set functions can measure every input used. Different functions are used in different logical operations like AND and OR. Some specified inference systems included in FL (Jang 1993) should have following features:

1. A FL interface should have the capacity to convert the inputs into linguistic values.
2. A number of fuzzy sets based on fuzzy rules.
3. A knowledge base.
4. A database which explains the functions of the fuzzy sets.
5. An unit which carries operations.
6. An interface which converts results into 0 and 1 (output).

16.2.2 Artificial Neural Networks

ANNs mimics the human brain, i.e. biological neurons to solve multifaceted problems. Since the last few years, scientists showed their great interest to understand the mechanism of working of the brain and its structure. As we know every neuron is self-directed and self-determining and has capacity to work asynchronously. Due to great processing power of biological neurons, it inspired the study of ANNs with respect to their organization and computing power for complex problems. In comparison to the conventional methods, ANNs offer an adaptive, fast, robust solution, aptitude to handle inaccurate and fuzzy data, along with the capability to generalize in a well-organized way which is easy to understand. ANN is a powerful data processing method which is capable enough to record process input as well as output. ANNs became more popular because of their capability to solve complex engineering problems, apart from providing a powerful method for accurate solutions which can be verified through experimental data.

ANNs are analogous to synapses in human body and composed of many interconnected processing elements; these processing units are associated with different weights. Feedforward and recurrent neural networks are considered as two major categories for ANNs. Feedforward neural networks are acyclic, whereas recurrent neural networks are cyclic. ANN is successfully applied to identify various genes of biological importance which might be associated with crop improvements in agriculture (Kesheri et al. 2016). A number of drawbacks are also associated with ANN.

16.2.3 Genetic Algorithms

The genetic algorithm is one of the popular data mining tools which was proposed by computer scientists J. H. Holland. Genetic algorithm is basically a heuristic search technique used for optimization. This is also known as evolutionary computing which is used to solve problems which need optimization. GAs generally use iterative method to find the fit individual in a selected population. GA uses parallel processing to reach the optimal solution. The processing technique used by GAs is based on theory of evolution proposed by Darwin (1859). GAs use inheritance, mutation, selection and crossover operators to provide a desired output of a particular input. GA works simultaneously on selected population with solutions to the problem related to individual. It starts with subpopulation to select a set of surviving individuals that will have the capacity of reproduction to carry out the life in that particular population. Then, the individuals will manage some changes using genetic mutation and crossover for their survival. A very careful selection of different approaches of genetic algorithm provides optimal solution after few iterations. GA is computationally simple, less expansive and very powerful method. It is powerful enough to do a number of iterative searches for difficult combinational problems. Due to these reasons, GAs became alternative tools for traditional optimization.

Genetic algorithms have advantage over other data mining methods in terms of movement which is very fast, from one generation to another generation. The major disadvantage of genetic algorithm is that this method is biased towards fit individuals.

16.2.4 Bayesian Inference

Another approach is Bayesian inference (BI); it is also known as probabilistic computing. This method is based on probability which can handle uncertainty by combining statistical methods and probability distributions. Usually, hypothesis (i.e. either true or false) testing is used to check which hypothesis is correct.

Bayesian networks are well known as one of the best decision-making tools in multifaceted conditions in a wide range of disciplines (Charniak 1991). Bayesian networks are mainly based on probabilistic graphical models which are defined in a set of variables. Bayesian networks provide the fine representation of a cognitive process. This cognitive process is based on two attributes: first is called link and second is called node. In this process the state of parent node helps in prediction of the state of the child node. Further, conditional probability tables, which also use Bayesian statistics, are used to explore the actual connection between the states of nodes for parent and child nodes.

16.2.5 Decision Tree

Decision tree (DT) was discovered by Magee (1964), and it is composed of a treelike structure. In DT, various attributes are considered to evaluate the problem, and these attributes are further used to predict the output. In DT, a recursive algorithm is used to find out the attribute having highest information, which is further used for the evaluation first. DTs are mostly used to identify a particular approach to achieve the desired goal. DT is also used for calculating conditional probabilities and predictive model in data mining/machine learning (Teorey 1999; Witten and Frank 2000). One of the major drawbacks of DT is that it cannot provide the final decision to any problem. The classification tree is used for decision-making. In DT, leaves describe classifications, whereas branches describe the common features. DTs are helpful in analyzing large quantity of data in a short period which may be applied for many plant biology and crop improvement programmes.

16.3 Applications of Soft Computing in Plant Biology and Agriculture

Nowadays, many applications which are having good accuracy are available which can be also used at industrial level. Problems in plant biology and agriculture related to soil management, crop management, water management and postharvesting have been resolved through soft computing. Problems related to precision agriculture, food safety and food processing can be solved through soft computing/machine learning methods based on prediction and optimization. It is quite valuable to employ ANNs, FL and GAs in different combination instead of using it alone. A list of hybrid soft computing techniques/machine learning tools (Table 16.1) is also being used in plant biology and agriculture. This list indicates that the fusion of FL and ANNs is the most commonly used for evolving the hybrid methods of machine learning techniques. A list of hybrid methods of soft computing/machine learning techniques which are being applied in plant biology and agriculture is shown in Fig. 16.2.

16.3.1 Crop Management

Pearson and Wicklow (2006) established a novel method of ANN to recognize fungal species that contaminate single kernels in place of input for crop protection.

Table 16.1 Major finding using combination of soft computing techniques for plant biology and agricultural improvement

SL. no.	Combination type	Applications	References
	ANN + GA	Snack foods identification	Jindal and Srisawas (2001)
1.	FL + GA	Optimization of design of threshing units	Miu and Perhinschi (2001)
2.	ANN + GA	Cold storage optimization for tomato storage	Morimoto et al. (2003)
3.	FCM + RBF	Performance of vegetative strips	Andriyas et al. (2003)
4.	ANFIS modelling	Prediction of various properties of soil	Lee et al. (2003)
5.	ANFIS classification	Detection of weed by segmentation of colour image of weed	Neto et al. (2003)
6.	SOM + FCM	Segmentation of coloured image of beans	Chtioui et al. (2003)
7.	Fuzzy–ANN	Classification of soil	
8.	ANFIS classification	Colour images of plant and their residues and soil	Meyer et al. (2004)
9.	Fuzzy c-means clustering + RBF	Determination of phosphorous movement	Goel et al. (2004)
10.	ANN + GA	Modelling of rainfall	Jain and Srinivasulu (2004)
11.	GA + ANN	Determination of sugarcane maturation curves	Madeiro et al. (2006)
12.	GA + ANN	Helps in determination of sugarcane harvest period	Oliveira et al. (2006)
13.	ANFIS classification	Monitoring hydraulic pump health	Hancock and Zhang (2006)
14.	ANN + GA	Automated calibration of watershed	Lakshmi et al. (2006)
15.	ANN + GA	For detection of fungi infection, automated classification of corn kernels	Pearson and Wicklow (2006)
16.	ANN + fuzzy	Applications in plant tissue culture	Prasad and Dutta Gupta (2008)
17.	Support vector machines	Classification and identification of plant diseases	Rumpfa et al. (2010)
18.	ANN modelling + fuzzy control	In vitro rhizogenesis	Gago et al. (2010a)

(continued)

Table 16.1 (continued)

SL. no.	Combination type	Applications	References
19.	ANN modelling + GA	Used in complex plant processes	Gago et al. (2010b)
20.	Neuro-fuzzy approach	"In vitro" direct rooting	Gago et al. (2010c)
21.	ANN with evolutionary preprocessing	Plant virus identification	Glezakos et al. (2010)
22.	Optimized hyperspectral spectral indices and partial least-squares regression	Estimation of nitrogen content	Li et al. (2014)
23.	ANN with SVM	Prediction of soil organic carbon	Werea et al. (2015)
24.	ANN with fuzzy	Evaluation of polyphenol oxidase (PPO) activity in lychee pericarp	Yang et al. (2015)
25.	ANN with fuzzy	Determination of influential weather parameters on reference evapotranspiration	Petkovića et al. (2015)
26.	GA with ANN	Estimating of soil temperature	Nahvia et al. (2016)

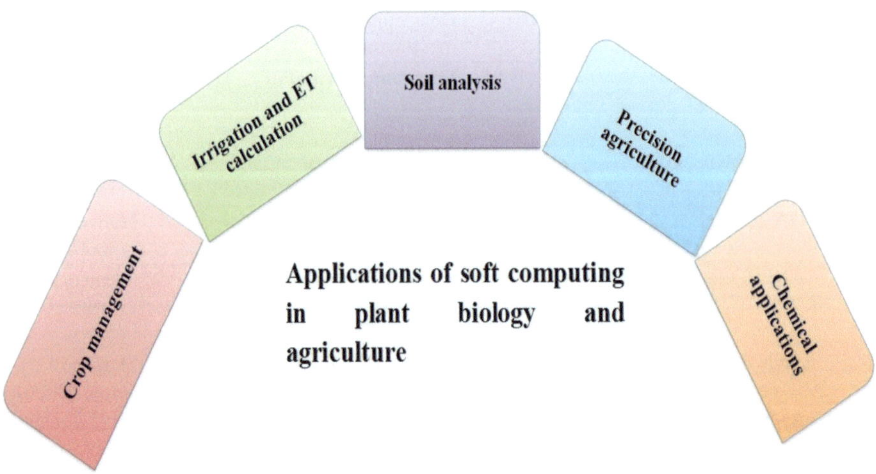

Fig. 16.2 Various applications of soft computing in plant biology and agricultural field

The ANN was trained using GAs where they observed that the GA training algorithm is good for training the data sets which doesn't over-fit the data (Lestander et al. 2003). Meyer and co-workers used uncertain inference structures in ANFIS for generation of colourful images of grass, cornstalk residue and wheat straw residue (Meyer et al. 2004). Neto et al. (2003) also used ANFIS for generation of colourful image which was used to distinguish weeds from the others plants in background. Odhiambo et al. (2001a, b) illustrated the means to remove trial and error in defining connection which are based on supervised learning which optimizes functions.

16.3.2 Soil Analysis

Soil profile studies involve collective use of ANN and fuzzy system for unsupported assembling and arrangement of soil profiles by means of ground-penetrating detector. ANN categorizes soil in a number of profile strips into an assured number of groups. The fuzzy members for each soil profiles are determined in the set of confidential groups. Residue and phosphorous movement were recognized by fuzzy K-mean clustering algorithms (Goel et al. 2004). Many researchers used partial least-squares method as well as adaptive network fuzzy inference arrangement methods for determination of water content and salinity in soil.

16.3.3 Precision Agriculture

Xiang and co-workers established an inclusive artificially intelligent controller which was based on ANN and fuzzy network system. In that they used a checker to spontaneously regulate camera for advantage of nonconformities during day hours (Xiang and Tian 2007). Oliveira et al. (2006) suggested a system which is integrated with ANN monitored for heuristic examination by gas, which is also used to approve suitable sugarcane regions to be harvested. It has also been confirmed that use of feedforward, completely related, BP-trained ANN may be imprecise for the nonlinear harvest determination connecting advance corn yield (Liu et al. 2001). Some of the data were used to train the database which is called as training set. This training set increases the accuracy for prediction of rainfall by ANN. Once the ANN is well trained, GA was integrated for optimization of the various input features so that maximum accuracy can be achieved.

16.4 Support Vector Machines

Support vector machines (SVMs) are known as best classifiers nowadays. SVMs can solve many complex problems and have fascinated more attention in recent times in plant biology and agricultural field. SVMs are also associated with ANNs many times which increases the prediction accuracy. In reality, SVM model is similar to a two-layer neural network. SVMs can also be used as substituting training tool for radial- and polynomial-based functions. SVMs have very sophisticated classification accurateness as compared to multilayer perceptron ANNs.

Table 16.2 List of various soft computing techniques used for plant biology and agricultural improvement

SL. no.	Application methods	Application area	References
1.	SVM approach	Plant disease classification	Tian et al. (2004)
2.	SVM approach	Identification of different varieties of tea	Chen et al. (2006a, b)
3.	SVM approach	Nitrogen stress identification and classification of various kinds of weeds	Karimi et al. 2006
4.	SVM approach	Immature hazelnut screening	Onaran et al. (2006)
5.	SVM approach	Analysis of compound feeds	Pierna et al. (2006)
6.	Least-squares SVM classification	Classification of wheat classes	Wang and Paliwal (2006)
7.	Gaussian kernel-based SVM classification	Meat classification	Jiang et al. (2007)
8.	SVM modelling and prediction	Sediment yield determination of a watershed	Oommen et al. (2007)
9.	Multi-class SVM with kernel of RBF neural network	Discrimination of individual fungal wheat kernels	Zhang et al. (2007)
10.	Least-squares SVM modelling	Vitamin C content determination in kiwifruit	Fu et al. (2008)
11.	SVM modelling	Determination of various concentration of instant coffee	Kovacs et al. (2008)
12.	SVM modelling	Evaluation of hydraulic properties of soil	Lamorski et al. (2008)
13.	Least-squares SVM classification	Identification of paddy seeds by their year of harvesting	Li et al. (2008)
14.	Least-squares SVM classification	Chinese cabbage variety seeds identification	Wu et al. (2008)
15.	Least-squares SVM modelling	Composition of wine made from rice	Yu et al. (2008)
16.	SVM classification	Class recognition of rice blast with multispectral imaging to supervise variable spray	Qi and Ma (2009)
17.	Least-squares SVM	Amylose and protein content determination in rice after gamma irradiation	Shao et al. (2009)
18.	SVM with laser-induced fluorescence	Accurate identification of nitrogen fertilizer application of paddy rice	Yang et al. (2015)
19.	Using support vector machine	Identification and classification of various diseases in *Brassica* sp.	Palak et al. (2015)
20.	SVM-based modelling	Used to detect evapotranspiration using hydro-climatic variables in a sub-tropical environment	Shukla et al. (2015)

(continued)

Table 16.2 (continued)

SL. no.	Application methods	Application area	References
21.	SVM approach	High-throughput stress phenotyping in plants	Singh et al. (2016)
22.	SVM	Soil quality assessment	Liua et al. (2016)

In plant biology and agricultural field, the growing interest in SVMs is increasing day by day. There are popular applications of SVMs (Table 16.2) which are used in plant biology and agricultural fields. The increasing attention in SVMs is because of (a) their essential success regarding traditional classifiers, consequently resulting in elevation of classification precisions in a simplified way, (b) some degree of determination is necessary for structural design and (c) the probability of elucidating problems rendering to linearly self-conscious quadratic user interface design (Melgani and Bruzzone 2004). In soil analysis, soil hydraulic factors affect dignified soil properties using SVMs (Lamorski et al. 2008). Various researches indicated that the importance of SVM was prominent for soil abilities, whereas comparative errors and the correlation were lower. Currently developed SVMs are used to calculate water retention and hydraulic conductivity of soil (Twarakavi et al. 2009).

It has been also proposed that SVMs act as a tool for categorizing airborne hyperspectral pictures in use for a cornfield (Karimi et al. 2006). The SVM technique ensured truncated misclassification proportions when it was compared with ANN. Uncovering stresses in initial crop development via the SVM could provide site-specific remedies. Tian et al. (2004) used SVM and designed for better consistency especially for coloured images for various plant diseases. The investigations also demonstrated that the SVM had outstanding arrangement and broad view in resolving erudition problem. This technique works well for classification of plant disease with a small amount of sample (Trebar and Steele 2008). The use of disseminated SVM structural design showed good results against optimization of large data.

16.5 Comparison and Limitations of Soft Computing Techniques

Soft computing techniques have some restrictions because of lack of conscious in theoretical study as well as practical application. Calculations on activities of many soft computing techniques were thoroughly discussed by Tikk et al. (2003). These soft computing techniques were mainly based on neural networks and fuzzy systems, and it was observed that the hybrid method increased the degree of accuracy dramatically. Nevertheless, ‘building blocks’ in large quantities may be required to accomplish the prescribed accuracy for the rough calculation.

Uncertainty may exist on the number of building blocks; if their numbers are restricted during calculation, they may lose their universal approximation property. Consequently, there should be a balance between accuracy and the number of the building blocks which is used to count the hidden layers (i.e. fuzzy sets or hidden neurons) in ANN which are used for training of the data sets and predictions. Similarly, fuzzy systems are involved in the time-consuming as well as difficult process knowledge acquisition and demonstration.

16.6 Future Prospects

Nowadays, a number of soft computing methods are being used in plant biology and agriculture; each of these methods has some limitations. Fusion of these computing methods contributes a lot in better generalization and predictions. In view of that, FL was fused with ANNs to include the advantages of these methods for better accuracy and predictions. Nowadays, due to wonderful development in electronics and information technology, global positioning system (GPS), remote sensing (RS), variable rate technology (VRT) and geographic information system (GIS) technologies have started being used for better agricultural production and irrigation. For good output due to use of the technologies, various kinds of information sources such as scientists, farmers, experts, engineers and system incorporation are highly required. Therefore, it delivers relevant information about location based on climatic data like soil, weather and water. Various metrological data are collected by different automated instruments on a daily basis for the measuring of wind, speed, solar radiation, rainfall, air temperature, etc. This information will help farmers to provide more precise information about seeding time, optimum use of fertilizers and irrigated water to enhance crop productivity.

For proper expansion of soft computing technologies, it is a must to frame proper guidelines for optimal artificial neural network structures and various algorithms responsible for training of the data sets used by ANN. These developments will allow the utilization of numerous soft computing techniques among which SVM has proved one of the most popular techniques. In practical, SVMs produced high accuracy specially in classifying objects, when it is compared to ANNs. Moreover, SVMs are involved in modelling the setup of control parameters especially for structural design.

In context of the soil and water, they are playing an essential role in crop management, precision agriculture and sensor-derived information. A number of soft computing technology-based applications might be involved in classification of agricultural soil and their distribution. Such applications might be also used for water resource optimization for irrigation planning. Now only in the above area of agriculture these applications might be also used for detection and classification of crop stress. These applications are also being used for detection and classification of pests which include diseases as well as insects. Moreover, they may also include detection, analysis of crop yield, remote sensing and field recommendations for variable rate of fertilization application. Modern science is targeting the fusion of soft computing methods with their integration with hardware/instruments and their application in plant biology and agriculture. The fusion of soft computing technologies with high-performance hardware may have the capability to produce most accurate results for the problems related with reliable computing systems in a cost-effective manner.

References

Andriyas S, Negi SC, Rudra RP, Yang SX (2003) Modelling total suspended solids in vegetative filter strips using artificial neural networks. Trans ASABE 032079. 10.13031/2013.13770

Burges CJC (1998) A tutorial on support vector machines for pattern recognition. Data Min Knowl Discov 2:121–167. https://doi.org/10.1023/A:1009715923555

Charniak E (1991) Bayesian networks without tears. AI Mag 12(4):50–63

Chen Q, Zhao J, Cai J, Wang X (2006a) Study on identification of tea using computer vision based on support vector machine. Chin J Scient Instru 27(12):1704–1706

Chen Y, Zheng J, Xiang H, Huang S (2006b) Study on an intelligent system for precision pesticide application based on fuzzy control and machine vision. Trans ASABE 061129. 10.13031/2013.20631

Chtioui Y, Panigrahi S, Backer LF (2003) Self-organizing map combined with a fuzzy clustering for color image segmentation of edible beans. Trans ASAE 46(3):831–838

Darwin C (1859) On the origin of species, vol 46. John Murray, London. 10.13031/2013.13577

Eerikäinen T, Linko P, Linko S, Siimes T, Zhu YH (1993) Fuzzy logic and neural networks applications in food science and technology. Trends Food Sci Tech 4:237–242. https://doi.org/10.1016/0924-2244(93)90137-Y

Fu X, Ying Y, Xu H, Yu H (2008) Support vector machines and near infrared spectroscopy for quantification of vitamin C content in kiwifruit. Trans ASABE 085204. 10.13031/2013.24721

Gago J, Landín M, Gallego PP (2010a) Artificial neural networks modeling the in vitro rhizogenesis and acclimatization of Vitis vinifera L. J Plant Physiol 167:1226–1231. https://doi.org/10.1016/j.jplph.2010.04.008

Gago J, Martínez-Núñez L, Landín M, Gallego PP (2010b) Strengths of artificial neural networks in modelling complex plant processes. Plant Signal Behav 5(6):1–3. https://doi.org/10.4161/psb.5.6.11702

Gago J, Landín M, Gallego PP (2010c) A neurofuzzy logic approach for modelling plant processes: a practical case of in vitro direct rooting and acclimatization of Vitis vinifera L. Plant Sci 179:241–249. https://doi.org/10.1016/j.plantsci.2010.05.009

Genetic Algorithms in Search, Optimization, and Machine Learning. Choice Reviews Online 27.02 (1989): 27–0936–27–0936. doi:https://doi.org/10.5860/choice.27-0936

Glezakos TJ, Moschopoulou G, Tsiligiridis TA, Kintzios S, Yialouris CP (2010) Plant virus identification based on neural networks with evolutionary preprocessing. Comput Electron Agric 70:263–275. https://doi.org/10.1016/j.compag.2009.09.007

Goel PK, Andriyas S, Rudra RP, Negi SC (2004) Modeling sediment and phosphorous movement through vegetative filter strips using artificial neural networks and GRAPH. Trans ASAE 042263. 10.13031/2013.17674

Hancock KM, Zhang Q (2006) A hybrid approach to hydraulic vane pump condition monitoring and fault detection. Trans ASABE 49(4):1203–1211. 10.13031/2013.21720

Horikawa S, Furuhashi T, Uchikaw Y (1992) On fuzzy modelling using fuzzy neural networks with back propagation algorithm. IEEE Trans Neural Netw 3(5):801–806. https://doi.org/10.1109/72.159069

Jain A, Srinivasulu S (2004) Development of effective and efficient rainfall-runoff models using integration of deterministic, real-coded genetic algorithms and artificial neural network techniques. Water Resour Res 40:w04302. https://doi.org/10.1029/2003wr002355

Jang RJS (1993) ANFIS: adaptive-network-based fuzzy inference system. IEEE Trans Syst Man Cybern 23(3):665–685. https://doi.org/10.1109/21.256541

Jiang L, Zhu B, Jing H, Chen X, Rao X, Tao Y (2007) Gaussian mixture model-based walnut shell and meat classification in hyperspectral fluorescence imagery. Trans ASABE 50(1):153–160. 10.13031/2013.22388

Jindal VK, Srisawas W (2001) Acoustic testing of snack food texture. Trans ASAE 016038. 10. 13031/2013.5541

Kanchan S, Mehrotra R, Chowdhury S (2014) Evolutionary pattern of four representative DNA repair proteins across six model organisms: an in silico analysis. Netw Model Anal Health Inform Bioinform 3(1):70. https://doi.org/10.1007/s13721-014-0070-1

Kanchan S, Mehrotra R, Chowdhury S (2015) *In silico* study of endonuclease III protein family identifies key residues and processes during evolution. J Mol Evol 81:54–67. https://doi.org/10. 1007/s00239-015-9689-5

Karimi Y, Prasher SO, Patel RM, Kim SH (2006) Application of support vector machine technology for weed and nitrogen stress detection in corn. Comput Electron Agric 51 (1–2):99–109. https://doi.org/10.1016/j.compag.2005.12.001

Kesheri M, Kanchan S, Richa SRP (2014) Isolation and in silico analysis of Fe-superoxide dismutase in *Nostoc commune*. Gene 553(2):117–125. https://doi.org/10.1016/j.gene.2014. 10.010

Kesheri M, Kanchan S, Chowdhury S, Sinha RP (2015a) Secondary and Tertiary Structure Prediction of Proteins: A Bioinformatic Approach. In: Zhu Q, Azar AT (eds.), Complex system modelling and control through intelligent soft computations, Studies in Fuzziness and Soft Computing. Vol 319, Springer-Verlag Germany, pp 541–569. doi:https://doi.org/10.1007/978-3-319-12883-2_19

Kesheri M, Kanchan S, Richa SRP (2015b) Computational methods and strategies for protein structure prediction. In: Sinha RP, Richa Rastogi RP (eds) Biological sciences: innovations and dynamics. New India Publishing Agency, New Delhi, pp 277–291

Kesheri M, Sinha RP, Kanchan S (2016) Advances in soft computing approaches for gene prediction: a bioinformatics approach. In: Dey N, Bhateja V, Hassanien AE (eds) Advancements in bio-medical sensing, imaging, measurements and instrumentation, vol 651. Springer, Berlin, pp 383–405

Kesheri M, Kanchan S, Sinha RP (2017) Exploring the potentials of antioxidants in retarding ageing. In: Benjamin S, Sarath Josh MK (eds) Examining the development, regulation, and consumption of functional foods. IGI Global, Hershey, pp 166–195. https://doi.org/10.4018/ 978-1-5225-0607-2.ch008

Kumari A, Kanchan S, Kesheri M (2016) Applications of bio-molecular databases in bioinformatics. In: Dey N, Bhateja V, Hassanien AE (eds) Advancements in bio-medical sensing, imaging, measurements and instrumentation, vol 651. Springer, Berlin, pp 329–351. https://doi.org/10. 1007/978-3-319-33793-7_15

Lakshmi G, Sudheer KP, Chaubey I (2006) Auto calibration of complex watershed models using simulation-optimization framework. Trans ASABE 062126. 10.13031/2013.20715

Lamorski K, Pachepsky Y, Slawinski C, Walczak RT (2008) Using support vector machines to develop pedotransfer functions for water retention of soils in Poland. Soil Sci Am J 72:1243–1247. https://doi.org/10.2136/sssaj2007.0280n

Lee KH, Zhang N, Das S (2003) Comparing adaptive neuro-fuzzy inference system (ANFIS) to partial least-squares (PLS) method for simultaneous prediction of multiple soil properties. Trans ASAE 033144. 10.13031/2013.15017

Lestander TA, Leardi R, Geladi P (2003) Selection of near-infrared wavelengths using genetic algorithms for the determination of seed moisture content. J Near Infrared Spec 11(4):433–446. https://doi.org/10.1255/jnirs.394

Li X, He Y, Wu C (2008) Least square support vector machine analysis for the classification of paddy seeds by harvest year. Trans ASABE 51(5):1793–1799. 10.13031/2013.25294

Li F, Mistele B, Hu Y, Chen X, Schmidhalter U (2014) Reflectance estimation of canopy nitrogen content in winter wheat using optimised hyperspectral spectral indices and partial least squares regression. Eur J Agron 52:198–209. https://doi.org/10.1016/j.eja.2013.09.006

Liu J, Goering CE, Tian L (2001) A neural network for setting target corn yields. Trans ASAE 44 (3):705–713. 10.13031/2013.6097

Liua Y, Wanga H, Zhanga H, Libera K (2016) A comprehensive support vector machine-based classification model for soil quality assessment. Soil Till Res 155:19–26. https://doi.org/10.1016/j.still.2015.07.006

Madeiro SS, Oliveira FR, Alexandre FBA, Neto FB (2006) Intelligent modelling of sugar-cane maturation. In: Proceedings of the 4th world congress conference on computers in agriculture and natural resources, Orlando 642–648. doi:10.13031/2013.21950

Magee JF (1964) Decision trees for decision making. Harv Bus Rev 42:126–138

Melgani F, Bruzzone L (2004) Classification of hyperspectral remote sensing images with support vector machines. IEEE Transac Geosci Remote Sens 42(8):1778–1790. https://doi.org/10.1109/TGRS.2004.831865

Meyer GE, Hindman TW, Jones DD, Mortensen DA (2004) Digital camera operation and fuzzy logic classification of uniform plant, soil, and residue color images. Appl Eng Agric 20 (4):519–529. 10.13031/2013.16482

Miu PI, Perhinschi MG (2001) Optimal design and process of threshing units based on a genetic algorithm. II. Application. Trans ASAE 013125. 10.13031/2013.7431

Morimoto T, Tu K, Hatou K, Hashimoto Y (2003) Dynamic optimization using neural networks and genetic algorithms for tomato cool storage to minimize water loss. Trans ASAE 46 (4):1151–1159. 10.13031/2013.13938

Nahvia B, Habibib J, Mohammadic K, Shamshir bandd S, Razgane OSA (2016) Using self-adaptive evolutionary algorithm to improve the performance of an extreme learning machine for estimating soil temperature. Comput Electron Agric 124:150–160. https://doi.org/10.1016/j.compag.2016.03.025

Neto JC, Meyer GE, Jones DD, Surkan AJ (2003) Adaptive image segmentation using a fuzzy neural network and genetic algorithm for weed detection. Trans ASAE 033088. 10.13031/2013.13854

Nie J, Linkens D (1992) Neural network–based approximate reasoning: principles and implementation. Int J Control 56(2):399–413. https://doi.org/10.1080/00207179208934320

Odhiambo LO, Yoder RE, Yoder D (2001a) Estimation of reference crop evapotranspiration using fuzzy state models. Trans ASAE 44(3):543–550. 10.13031/2013.6114

Odhiambo LO, Yoder RE, Yoder DC, Hines JW (2001b) Optimization of fuzzy evapotranspiration model through neural training with input–output examples. Trans ASAE 44(6):1625–1633. 10.13031/2013.7049

Oliveira FR, Pacheco DF, Leonel A, Neto FB (2006) Intelligent support decision in sugarcane harvest. In: Proceedings of the 4th world congress conference on computers in agriculture and natural resources, Orlando, FL, pp 456–462. 10.13031/2013.21917

Onaran I, Pearson TC, Yardimci Y, Cetin AE (2006) Detection of under developed hazelnuts from fully developed nuts by impact acoustics. Trans ASABE 49(6):1971–1976. 10.13031/2013.22277

Oommen T, Misra D, Agarwal A, Mishra SK (2007) Analysis and application of support vector machine based simulation for runoff and sediment yield. Trans ASABE 073019. https://doi.org/10.1016/j.biosystemseng.2009.04.017

Ovaska SJ, Vanlandingham HF, Kamiya A (2002) Fusion of soft computing and hard computing in industrial applications: an overview. Ieee T Syst Man Cyb 32(2):72–79

Pearson TC, Wicklow DT (2006) Detection of corn kernels infected by fungi. Trans ASABE 49 (4):1235–1245. 10.13031/2013.21723

Petkovića D, Gocicb M, Trajkovicb S, Shamshirbandc S, Motamedid S, Hashimd R, Bonakdari H (2015) Determination of the most influential weather parameters on reference evapotranspiration by adaptive neuro-fuzzy methodology. Comput Electron Agric 114:277–284. https://doi.org/10.1016/j.compag.2015.04.012

Pierna JAF, Baeten V, Dardenne P (2006) Screening of compound feeds using NIR hyperspectral data. Chemometr Intell Lab Syst 84:114–118. https://doi.org/10.1016/j.chemolab.2006.03.012

Prasad VSS, Dutta Gupta S (2008) Applications and potentials of artificial neural networks in plant tissue culture. In: Gupta D, Gupta S, Ibaraki Y (eds) Plant tissue culture engineering. Springer-Verlag, Berlin, pp 47–67. https://doi.org/10.1007/1-4020-3694-9_3

Priya P, Kesheri M, Sinha RP, Kanchan S (2016) Molecular dynamics simulations for biological systems. In: Karâa W. B. A., Dey N. (eds.), Biomedical image analysis and mining techniques for improved health outcomes, advances in bioinformatics and biomedical engineering (ABBE) series. IGI Global, USA 286–313. doi:https://doi.org/10.4018/978-1-5225-1762-7.ch040

Qi L, Ma X (2009) Rice blast detection using multispectral imaging sensor and support vector machine. Trans ASABE 095891. 10.13031/2013.26985

Rumelhart DE, McClelland JL (1986) Parallel distributed processing: explorations in the microstructures of cognition, vol I. MIT Press, Cambridge, MA

Rumpfa T, Mahleinb A-K, Steinerb U, Oerkeb E-C, Dehneb H-W, Plümera L (2010) Early detection and classification of plant diseases with support vector machines based on hyperspectral reflectance. Comput Electron Agric 74(1):91–99. https://doi.org/10.1016/j.compag.2010.06.009

Shao Y, Zhao C, He Y, Bao Y (2009) Application of infrared spectroscopy technique and chemometrics for measurement of components in rice after radiation. Appl Eng Agric 52 (1):187–192. 10.13031/2013.25929

Simpson PK, Jahns G (1993) Fuzzy min–max neural networks for function approximation. In: proc. IEEE Int Conf Neural Netw 3:1967–1972. https://doi.org/10.1109/ICNN.1993.298858

Singh A, Ganapathysubramanian B, Singh AK, Sarkar S (2016) Machine learning for high throughput stress phenotyping in plants. Trends Plant Sci 21(2):110–124. https://doi.org/10.1016/j.tplants.2015.10.015

Takagi T, Hayashi I (1991) NN-driven fuzzy reasoning. Int J Approx Reason 5(3):191–212

Teorey TJ (1999) Database modeling and design. Morgan Kaufmann Publishers, San Francisco

Tian Y, Zhang C, Li C (2004) Study on plant disease recognition using support vector machine and chromaticity moments. Trans Chi Soci Agric Machin 35(3):95–98

Tikk D, Koczy LT, Gedeon TD (2003) A survey on universal approximation and its limits in soft computing techniques. Int J Approx Reason 33(2):185–202. https://doi.org/10.1016/s0888-613x(03)00021-5

Trebar M, Steele M (2008) Application of distributed SVM architectures in classifying forest data cover types. Comput Electron Agric 63(2):119–130. https://doi.org/10.1016/j.compag.2008.02.001

Twarakavi NKC, Simune k J, Schaap MG (2009) Development of pedotransfer functions for estimation of soil hydraulic parameters using support vector machines. Soil Sci Am J 73:1443–1452. https://doi.org/10.2136/sssaj2008.0021

Wang W, Paliwal J (2006) Spectral data compression and analyses techniques to discriminate wheat classes. Trans ASABE 49(5):1607–1612. 10.13031/2013.22035

Werea KB, Buic DT, Dicka ØB, Singh BR (2015) A comparative assessment of support vector regression, artificial neural networks, and random forests for predicting and mapping soil organic carbon stocks across an Afromontane landscape. Ecol Indic 52:394–403. https://doi.org/10.1016/j.ecolind.2014.12.028

Whittaker A D, Park B S, McCauley J D, Huang Y (1991) Ultrasonic signal classification for beef quality grading through neural networks. In: Automated agriculture for the 21st century Trans ASAE, pp 116–125

Witten IH, Frank E (2000) Data mining: practical machine learning tools and techniques with JAVA implementations. Morgan Kaufmann Publishers, San Francisco

Wu D, Feng L, He Y, Bao Y (2008) Variety identification of Chinese cabbage seeds using visible and near-infrared spectroscopy. Trans ASABE 51(6):2193–2199. 10.13031/2013.25382

Xiang H, Tian LF (2007) Artificial intelligence controller for automatic multispectral camera parameter adjustment. Trans ASABE 50(5):1873–1881. 10.13031/2013.23939

Yang J, Gong W, Shi S, Du L, Sun J, Ma Y-Y, Song S-L (2015) Accurate identification of nitrogen fertilizer application of paddy rice using laser-induced fluorescence combined with support vector machine. Plant Soil Environ 61(11):501–506. 10.17221/496/2015-PSE

Yu H, Niu X, Ying Y, Pai X (2008) Non-invasive determination of enological parameters of rice wine by Vis-NIR spectroscopy and least squares support vector machines. ASABE 084875. 10.13031/2013.24669

Zadeh LA (1965) Fuzzy sets. Inf Control 8:338–353. https://doi.org/10.1016/S0019-9958(65)90241-X

Zadeh LA (1973) Outline of a new approach to the analysis of complex systems and decision processes. IEEE Trans Syst Man Cybern SMC-3:28–44. https://doi.org/10.1109/tsmc.1973.5408575

Zadeh LA (1981) Possibility theory and soft data analysis. In: Cobb L, Thrall RM (eds) Mathematical frontiers of the social and policy sciences. Westview Press, Boulder, pp 69–129

Zhang Q, Litchfield JB (1992) Advanced process controls: applications of adaptive, fuzzy and neural control to the food industry. In: Food processing automation II. Trans ASAE, pp 169–176

Zhang H, Paliwal J, Jayas DS, White NDG (2007) Classification of fungal infected wheat kernels using near-infrared reflectance hyperspectral imaging and support vector machine. Trans ASABE 50(5):1779–1785. 10.13031/2013.23935

Future Perspectives of Computational Biology: Demanding Shifts in Analytical Thinking to Unfold Biological Complexities

17

Hemant J. Purohit, Hitesh Tikariha, and Vipin Chandra Kalia

Abstract

With available knowledge and databases, the mining of more information has driven the last decade of computational biology. We validated the existing known information with omics data. There is need in overall shift in our approach; instead of understanding the architecture of hierarchical gene network, we should work on condition-specific shift in hierarchies or partnerships of gene to manage plasticity. We assume that there will be a great shift in metabolomics approach to understand how cell manages to perform at its minimum driving energy level. Transformation of decision-making system with systematic mathematical and multiple soft computing modeling platform will be crucial to untangle the thread of pattern with which the nature follows for the process of evolution, expression, and engineering the cellular machinery.

Keywords

Functional element · Machine learning · Fractal analysis · Pan-genome · System biology

H. J. Purohit (✉) · H. Tikariha
Environmental Biotechnology and Genomics Division, CSIR-National Environmental Engineering Research Institute (NEERI), Nagpur, Maharashtra, India
e-mail: hj_purohit@neeri.res.in; hemantdrd@hotmail.com

V. C. Kalia
Microbial Biotechnology and Genomics, CSIR-Institute of Genomics and Integrative Biology (IGIB), Delhi University Campus, Delhi, India
e-mail: vckalia@igib.in; vc_kalia@yahoo.co.in

H. J. Purohit et al. (eds.), *Soft Computing for Biological Systems*,
https://doi.org/10.1007/978-981-10-7455-4_17

17.1 Introduction

A surge in availability of information has resulted in the emergence of new challenges in understanding the biological systems. The observed data is being analyzed and used for predicting its behavior using appropriate mathematical modeling. This is complemented by the statistical tools, which ensure the accuracies of the recorded data (Bassalo et al. 2016; Wright et al. 2016). The application of soft computing and support from fuzzy logics includes the explorations of diverse hypothesis in different models (Liu et al. 2016; Mahata et al. 2017). The book presents the current trends in soft computing tools and techniques in various domains of biological sciences (Meza-Lucas et al. 2016).

17.2 DNA, RNA Sequence, and Proteins to Functional Element Prediction

The question is whether the biological system is very complex or it has the great architecture of simple system components. Due to inherent plasticity of the biological systems, the simple architectures may give an overall projection as a most complicated system. If we look at the most simple form of biological intelligence, then its most organized and least understood molecule, i.e., DNA, is sometimes considered as junk DNA or non-transcribed DNA (Palazzo and Gregory 2014). Other than that, DNA sequences also have an occurrence of nucleotide patterns either following the ordered nature of distribution or with some insertions and deletions (Bhushan et al. 2013, 2015; Puri et al. 2016). With the concept of the epigenome, this information gains more interest, which to an extent is translated via mRNA to proteins. If we closely look into the biochemical reactions, then from a cell to a multicellular system, we can identify the functional units required to bring out a specific biological event. We consider these participating functional units as enzymes, cofactors, signaling molecules or transport system, and so on. Different functional biomolecules have an active site, modulation, or regulatory elements. This demands massive mining exercise for functional elements and generating the correlations at DNA, RNA, and at protein levels with their primary structures. These elements could be nucleotide patterns and their organization or organized distributions of amino acids as motifs (Bohlin et al. 2017; Kalia et al. 2017). At the end of this exercise, we shall end up with a database of functional elements that participates in different biochemical events in the cells. With this fact, it will be interesting to generate new organization of functional elements or predict new functions for defined biomolecules. These new functions may not work with expected efficiency but definitely could help stress cells with the desired plasticity.

NGS has changed our data generation capacities for any environment. This has revolutionized and provided the depth of information for DNA and RNA transcripts. Now, we are trying to correlate available microbiome with every living environment. This has created a new avenue for metagenomics with depth of the sequence data (Qiu et al. 2017). The generated data may be up to 70% or more cannot be even annotated (Spetale et al. 2016). The conventional approach based on the similarity index and known information cannot correlate the expected biological phenomenon associated with that environmental DNA. The emerging area of gut microbiome (Pooja et al. 2015), where the shift in community with a clinical scenario, will be a very challenging task in the future; that will be further superimposed by the SNPs of that host. The approach of understanding of functional elements and their organization might be the way to dig into this massive data. This will suggest new ways of bacterial survival and may be new pathways, either to help and survive in the host with options of symbiosis or to dominate the community, which is associated with the host.

17.3 Protein Network and Docking

With the development of so many soft computational tools and algorithms for a given sequence, its protein 3D structure can be determined and so its nature and a predictable function based on the domain it possesses. Now there is a necessity of algorithms, which can predict the possible networks/ direct interactions for a protein based on functional elements. This should further follow the movement of this protein in total cell metabolic machinery and suggest possible participation in different cellular events (Li et al. 2014). All this demands more efficient docking and simulation platform for protein/receptors to generate 3D structure and an automatic pipeline, which sort out the possible interaction between proteins and ligands. Another area of interest is protein structure elasticity. A protein interacts with a molecule with specificity, we can predict different possible energy levels and possible outcome, but we need comprehensive algorithms to understand in the milieu of cytoplasm with a pool of intermediates and their interactions with a differential specificity of chemically similar molecules/analogs. How this titration reaction segregates and aligns with a particular time-specific decision-making system will be the next step in our biological comprehension.

Fabricating new molecules with protein engineering still needs a lot of iteration in the understanding of functional elements associated with enzymatic/receptor-mediated processes. There are few databases like STRING, Gene MANIA, and I2D which generate a network of protein, but a lot more has to be done (Mostafavi and Morris 2012). Construction of new molecule needs to be more accurate and requires in silico testing with docking dynamic study for all its possible types of action inside a cell milieu with competing metabolites. Exploration of a transported drug inside a cell with simulated gene network, protein expression profile, and related metabolic potential/ metabolite spacing will provide a pathway for uncovering of all the actions a drug molecule can encounter in a cell. This might open up a way to understand side effects of the drug, which is essentially a mismanaged metabolic event.

17.4 Soft-Tool Development

Computational tools in the sphere of omics will require the development of different options of data mining and organization. The future interpretation drawn from such data set will be like QR code where a set of same functional elements organized in different ways in two- or three-dimensional scale will suggest different cellular events. The information across different data set has already started merging, but now it's the time to generate a more stringent rule for data merger. Not only integration of tools is required but a concatenation of tools in a systematic pipeline is also in the stipulation. Working with different levels of cell organization and making inferences based on selected signatures of the biochemical events could help in predicting the final course of cell decision-making framework. This can be possible if deep machine learning and artificial intelligence can be utilized for such process, which will require strong hypothesis generation with a team of experts. The idea such as bioboxes can be a solution to cater these processes, which proposes the concept of containerization of software and make it interchangeable thus increasing its accessibility (Belmann et al. 2015).

17.5 Epigenomics and Pan-Genomics

The idea of epigenetics a decade back was limited to eukaryotes only, but now it has been associated with gene regulatory network in prokaryotes also. Bacterial epigenetics similar to eukaryotes unfolds the methylation pattern associated with regulation of genes in bacteria. Histone modification system is absent in bacteria, and epigenetics is centered on discovering of distribution of DNA methylation pattern. Study of bacterial epigenetics is mostly focused on pathogenic ones like deciphering cystic fibrosis epigenetic control by *Pseudomonas aeruginosa* (Madhavaram 2016) and conception of ciprofloxacin resistance in *E.coli* due to rise in methylation level (Yugendran and Harish 2016). The epigenetic role is extensively studied in bacterial restriction-modification system only. But now it has approached a wider depth and so requires new tools to unfold the circuit of epigenetic regulation in bacteria along with NGS analysis. Need for exploration of methyltransferases, mechanism of transfer, and process regulation by cytosine methylation are demanding ones. This will surely lead to a surge in the generation of more software and algorithms and a new type of networking system to connect metagenome and metabolome with epigenome.

When the understanding of bacterial diversity and filling of a gap in ecological interaction begins with metagenome analysis, pan-genome emerges as a way to bridge the gap. Pan-genome is simply just an analysis plot to extrapolate the genetic diversity and classification of the genes and their level of sharing in different life forms (Dumas et al. 2016; Monat et al. 2017). The application of pan-genome can contribute to increasing the gene pool in a particular system and interaction. It can guide us to understand the complex intercommunication and relay network existing between a group of organisms and with organisms themselves. To fulfill such promising concept, NGS technologies can surely help along with metagenomic, gene regulatory network and pan-genomic software. There are already computing techniques like GET_HOMOLOGUES, PANNOTATOR, SplitMEM, PanGP, Roary, etc. to carry out pan-genome analysis. But the problems of identifying core genomes, describing pan-genome as open or closed and size estimation limits the precision of the result generated by this software (Vernikos et al. 2015). Hence, it opens a challenge for bioinformaticians to architect stronger algorithm to overcome this problem.

17.6 Genomic Plasticity and Evolution

Understanding the trend in evolution has been always a key to understanding the biological phenomenon. It has broadened its way from simple evolutionary genetics to different branches of omics. This property of genomic plasticity creates a difficult task of interpreting functional and metabolic changes in the cell or organism itself (Ricker et al. 2016; Vandecraen et al. 2017). There is a need for algorithms, which can capture subtle changes in overall DNA information. The challenges include the study of adaptations in the microbial community or stress-mediated shift in community dynamics. The scenario generates the specific community structure and redefines the overall community intelligence that drives the required function to meet the enforced changes by the environment. The other scenario could be the changes in genome structure. The changes in genetic composition are of special concern in disease diagnosis and construction of an ideal prevention strategy. The host-pathogen interaction is another area where prediction of host-mediated changes in pathogen physiology will suggest new strategies to prevent overall pathogenesis (Goh and Knight 2017). There can be the construction of a sophisticated evolutionary model that in a biological system predicts the course of gene shuffling, transversion or translational changes, and events like gene transfer. This part can really be helpful for combating epidemic diseases and sudden outburst of some pathogens in an area.

17.7 Biological Modeling

In the domain of biological modeling, new tools such as "fractal analysis" will help in computing the changes using data generated as signal, networks, or even predicted molecular motions. Future modeling will require a lot of unsupervised validation of model where the wet laboratory data with different hypothesis will make the base (Tsigkinopoulou et al. 2017). This will help in capturing the subtle variants that might be influencing the overall behavior of the system. There are a broad range of areas where these types of studies could be essential, which includes the survival of microbes in wastewater to colonization in human gut or a pathogenesis scenario. Mathematical models are applied to study the ecological functioning and biodiversity (Fitzpatrick and Keller 2015), process efficiency of a wastewater treatment plant (Pan et al. 2015), immunological processes and response (Cappuccio et al. 2016), and even the exploration of metabolic regulation in aging conditions (Mc Auley et al. 2015). The understanding in gene regulation finds a new paradigm with siRNA (Dar et al. 2016; Alkan et al. 2017; He et al. 2017) or CRISPR-Cas systems (Mohanraju et al. 2016; Cloney 2017) and will demand new tools to explore their interactions in gene silencing.

17.8 Pattern-Based Expression and Organization

Mining of data set from the independent experiment for a typical pattern gives a trend associated with that particular system. This data set could be DNA sequence data, RNA expression profile of siRNA associated with the atypical condition. Finding the pattern and deducing out the core function are tasks that really project out the machinery of the system. There are many ways by which this expression pattern can solve many of the biological problems. Finding the heterogeneity in colorectal cancer with the help of exploration of gene expression network was a way by which pattern-based study can assist in expanding deep knowledge about a disease (Budinska et al. 2013). Another is that this expression pattern can reveal niche diversification in a microbial community (Gifford et al. 2013).

DNA pattern-based study at genome level can also help to deduce evolutionary lineages and polymorphisms associated with the genes (Cornejo et al. 2015; Yu et al. 2015; Ambardar et al. 2016; Kumar et al. 2017). Apart from revealing genetic alterations, this genome level pattern analysis can be used for mining gene cluster of interest (Duncan et al. 2015; Kumar et al. 2016). Moreover, this genome-wide pattern analysis at large extent can really provide us the solution of trait selection during evolution (Kalia et al. 2015, 2016; Kekre et al. 2015; Mathieson et al. 2015; Koul and Kalia 2016; Lee and Rho 2016; Lee et al. 2016). These examples grant ample opportunities for further research by using soft tools for pattern-based expression or genome-wide studies. It also opens multiple doors for the application developer to tackle problems of pattern-based study (Kumar et al. 2015). This also enforced the addition of pipeline system for pattern-based study because one has to deal with different types of data set.

17.9 System Biology and Decision-Making in Biological Sciences

System biology was evolved to understand complex interactions occurring in the biological matrix. Now with the evolution of huge computational power, the framework shifted to integrate the separate data generated by omics to completely new interface and suggest an output through system biology (Batchelor and Loewer 2017). An array of tools and networking system is being built to understand the biological phenomenon as a whole (Bartocci and Lió 2016); for example, to know what physiological, immunological, gut bacterial responses are generated when a new drug molecule enters the system (Nickerson et al. 2017). Most of the investigations consider response/signaling in a cell and study as a separate, but its effect on peripheral and housekeeping pathways is generally ignored; a holistic conclusion through systematic interconnecting network could provide more insight into overall physiology. A model scheme has to be laid out for such complex interactive system. Work has already been initiated in this direction; one such example is pathway tools which integrate genomic, metabolic, and regulatory data and thus assist in the investigation of the biological network in question (Karp et al. 2016). It has also included the energy and flux modeling system to estimate the possible interaction and reactions occurring for an accurate metabolic pathway prediction.

17.10 Conclusion

The understanding in biological sciences started with one gene and a physiological scenario as a concept (Risch and Merikangas 1996). Time has changed with more information and analytical tools; now we are considering even one gene and many scenarios (Zhu et al. 2014). Investigations are supported by omics as an option of data generation and system biology tools with their analytical capabilities are integrating our understanding to a more complex outcome (Shaik and Ramakrishna 2014; Kalia and Kumar 2015; Puranik and Purohit 2015; Bracken et al. 2016; Ram et al. 2016). This way we can conclude that we are going away from reaching any conclusion soon for a defined biological system (Wu et al. 2016).

17.11 Opinion

We need to rethink and change our strategy of analysis, which is mostly driven by statistics for selecting the most appropriate data. The system may not require too many copies of a regulatory protein, but its influence could lead to many copies of a functioning protein, which will provide a typical phenotype. Similarly, in a metagenome data, we select a cutoff to understand a dominating population, but a time series data or a stress condition suddenly picks a bacterium that was otherwise removed from the statistics. Maybe in few years, we shall be correlating every single disease condition with the signature gut microbiome. This suggests that for every scenario, irrespective of omic tools applied, the data needs better mining, binning, and discriminating approach to understand the cause of the key events and the metabolic status of the cell supporting the key event, and new emerging biochemical perturbances in cell expected in different compartments need to be understood. The time series data supported with compartmentalization will add to our understanding.

References

Alkan F, Wenzel A, Palasca O, Kerpedjiev P, Rudebeck AF, Stadler PF, Hofacker IL, Gorodkin J (2017) RIsearch2: suffix array-based large-scale prediction of RNA–RNA interactions and siRNA off-targets. Nucleic Acids Res 45:e60–e60. https://doi.org/10.1093/nar/gkw1325

Ambardar S, Gupta R, Trakroo D, Lal R, Vakhlu J (2016) High throughput sequencing: an overview of sequencing chemistry. Indian J Microbiol 56:394–404. https://doi.org/10.1007/s12088-016-0606-4

Bartocci E, Lió P (2016) Computational modeling, formal analysis, and tools for systems biology. PLoS Comput Biol 12:e1004591. https://doi.org/10.1371/journal.pcbi.1004591

Bassalo MC, Liu R, Gill RT (2016) Directed evolution and synthetic biology applications to microbial systems. Curr Opin Biotechnol 39:126–133. https://doi.org/10.1016/j.copbio.2016.03.016

Batchelor E, Loewer A (2017) Recent progress and open challenges in modeling p53 dynamics in single cells. Curr Opin Sys Biol 3:54–59. https://doi.org/10.1016/j.coisb.2017.04.007

Belmann P, Dröge J, Bremges A, McHardy AC, Sczyrba A, Barton MD (2015) Bioboxes: standardised containers for interchangeable bioinformatics software. Gigascience 4:47. https://doi.org/10.1186/s13742-015-0087-0

Bhushan A, Joshi J, Shankar P, Kushwah J, Raju SC, Purohit HJ, Kalia VC (2013) Development of genomic tools for the identification of certain *Pseudomonas* up to species level. Indian J Microbiol 53:253–263. https://doi.org/10.1007/s12088-013-0412-1

Bhushan A, Mukherjee T, Joshi J, Shankar P, Kalia VC (2015) Insights into the origin of *Clostridium botulinum* strains: evolution of distinct restriction endonuclease sites in *rrs* (16S rRNA gene). Indian J Microbiol 55:140–150. https://doi.org/10.1007/s12088-015-0514-z

Bohlin J, Eldholm V, Pettersson JH, Brynildsrud O, Snipen L (2017) The nucleotide composition of microbial genomes indicates differential patterns of selection on core and accessory genomes. BMC Genomics 18:151. https://doi.org/10.1186/s12864-017-3543-7

Bracken CP, Scott HS, Goodall GJ (2016) A network-biology perspective of microRNA function and dysfunction in cancer. Nat Rev Genet 17:719–732. https://doi.org/10.1038/nrg.2016.134

Budinska E, Popovici V, Tejpar S, D'ario G, Lapique N, Sikora KO, Di Narzo AF, Yan P, Hodgson JG, Weinrich S, Bosman F, Roth A, Delorenzi M (2013) Gene expression patterns unveil a new level of molecular heterogeneity in colorectal cancer. J Pathol 231:63–76. https://doi.org/10.1002/path.4212

Cappuccio A, Tieri P, Castiglione F (2016) Multiscale modelling in immunology: a review. Brief Bioinform 17:408–418. https://doi.org/10.1093/bib/bbv012

Cloney R (2017) Metagenomics: uncultivated microbes reveal new CRISPR-Cas systems. Nat Rev Genet 18:146. https://doi.org/10.1038/nrg.2017.1

Cornejo OE, Fisher D, Escalante AA (2015) Genome-wide patterns of genetic polymorphism and signatures of selection in *Plasmodium vivax*. Genome Biol Evol 7:106–119. https://doi.org/10.1093/gbe/evu267

Dar SA, Gupta AK, Thakur A, Kumar M (2016) SMEpred workbench: a web server for predicting efficacy of chemically modified siRNAs. RNA Biol 13:1144–1151. https://doi.org/10.1080/15476286.2016.1229733

Dumas E, Christina Boritsch E, Vandenbogaert M, Rodríguez de la Vega RC, Thiberge JM, Caro V, Gaillard JL, Heym B, Girad-Misquich F, Brosch R, Sapriel G (2016) Mycobacterial pan-genome analysis suggests important role of plasmids in the radiation of type VII secretion systems. Genome Biol Evol 8:387–402. https://doi.org/10.1093/gbe/evw001

Duncan KR, Crüsemann M, Lechner A, Sarkar A, Li J, Ziemert N, Wang M, Bandeira N, Moore BS, Dorrestein PC, Jensen PR (2015) Molecular networking and pattern-based genome mining improves discovery of biosynthetic gene clusters and their products from *Salinispora* species. Chem Biol 22:460–471. https://doi.org/10.1016/j.chembiol.2015.03.010

Fitzpatrick MC, Keller SR (2015) Ecological genomics meets community-level modelling of biodiversity: mapping the genomic landscape of current and future environmental adaptation. Ecol Lett 18:1–16. https://doi.org/10.1111/ele.12376

Gifford SM, Sharma S, Booth M, Moran MA (2013) Expression patterns reveal niche diversification in a marine microbial assemblage. The ISME J 7:281–298. https://doi.org/10.1038/ismej.2012.96

Goh C, Knight JC (2017) Enhanced understanding of the host–pathogen interaction in sepsis: new opportunities for omic approaches. The Lancet Resp Med 5:212–223. https://doi.org/10.1016/S2213-2600(17)30045-0

He F, Han Y, Gong J, Song J, Wang H, Li Y (2017) Predicting siRNA efficacy based on multiple selective siRNA representations and their combination at score level. Sci Rep 7:44836. https://doi.org/10.1038/srep44836

Kalia VC, Kumar P (2015) Genome wide search for biomarkers to diagnose *Yersinia* infections. Indian J Microbiol 55:366–374. https://doi.org/10.1007/s12088-015-0552-6

Kalia VC, Kumar P, Kumar R, Mishra A, Koul S (2015) Genome wide analysis for rapid identification of *Vibrio* species. Indian J Microbiol 55:375–383. https://doi.org/10.1007/s12088-015-0553-5

Kalia VC, Kumar R, Kumar P, Koul S (2016) A genome-wide profiling strategy as an aid for searching unique identification biomarkers for *Streptococcus*. Indian J Microbiol 56:46–58. https://doi.org/10.1007/s12088-015-0561-5

Kalia VC, Kumar R, Koul S (2017) *In silico* analytical tools for phylogenetic and functional bacterial genomics. In: Arora G, Sajid A, Kalia VC (eds) Drug resistance in bacteria, fungi, malaria and cancer. Springer Nature, Cham, pp 339–355. ISBN 978-3-319-48682-6. https://doi.org/10.1007/978-3-319-48683-3_15

Karp PD, Krummenacker M, Paley S, Wagg J (1999) Integrated pathway – genome databases and their role in drug discovery. Trends Biotechnol 17:275–81. https://doi.org/10.1016/S0167-7799(99)01316-5

Kekre A, Bhushan A, Kumar P, Kalia VC (2015) Genome wide analysis for searching novel markers to rapidly identify *Clostridium* strains. Indian J Microbiol 55:250–257. https://doi.org/10.1007/s12088-015-0535-7

Koul S, Kalia VC (2016) Comparative genomics reveals biomarkers to identify *Lactobacillus* species. Indian J Microbiol 56:253–263. https://doi.org/10.1007/s12088-016-0605-5

Kumar A, Mohanty NN, Chacko N, Yogisharadhya R, Shivachandra SB (2015) Structural features of a highly conserved Omp16 protein of *Pasteurella multocida* strains and comparison with related peptidoglycan-associated lipoproteins (PAL). Indian J Microbiol 55:50–56. https://doi.org/10.1007/s12088-014-04896-1

Kumar R, Koul S, Kumar P, Kalia VC (2016) Searching biomarkers in the sequenced genomes of *Staphylococcus* for their rapid identification. Indian J Microbiol 56:64–71. https://doi.org/10.1007/s12088-016-0565-9

Kumar R, Koul S, Kalia VC (2017) Exploiting bacterial genomes to develop biomarkers for identification. In: Arora G, Sajid A, Kalia VC (eds) Drug resistance in bacteria, fungi, malaria and cancer. Springer Nature, Cham, pp 357–370. ISBN 978-3-319-48682-6. https://doi.org/10.1007/978-3-319-48683-3_16

Lee S, Rho JY (2016) Development of a specific diagnostic system for detecting *Turnip Yellow Mosaic Virus* from Chinese cabbage in Korea. Indian J Microbiol 56:103–107. https://doi.org/10.1007/s12088-015-0557-1

Lee S, Kim CS, Shin YG, Kim JH, Kim YS, Jheong WH (2016) Development of nested PCR-based specific markers for detection of peach rosette mosaic virus in plant quarantine. Indian J Microbiol 56:108–111. https://doi.org/10.1007/s12088-015-0548-2

Li GW, Burkhardt D, Gross C, Weissman JS (2014) Quantifying absolute protein synthesis rates reveals principles underlying allocation of cellular resources. Cell 157:624–635. https://doi.org/10.1016/j.cell.2014.02.033

Liu F, Heiner M, Yang M (2016) Fuzzy stochastic petri nets for modeling biological systems with uncertain kinetic parameters. PLoS One 11:e0149674. https://doi.org/10.1371/journal.pone.0149674

Madhavaram A (2016) Biofilm production in response to DNA methylation in *Pseudomonas aeruginosa*. Int J Sci Eng Res 7:1183–1188

Mahata A, Mondal SP, Alam S, Roy B (2017) Mathematical model of glucose-insulin regulatory system on diabetes mellitus in fuzzy and crisp environment. Ecol Genet Genom 2:25–34

Mathieson I, Lazaridis I, Rohland N, Mallick S, Patterson N, Roodenberg SA, Harney E, Stewardson K, Fernandes D, Novak M, Sirak K, Gamba C, Jones ER, Llamas B, Dryomov S, Pickrell J, Arsuaga JL, de Castro JM, Carbonell E, Gerritsen F, Khokhlov A, Kuznetsov P, Lozano M, Meller H, Mochalov O, Moiseyev V, Guerra MA, Roodenberg J, Vergès JM, Krause J, Cooper A, Alt KW, Brown D, Anthony D, Lalueza-Fox C, Haak W, Pinhasi R, Reich D (2015) Genome-wide patterns of selection in 230 ancient Eurasians. Nature 528:499–503. https://doi.org/10.1038/nature16152

Mc Auley MT, Mooney KM, Angell PJ, Wilkinson SJ (2015) Mathematical modelling of metabolic regulation in aging. Meta 5:232–251. https://doi.org/10.3390/metabo5020232

Meza-Lucas A, Pérez-Villagómez M, Martínez-López JP, García-Rodea R, Martínez-Castelán MG, Escobar-Gutiérrez A, de la Rosa-Arana JL, Villanueva-Zamudio A (2016) Comparison of DOT-ELISA and Standard-ELISA for detection of the *Vibrio cholerae* toxin in culture supernatants of bacteria isolated from human and environmental samples. Indian J Microbiol 56:379–382. https://doi.org/10.1007/s12088-016-0596-2

Mohanraju P, Makarova KS, Zetsche B, Zhang F, Koonin EV, van der Oost J (2016) Diverse evolutionary roots and mechanistic variations of the CRISPR-Cas systems. Science 353: aad5147. https://doi.org/10.1126/science.aad5147

Monat C, Pera B, Ndjiondjop MN, Sow M, Tranchant-Dubreuil C, Bastianelli L, Ghesquiere A, Sabot F (2017) De novo assemblies of three *Oryza glaberrima* accessions provide first insights about pan-genome of African rices. Genome Biol Evol 9:1–6. https://doi.org/10.1093/gbe/evw253

Mostafavi S, Morris Q (2012) Combining many interaction networks to predict gene function and analyze gene lists. Proteomics 12:1687–1696. https://doi.org/10.1002/pmic.201100607

Nickerson ML, Witte N, Im KM, Turan S, Owens C, Misner K, Tsang SX, Cai Z, Wu S, Dean M, Costello JC, Theodorescu D (2017) Molecular analysis of urothelial cancer cell lines for modeling tumor biology and drug response. Oncogene 36:35–46. https://doi.org/10.1038/onc.2016.172

Palazzo AF, Gregory TR (2014) The case for junk DNA. PLoS Genet 10:e1004351. https://doi.org/10.1371/journal.pgen.1004351

Pan Y, Ni BJ, Lu H, Chandran K, Richardson D, Yuan Z (2015) Evaluating two concepts for the modelling of intermediates accumulation during biological denitrification in wastewater treatment. Wat Res 71:21–31. https://doi.org/10.1016/j.watres.2014.12.029

Pooja S, Pushpanathan M, Jayashree S, Gunasekaran P, Rajendhran J (2015) Identification of periplasmic a-amylase from cow dung metagenome by product induced gene expression profiling (Pigex). Indian J Microbiol 55:57–65. https://doi.org/10.1007/s12088-014-0487-3

Puranik S, Purohit HJ (2015) Dynamic interactive events in gene regulation using E. Coli dehydrogenase as a model. Funct Integ Genom 15:175–188. https://doi.org/10.1007/s10142-014-0418-8

Puri A, Rai A, Dhanaraj PS, Lal R, Patel DD, Kaicker A, Verma M (2016) An *in silico* approach for identification of the pathogenic species, *Helicobacter pylori* and its relatives. Indian J Microbiol 56:277–286. https://doi.org/10.1007/s12088-016-0575-7

Qiu Z, Yang H, Rong L, Ding W, Chen J, Zhong L (2017) Targeted metagenome based analyses show gut microbial diversity of inflammatory bowel disease patients. Indian J Microbiol 57. https://doi.org/10.1007/s12088-017-0652-6

Ram R, Mehta M, Nguyen QT, Larma I, Boehm BO, Pociot F, Concannon P, Morahan G (2016) Systematic evaluation of genes and genetic variants associated with type 1 diabetes susceptibility. J Immunol 196:3043–3053. https://doi.org/10.4049/jimmunol.1502056

Ricker N, Shen SY, Goordial J, Jin S, Fulthorpe RR (2016) PacBio SMRT assembly of a complex multi-replicon genome reveals chlorocatechol degradative operon in a region of genome plasticity. Gene 586:239–247. https://doi.org/10.1016/j.gene.2016.04.018

Risch N, Merikangas K (1996) The future of genetic studies of complex human diseases. Science 273:1516–1517

Shaik R, Ramakrishna W (2014) Machine learning approaches distinguish multiple stress conditions using stress-responsive genes and identify candidate genes for broad resistance in rice. Pl Physiol 164:481–495. https://doi.org/10.1104/pp.113.225862

Spetale FE, Tapia E, Krsticevic F, Roda F, Bulacio P (2016) A factor graph approach to automated GO annotation. PLoS One 11:e0146986. https://doi.org/10.1371/journal.pone.0146986

Tsigkinopoulou A, Baker SM, Breitling R (2017) Respectful modeling: addressing uncertainty in dynamic system models for molecular biology. Trends Biotechnol 35:518–529. https://doi.org/10.1016/j.tibtech.2016.12.008

Vandecraen J, Chandler M, Aertsen A, Van Houdt R (2017) The impact of insertion sequences on bacterial genome plasticity and adaptability. Crit Rev Microbiol 43; https://doi.org/10.1080/1040841X.2017

Vernikos G, Medini D, Riley DR, Tettelin H (2015) Ten years of pan-genome analyses. Curr Opin Microbiol 23:148–154. https://doi.org/10.1016/j.mib.2014.11.016

Wright AV, Nuñez JK, Doudna JA (2016) Biology and applications of CRISPR systems: harnessing nature's toolbox for genome engineering. Cell 164:29–44. https://doi.org/10.1016/j.cell.2015.12.035

Wu S, Joseph A, Hammonds AS, Celniker SE, Yu B, Frise E (2016) Stability-driven nonnegative matrix factorization to interpret spatial gene expression and build local gene networks. Proc Nat Acad Sci 113:4290–4295. https://doi.org/10.1073/pnas.1521171113

Yu S, Peng Y, Zheng Y, Chen W (2015) Comparative genome analysis of *Lactobacillus casei*: insights into genomic diversification for niche expansion. Indian J Microbiol 55:102–107. https://doi.org/10.1007/s12088-014-0496-2

Yugendran T, Harish BN (2016) Global DNA methylation level among ciprofloxacin-resistant clinical isolates of *Escherichia coli*. J Clin Diagnos Res 10:DC27. https://doi.org/10.7860/JCDR/2016/19034.7830

Zhu X, Need AC, Petrovski S, Goldstein DB (2014) One gene, many neuropsychiatric disorders: lessons from Mendelian diseases. Nat Neuroscience 17:773–781. https://doi.org/10.1038/nn.3713

Index

H. J. Purohit et al. (eds.), *Soft Computing for Biological Systems*,
https://doi.org/10.1007/978-981-10-7455-4

MIX
Papier aus verantwortungsvollen Quellen
Paper from responsible sources
FSC® C105338

If you have any concerns about our products,
you can contact us on
ProductSafety@springernature.com

In case Publisher is established outside the EU,
the EU authorized representative is:
Springer Nature Customer Service Center GmbH
Europaplatz 3, 69115 Heidelberg, Germany

Printed by Libri Plureos GmbH
in Hamburg, Germany